OXFORD RESEARCH STUDIES IN GEOGRAPHY

General Editors

J. Gottmann   J.A. Steers

F.V. Emery   C. D. Harris

The American Steel Industry

1850 - 1970

A GEOGRAPHICAL INTERPRETATION

Kenneth Warren

UNIVERSITY OF PITTSBURGH PRESS

Published by the University of Pittsburgh Press, Pittsburgh, Pa. 15260
© 1973 Oxford University Press. This reprint has been authorized by
Oxford University Press
All rights reserved
Manufactured in the United States of America

Library of Congress Cataloging-in-Publication Data

Warren, Kenneth.
  The American steel industry, 1850–1970: a geographical
interpretation/Kenneth Warren.
    p. cm.
  Reprint. Originally published: Oxford: Clarendon Press, 1973. (Oxford
research studies in geography)
  Includes bibliographies and index.
  ISBN 0-8229-3597-X
  1. Steel industry and trade — United States — Location — History.
I. Title.
HD9515.W33 1988
338.4′7669142′09 — dc19   88-14687
                                                                 CIP

EDITORIAL PREFACE

Geography has recently expanded rapidly as a field of scientific research. The subdivisions of geographical study are steadily growing more numerous and diverse as the concerns of geographers and the data made available to them expand. Similar trends have developed in many other fields of science; they testify to the greater breadth and depth of the subject as knowledge of it advances in various directions.

It seemed desirable in the circumstances to improve communications among scholars concerned with geographical questions. Between the comprehensive textbook, the article in a learned journal, and the 'popular' book on geography there is a place for the research study: the book which treats a limited subject in depth.

The Oxford Research Studies will seek to publish the results of original and specialized research on a wide variety of geographical subjects. Titles now under consideration range in subject from the physical aspects of geography to historical, social, economic, and political phenomena of a geographical nature. The first volumes in the series happen to be written by members of the staff of the School of Geography at Oxford but the editors welcome manuscripts from any source if they are consistent with the purposes of this series.

The American Steel Industry 1850–1970: A Geographical Interpretation is an original research study in historical and economic geography. It treats a basic industry, steel, in a country of continental dimensions, the United States, in historical perspective, from 1850, when it lagged far behind British industry, to 1970, when after a long period of international leadership, it was in turn challenged by rising production in the Soviet Union and Japan. The industry is viewed in its historical and geographical setting with both physical factors of resources and human factors of market, organization, technology, and governmental policy. The study illustrates admirably the changing weights of locational factors such as markets, fuel, and ore with changing technology: within the period studied production changed from pig iron to steel; the fuel, from charcoal to anthracite, to bituminous coking coal, to complex combinations of fuels; the ore from local widespread low-grade deposits of the Eastern United States, to high-quality but highly localized deposits of the Upper Great Lakes, to imported ores, and to pelletized concentrates made from large bodies of low-grade ore, involving massive capital outlays but making possible large savings in transportation and in the costs of smelting itself.

Kenneth Warren, of the staff of the School of Geography in Oxford, presents in rich detail the drama of the shifting location of the American steel industry. He traces the radical innovations in technology, alterations of locational factors, the relation of the industry to general economic development and particularly to railroad transportation as a carrier of its raw materials and products and indeed as a major market itself. Nor does he neglect the role of leadership and management, and the organization of the industry. Because of heavy capital investment the steel industry is characterized by considerable geographical and economic inertia, yet, viewed over the period of 120 years, the industry is seen as having undergone revolutionary transformations.

J.G.   J.A.S.

Oxford, Autumn 1972   F.V.E.   C.D.H.

PREFACE

In the period examined here United States production of pig iron rose from well under 750,000 tons a year to over 95 million tons, and steel output, a few thousand tons only at the beginning, increased to record levels in excess of 140 million tons. Although concerned with development in time, my focus has been primarily on spatial change—concentrating on evolving patterns of location, examining the ways in which differentiation in the course of development brought about localization of particular lines of activity in certain areas, attempting to explain differences in the type and rate of growth and the competitive power of various iron and steel districts. This I take to be the distinctive perspective of the economic geographer in his study of location. I should add that the basic work on my manuscript was complete before the publication of Father Hogan's great work on the development of the American steel industry and that I have therefore made no reference to it at all in preparing this text.[1] The reader should most certainly refer to it, not only for its intrinsic merits or as a supplement to this survey, but also as an indication of the very different academic perspective and approach of an economist and economic historian as compared with an economic geographer. As a geographer I am conscious that I have neglected two major aspects of study in order to focus on locational change itself. One is the landscape-shaping and region-forming consequence of growth and change in the spatial patterns of basic industry, an essential of full geographical analysis. At a later stage I hope to combine such a study with a more direct analysis of the role of individual business genius in locational change, in other words, from the locational point of view, to undertake a case study of the behavioural and decision-taking factors and their implications. The second field which I have even more completely ignored here is the comparative one, contrasts with other nations' steel industries, and more particularly the varying ideological frameworks for locational studies. I refer not to the different business ethos or industrial psychology of late-nineteenth-century Britain and the U.S.A., but to much more radical differences in the set of rules under which the locational game can be played in a Marxist as opposed to a capitalist society. Again I hope that some time in the future it may be possible to make a comparative study of the development of basic industry under such widely different frameworks of believed economic rationality.

In the present pattern of American steel-making there are innumerable anomalies, unsuitable and apparently irrational locations. This affects both individual works and whole districts. Why for instance was western steel-making so small until such a late date? Why do the Pittsburgh and Youngstown districts have capacity so greatly beyond the needs of their metal fabricating industries, while New England, with all its metal-working trades, has no integrated works, and the largest single market for steel in the nation, Southern Michigan, which also lies in a key position in the Great Lakes flows of iron ore and coal, still has such a relatively small steel industry? How is it that there are such major isolated works as Johnstown, Steelton, or Bethlehem on inadequate river sites in landbound locations? Why is south-eastern steel-making so heavily centred in Birmingham, Alabama? Only a study of changing location factors and evolving patterns of location will explain these anomalies of of the present day geography of steel.

Until after the mid-nineteenth century the American iron industry lagged far behind that of Britain both in scale of operation and in technical and organizational efficiency. Plants still used local materials and supplied local markets. With a few exceptions, they were scattered rather than concentrated into iron-making districts, and most of them were still east of the Appalachians. Later, as settlement spread westwards, rich new mineral resources were proved and opened, notably in the case of the prime coking coalfields of south-western Pennsylvania and the iron ranges of the upper Great Lakes. Efficient bulk transport was developed, first by river and then and more important by lake vessel and by rail. As this happened, iron and later steel manufacture was concentrated in fewer, larger units and in major producing districts. Output in the areas between the Ohio River and the shores of the Great Lakes grew to equal and, by the end of the nineteenth century, far to exceed that of the east. In the process wholly new metallurgical districts emerged, while some of the most renowned old ones stagnated, declined, or in exceptional cases disappeared from the production lists altogether. In the twentieth century changes in the raw material and marketing context of steel production continued. New demand lines, the technical breakthrough of the continuous wide hot strip mill, and the near collapse of some old trades had important and varying repercussions in the various districts in the interwar years. Since World War II material supply sources have been largely changed, especially in the case of iron ore. Within the last decade there have been far-reaching technical changes, above all the introduction and rapid spread of oxygen steel-making processes. Industrial growth and, with it, metal-working industry are spreading much more widely throughout the United States, but the great bulk of steel consumption and still more of production still lies in the manufacturing belt of the north-east.

In surveying the evolving economic geography of American steel since bulk production began 120 years ago, my aim has been to account for the present spatial pattern and to provide a case study of locational change in a major manufacturing industry. Throughout, I have had help, guidance, and invaluable criticism from firms and individuals. Only in a very few cases was help refused or when offered and accepted, not in the end provided. For obvious reasons it is impossible here to identify those in the industry who have helped me, but I would like them to know that my appreciation is no less sincere. However, I can acknowledge the stimulus provided by the friendly encouragement and criticism of two retired steelmen, the late Mr. G.A.V. Russell of Tunbridge Wells, England, and Mr. Joseph Malborn of Cleveland, Ohio. Most of the maps, which I regard as the basis of the whole analysis, were produced in their finished form by the patience and tolerance of the cartographers of my former department in the University of Leicester, and by Miss Helen Bromley of the School of Geography, Oxford. I also wish to acknowledge permission to reproduce illustrative material: details of sources are given in the underlines.

Kenneth Warren,

September 1972. Witney, Oxfordshire.

[1] W.T. Hogan, *An Economic History of the Iron and Steel Industry in the United States,* 5 vols., 1972.

CONTENTS

LIST OF MAPS AND DIAGRAMS

ABBREVIATIONS

The following abbreviations are used throughout in reference to sources:

A.I.S.A. American Iron and Steel Association. *Statistics of the American and Foreign Iron Trades.*

A.I.S.I. American Iron and Steel Institute. *Annual Statistical Report.*

A.M.M. *American Metal Market.*

B.A.I.S.A. *Bulletin of the American Iron and Steel Association.*

B.F.S.P. *Blast Furnace and Steel Plant.*

Eng. *Engineering.*

I.A. *Iron Age.*

I.C.T.R. *Iron and Coal Trades Review.*

I.S. Eng. *Iron and Steel Engineer.*

I.T.R. *Iron Trade Review.*

J.F.I. *Journal of the Franklin Institute.*

J.I.S.I. *Journal of the Iron and Steel Institute.*

T.A.I.M.E. *Transactions of the American Institute of Mining Engineers.* (later *Transactions of the American Institute of Mining and Metallurgical Engineers).*

CHAPTER 1

Location of Heavy Industry

The briefest consideration of the location pattern of manufacturing in any advanced economy reveals a highly complex situation. The older and more highly developed the economy the more difficult is analysis of the pattern. To make some order out of the apparent chaos appropriate tools and a plan of action are essential. It is useful to start with certain simplifying assumptions, but as these yield their crop of results they must be modified, more variables being introduced in order to tackle some of the remaining problems. Bit by bit a more complete insight into the reality of location patterns is achieved and in the process techniques of analysis and the general conceptual framework are refined. This study is concerned with the location of American iron and steel manufacture. First, and briefly, the situation is examined under general simplified assumptions, and then the processes of locational change are analysed. In the first stage the discussion is in general terms, specific reference to the American industry being kept to a minimum. It should be stressed, however, that the models which illustrate possible patterns of growth have been devised to mirror some of the broad influences affecting that industry over the last century; they are themes in relation to which the actual locational patterns represent a host of variations.

BUSINESS AIMS AND LOCATIONAL CHOICE

A strictly economic locational choice for any manufacturing concern should be a rational process of evaluating the factors involved in the whole production process, aimed at ensuring that the value of all saleable outputs exceeds the cost of all inputs at the intended scale of operation by the maximum amount. Non-economic considerations, perhaps based on social aims or representing government interference, may influence or even determine the locational decision, but even in these circumstances a strictly commercial assessment is essential to provide a measure of the true cost to the firm, community, or nation. The locational problem for the individual concern involves the recognition of all the variables and their manipulation in order to maximize profits.[1] The situation of any firm considering a new plant or extension of an old one, however, is still more complicated. In the first place the business unit operates within an industry, a setting of other concerns taking independent and perhaps unknown decisions. Secondly, the plant operates in time as well as in space, and during its operating life there will inevitably be changes both in the variables it can manipulate and in the rank and stance of its competitors. In some conditions these changes may be almost revolutionary. Recognizing these complications from the start, one may profitably approach them by easy stages, first by looking at the way in which individual factors affect the situation.

A LOCATIONAL MODEL FOR A DEVELOPING IRON INDUSTRY

A time-honoured simplification in locational analysis is to start with an isolated uniform plain on which patterns of economic activity gradually evolve. For the American iron industry this is perhaps as good a starting-point as any. If we assume settlement of such an area by an agricultural population bringing tools, implements, and household hardware with them, we have the simplest datum surface possible. As August Lösch pointed out, a hierachy of settlements and of manufacturing activity will emerge to serve the population, whose patterns of farming activity soon diversify the plain.[2] Eventually iron manufacture will be developed to supply their needs. Scale of production will be small because of difficulties of transport, so that local, small supplies of ore and fuel will be used. Even at this stage, however, there are limits to the subdivision of operations and availability of materials so that iron production will be much less widely spread than, for instance, baking or the work of blacksmiths, and will possibly be associated with centres of intermediate or larger size in the hierarchy of service centres in the plain, or, more probably, be at new locations specially related to mineral working.

In stage two of analysis it is necessary to introduce certain complications which modify the cost situation and the location pattern of the industry. These upset the assumption that costs are more or less equal over the whole area. For instance, even though in a pre-railway age land transport costs may be more or less equal in all directions by packhorse or cart the relatively small rivers navigable at this stage of economic growth extend market range and so favour the concentration of production units and their subsequent growth. Such larger agglomerations of iron-making capacity require bigger mineral supplies and the new scale of working and the use of water carriage will permit these to come from greater distances. At an early stage of production charcoal fuel, widely available in a newly settled, moist temperate country, is perhaps employed but this widely spread supply becomes limited as farmers clear the original vegetation from the better land and the furnaces make their own mark, so that costs of fuel change unequally throughout the area. Rising wood fuel costs and the opportunity of increased production to serve bigger market areas may cause a transfer to mineral fuel, but at this point the vagaries of mineral distribution, always important modifying influences for the iron ore part of the furnace charge, become even more significant. Depth, water conditions, richness, and accessibility of coal and iron ore become factors which will favour one old-established centre and penalize another. They call into being wholly new investments in iron-making plant, which cause the elimination of others, favoured in a previous era. Apart from iron content, the other elements present in an ore deposit may be of significance, for iron of a particular chemical composition suitable for manufacture of one product is unsuitable for production of another. At an early stage of development, however, when specifications are less exacting and particularly when puddled iron, not steel, is the material employed in the rolling mills, differences in the character of raw materials are less important factors in the areal differentiation of iron manufacture than the existence of specialized markets. A favoured point on a sea coast will become a shipbuilding centre where boiler plate and thinner grades of finished flat-rolled iron, ranging downwards into sheet, will be turned out, and later, as wooden shipbuilding declines, iron ship plate and later steel plate will also be produced. A major river-crossing point for the roads which replace the earlier haphazard pathways across country, and whose course is shaped in turn by favourable topography, may also build boats and, if it is an embarkation point for the settler, will supply him with implements, domestic utensils, nails for his cabin and fencing materials which the difficulties of overland transport prevent him from bringing. Such patterns of early specialization tend to be perpetuated.

Into this already dynamic system of manufacturing locations there will be thrown at certain stages new factors of fundamental importance. One is the railway, for the first time providing bulk transport over land, lessening still further the dependence on local supplies of minerals and widening the market. Itself a great consumer of iron, it enhances the status of existing river-based concentrations of iron manufacture, especially as these are early targets of the railway promoters. At this early stage, too, the railway system is only skeletal and while plants away from the main lines are protected from direct competition in their local area they do not share in the growth that enables others to extend production, install new plant and technologies and exploit new raw materials. Eventually as the railway network thickens new growth points emerge, points which previously have no particular attraction but which now acquire an artificial nodality. Railway costing and pricing policy becomes important. Terminal costs are relatively more important in relation to movement costs than with packhorses and wagons. Reflecting this, the rates charged per ton mile decline or taper with increasing distance. One long rail haul is cheaper than two of half the distance, and heavy industry is, therefore, encouraged to locate itself either near a major source of its raw materials or perhaps close to a major market centre rather than at some intermediate point. Minimization of transhipment costs also explains why heavy bulk-reducing trades are frequently found at transhipment or 'break-of-bulk' points where if two raw materials are employed the number of handlings involved may be reduced from six to four, or, if the transfer had not been well organized, effectively from eight to four. Alternatively, the desire to reduce the incidence of haulage costs is behind the introduction of ballast rate/freight charges, a low back-haul rate in wagons which have carried one bulk material being a means to secure a return cargo giving some additional revenue. Naturally the ballast rate principle contributes to polarization of development at the originating, terminating, or transhipment point of a bulk transport system. Tapering, break-of-bulk, and ballast rate advantages apply in the movement of low-grade raw materials or bulky products by water as well as by rail.

Another important result of the new techniques of mass movement involves increasing economies of scale. With high-volume output each unit can be turned out with a smaller input of the factors of production, until, beyond a certain point—which may differ for each plant—diseconomies may begin to appear. The plant located near a major market is favoured by the growth prospect this provides. Alternatively, highly efficient company organization, at a time when there are still wide differences between firms in this respect, may so lower production costs that the market area is widened and the scale of operations of a plant can thereby be increased. The new means of transport extend marketing advantages from the local to the regional scale and by lowering transfer costs increase the effect that meticulous attention to efficiency may have. Plants previously isolated now come into competition, and the ranks of producers may be thinned as a result, while new high levels of capacity are reached in the plants of the victors.

CHANGES IN TECHNOLOGY, MATERIALS, AND MARKETS

The change in scale that is assisted by rail or bulk water movement will go ahead most rapidly at a time of sweeping technical change, those units which receive the new equipment being favoured in the competitive struggle. The change from puddled iron to steel was an especially important step. The new material came from equipment of much higher productivity, requiring bigger investment than the old. Fewer firms could afford to build a Bessemer shop and fewer were needed to meet even a rapidly growing overall demand for iron and steel. The early locations of steel-making in any old industrial country were largely chosen from among the old puddled iron plants, but the new material had rather different material and market conditions. It was in particular less fuel intensive, so that major new growth points did emerge.

Changes in the range of practicable transport, in scale of operation, in techniques of manufacture, and in type of product affected the pattern of mineral supply. Low-grade distant deposits, worked in bulk and efficiently transported, now replaced small, and even rich local sources of supply. The geography of markets also changed, largely as a result of independent influences, the evolution of settlement patterns, the progress of urbanization, and the development of industries which were consumers of iron and steel, and for which, although proximity to this material was desirable, other locational influences were more important. In developing countries there is another element in both raw material supply and marketing, the opening of wholly new sections. This involves both the discovery of new mineral supplies, the provision of transport, capital, and labour to exploit them, and the emergence of rural and urban outlets which could not have been foreseen when the older growth points were established. In some parts of such a country explanation of the 'locational relics' poking through the pattern of works is to be found half a century or more before that which applies elsewhere, in a period when the effective national economy was spread over less than half the area.

THE ROLE OF ENTERPRISE

Initiative in economic growth and, therefore, in the shaping of industrial location patterns lies with the individual firm. Not all respond readily or equally to opportunities such as those represented by technical change, the arrival of the railway, or the growth of fresh markets. The form and scale of innovation are important to the success of the firm and to the future stature of a producing point. Some decisions are taken by outsiders but affect the location pattern for the iron industry, as is clearly the case with the policy of suppliers of minerals, consuming trades, railway or shipping companies. As the scale of operations increases so monopolistic conditions may appear in the provision of supplies and services, and the more enterprising company will be induced to take over more of these operations, acquiring its own mines, bargaining with transport companies for favourable rates, buying or building its own facilities to knit its organization into a fully integrated supply chain and even perhaps working up some of its own finished products as in constructional work or shipbuilding. In the nineteenth century the vigour of an organization frequently depended on a very few men, perhaps on an outstanding individual or partnership of which one member was a man of broad vision, the other a practical man, equipped with a knowledge of things or distinguished as a leader of men. This age passed into that of the giant concern where the professional manager was more prominent and the limited company the usual form of organization. In exceptional cases, however, the earlier stage shaped the conditions for future development, so that the outstanding firm of the late nineteenth century could build up plant capacity which with subsequent modifications would survive as a leading centre of production fifty years later. Moreover, then and later the business unit seems to have been subject to a built-in stagnation factor, which, along the lines of W.W. Rostow's 'secular stagnation', would follow a period of favourable growth conditions. Only with difficulty does the son of the founder retain the same drive as his father, or the changing board of directors of a successful company continue to exhibit the willingness to innovate which its predecessors showed when making the first aggressive thrusts.[3] The old concern has plant which has become out of date, with costs of operation probably higher than those of a newcomer. This principle, however, does not apply when the cost of building a wholly new plant is much greater than the cost of re-equipping an old one. In so far as firms and even districts develop at widely differing dates such a stagnation principle has locational significance.

OUTSIDE CONSTRAINTS ON LOCATIONAL CHANGE

Locational change in general proceeds more slowly as an industry grows. Old plants develop their internal and external economies and are modernized and extended. As the scale of operations grows and integration increases, so depression and the possibility of cut-throat competition for business threaten the success of the heavily capitalized firms. This encourages co-operative action among the industry's leaders to provide conditions in which all may survive and prosper. Industry-wide devices to this end are limitations of make or a stabilizing pricing policy. In these conditions old locations survive longer than they would do under unbridled competition, but secure prices encourage fresh entrants, so that new plants are established in better locations. Eventually the new firms may become so important as to secure a modification of the old stabilizing policies. Such a process will be accelerated by technical changes.

Government policy provides another constraint in the working out of any locational pattern. A protectionist policy or a trend to free trade differentially affects the competitive position of various works by opening or excluding foreign raw materials, markets, or competition. Another aspect of government intervention is its indirect role via the work of various branches of the executive arm—improvement of waterways and port channels, regulation of railways, policies concerned with the development of agriculture or regional economic support. Finally, government may interfere directly, as for instance in trust or pricing policies, in supporting construction of new plants, and more specifically in laying down rules as to their location for either strategic or socio-economic reasons.

Shaped by a host of factors, some quantifiable, others said to be imponderable—because they either cannot be or have not yet been quantified—the end product is the location pattern of the industry today. This in turn may provide another starting point for study. It is measurable, plants and flows of material and finished products can be mapped, and the pattern will pose many of the questions for which broad answers may be sought in the hypothetical situations sketched above. These may now be summarized from a slightly different viewpoint.

Models of industrial location may be considered in different dimensions and activated under various sets of rules. The simplest location model is two-dimensional, the only variables being cost/distance. A third dimension involves differences in what may be described as the intensity of factors previously accepted at face value. Variations in scale and technical excellence of plant result in a range of costs of production, and freight charges are similarly affected by the economics of bulk transport, which in turn reflect the ease of reaching and serving supply and market areas. The time element is a fourth dimension, which transforms the box-like three-dimensional figure into a series of boxes overlapping, or more accurately with each one growing out of its predecessors. The effects of changes of scale and technology, the role of enterprise and stagnation, alterations in the patterns of material supply and consumption are among the multitude of considerations at this level of examination. The locational boxes may be made up of a range of possible materials—the mineral supply situation, the structure of demand, and the particular social set-up of individual countries. There were common technologies but widely varying patterns of growth in the metallurgical industries of Britain, Germany, and the United States in the nineteenth century. All these were capitalist economies, but, as the Soviet Union has shown, the box may be constructed according to another set of rules. The resulting location pattern differs in principle, not merely in kind, from that of a capitalist society.

Finally, it may be recalled that the image of the box, of an empty box, was used by Clapham to show the need for realistic studies, dealing with particular circumstances, processes, and firms, to balance the work of the economic theorist.[4] Locational models or locational theories in geography, too, need to be brought down to earth. The complications of reality have to be examined. It is true they provide no substitute for the approach of the theoreticians, for without this the result is likely to be an ill-sorted heap of locations, events, or personalities. On the other hand they are the essential counterpart, providing the test of experience. The physical, economic, social, and political conditions of the United States constituted a testing and then for many decades a uniquely favourable framework for the development of iron and steel manufacture.

THE AMERICAN CONTEXT FOR THE GROWTH OF IRON AND STEEL MANUFACTURE

United States conditions for iron and steel manufacture have always been very different from those in Europe. These differences may be summarized under four headings, although they are all interconnected—the institutional framework, time of development, spatial setting, and scale of growth. Throughout its history the industry has grown with a minimum of direct government intervention. In other countries government action has been much more important, sometimes in the initial stages of growth, as with Japan, more frequently as the economy has matured, as with Italy, or since the mid-thirties, Britain. There have been exceptions even in the United States. Since just before the start of World War II, there has been some intervention for strategic purposes in the location or financing of new plants. In pricing policy, whether in the industry, as with basing point pricing, in the transport of products, or in antitrust operations, government has intervened over a longer period. Less directly, government action has been of great importance throughout the period since the Revolution. As opposed to the free-trade policy under which Britain's archetypal capitalism grew, American manufacturing was built behind tariff barriers. Insulation was never completely effective, but locational change in much of the country and for much of the time has not been complicated by foreign trade. In the west and on the Gulf coast, and more generally in the years to the 1870s and since the late 1950s, foreign competition has been significant.

The time of development of major iron and steel capacity in the United States paralleled or even preceded that of Germany but came long after the first flush of British development. American pig iron production in 1850 was one fifth that of Great Britain although the populations of the two countries were approximately equal. In 1874 output was still no more than 41 per cent of Britain's; by 1895 it was 19 per cent greater. Britain was passed in steel only in 1890, but by 1906 American production was almost four times as great. Growth on this scale allowed a greater choice of new techniques, an opportunity to enjoy all the scale economies they permitted, and development of locations more suited to changes in raw material supply, marketing, and process conditions than in Britain. Late development allowed the unique American physical setting to have its full effect on the pattern of steel-making.

At the time of settlement the eastern United States had a broad cover of woodland which provided most sections with charcoal fuel. Charcoal prices rose with land clearance and competition from other fuel users. East of the Allegheny Front large coal resources are confined to the anthracite basins of eastern Pennsylvania. In the Appalachian plateau and westwards to the Mississippi are the much bigger reserves of bituminous coal, some of which later proved of great excellence as coking coals. The east has scattered iron ore deposits and a few bigger reserves like those of the Adirondacks or the Cornwall ore banks of eastern Pennsylvania. To the south the bigger ore fields of Alabama were unknown until the mid-nineteenth century. The rich iron ores fringing the Ozarks were proved in the 1840s at about the same time as the Upper Great Lakes ranges were discovered, though the latter were then beyond the fringe of settlement. Already the United States was recognized by outsiders as uniquely well endowed with minerals. As Sir S. Morton Peto put it in 1866, '... I could not fail to be struck, as a practical man, with the extraordinary and wonderful character of American resources, surpassing by far anything of which we have the slightest experience in the old world, great as are our own products, and remarkable as is the industry of our teeming population.[5]

The obstacles to the full development of these resources were serious. The most important was the physical problem of distance, and intimately linked with it was the vital psychological problem of a willingness to think in big terms. One unique material asset in overcoming the distance problem was the existence of the Great Lakes, providing a potential link between the lake ranges and the developing agricultural zone and the coalfields. Development of this resource required immense investment of ingenuity and finance, in establishing harbour works, specialized shipping, canal links, navigation aids, and linking land transport. In the agricultural belt itself economic growth justified the filling out of the railway system into a true network in the 1850s. Finally, westward movement of settlers and the growth of new communities provided the demand that encouraged ironmasters to seize these opportunities. A few years later the arrival of bulk steel-making with its more concentrated operations provided another incentive for the growth of large-scale organization both in production and in supply. In this context there gradually emerged a distinctive American production psychology.

There is little doubt that this attitude to output and productivity was unique but it remains questionable whether it was not merely the result of a willingness to grapple with equally unique challenges and opportunities. Earlier the British pioneers of puddling, of rolling, of bulk steel-making had dealt with situations similar in nature but different in scale, and as late as the Paris Exposition of 1867 Abram Hewitt emphasized the backwardness of American techniques in many respects. Yet already *The Economist* was foreseeing the way things would go and some of the wider implications too: 'As soon as America is densely peopled, to America must both our iron and our coal supremacy—and all involved therein—be transferred, for the United States are in these respects immeasurably richer even than Great Britain.'[6] Gradually the British recognized their own backwardness, although they did not always identify its true cause in the situation of a maturing, relatively slow-growing industrial economy. American writers in turn by the end of the nineteenth century frequently overlooked the differences of circumstance that helped to account for British tardiness, not only in material circumstances but also in attitudes. At this time Henry Campbell of the Pennsylvania Steel Company described an outdated industrial structure in the former home of capitalism: 'In England there is a tendency for the management of an enterprise to descend from father to son, and this must retard the advancement of progressive young men. There is also an opposition to change, a magnifying of every tradition into a law of nature, and a disinclination to be different from others.'[7] The generalization was too sweeping but there was much truth in it. One result was a change in the wider international pattern of steel location with British output growing much more slowly than that of its chief rivals.

Nurtured in its unique institutional and physical setting the American industry avoided any taint of the same disease for another half-century. In 1945, at the peak of their pre-eminence, American mills made over 62 per cent of the world's steel. By the 1960s, however, it was clear that a new era in the world steel situation had arrived. Carriage of minerals in bulk carriers over distances now almost global, revolutionary new steel-making techniques and new patterns of growth in the general manufacturing economy of the world again threatened the hegemony of old leaders. In the newly industrializing economy of nineteenth century America iron and steel made a most successful adjustment to the challenges presented by economic geology and economic geography. Now that industrial economy has in turn matured and steel-making must adjust again. Massive investments in new steel plant and in mineral processing, and vocal protests about unfair foreign competition are different aspects of the process of adjustment. Internal location change is taking place though much more slowly than in the nineteenth century. On the world scale the change is occurring more rapidly. By 1968 the United States made only 22·5 per cent of the world's steel. In 1971 for the first time since 1890 her output of steel was exceeded by another nation when the U.S.S.R. pulled ahead.[8]

[1] A. Marshall, *Economics of Industry*, 1919, pp. 166–7.

[2] A Lösch *The Economics of Location*, 1954, Part II, Economic Regions.

[3] W. Rostow, *The Process of Economic Growth*, 1953, pp. 71, 107, 108.

[4] J.H. Clapham, 'On Empty Boxes', *Economic Journal*, 1922.

[5] Sir S.M. Peto, *The Resources and Prospects of America*, 1866, p.6.

[6] *The Economist*, 6 Jan. 1866.

[7] H.H. Campbell, *The Manufacture and Properties of Iron and Steel*, Hill Publishing Company, New York, 1907, p.422.

[8] See also W. Isard, 'Some Locational Factors in the Iron and Steel Industry since the early Nineteenth Century', *Journal of Political Economy*, 56, June 1948; W. Isard and W.M. Capron, 'The Future Locational Pattern of Iron and Steel Production in the United States', ibid., 57, Apr. 1949, P.G. Craig, 'Location Factors in the Development of Steel Centres', *Papers and Proceedings of the Regional Science Association*, 3, 1957.

CHAPTER 2

American Iron Production to the Mid-Nineteenth Century

DEVELOPMENT TO 1850

In 1850 United States population was 23 millions. The population of the British Isles was 27 millions, of whom 6·5 millions lived in a still united but overwhelmingly rural Ireland. American pig iron output was 0·62 million net tons, that of British 2·75 million. Most British iron had been made with mineral fuel for three quarters of a century, and Cort's process of puddling and rolling with coal had focused the industry in big units on the coalfields and had reduced costs of production, so that British iron was competitive in all coastal markets of the world. About half of American pig iron was still made with charcoal. Puddling had begun in Pennsylvania only in 1817 and in New England in 1835. American development was still 'a tale of local markets supplied by comparatively small furnaces', and as entry to such a trade was easy so, too, elimination was frequent. 'Tiny ironworks every-where, but particularly in Pennsylvania, with poor equipment and an un-economic force of men, passed rapidly from birth to death; they rose fluttered and fell like May flies.'[1] (Fig. 1) There were important concentrations of iron manufacture near big markets, as in the immediate hinterland of Philadelphia, or near high-grade ore, as in south-western New England, but in general both the materials and market situations pointed to the advantages of scatter. Wood fuel made smelting dependent on extensive use of the land so that, as F.W. Taussig later said, the industry was then much more analogous to agriculture. Whereas supply of charcoal for a blast furnace might annually require the produce of 2,000 to 5,000 acres of woodland, a six-foot coal seam covering half an acre could provide an adequate substitute.[2] On the demand side most of the outlets were scattered, either in rural communities or in small agglomerations, for, as Zachariah Allen had put it in 1829, 'the manufacturing operations in the United States are all carried on in little hamlets, which often appear to spring up in the bosom of some forest, gathered around the waterfall that serves to turn the mill-wheel. These villages are scattered over a vast extent of country, from Indiana to the Atlantic and from Maine to North Carolina, instead of being collected together, as they are in England in great manufacturing districts.'[3]

There were various explanations of the smallness, slow growth, and technical backwardness of the American iron industry. They may be summarized under four headings, competition from Britain, high labour costs, lack of mineral fuel and transport difficulties in a widely dispersed economy. Much has been written about the tariff issue.[4] Coke smelting and puddling were already cutting British production costs when Hamilton's Report on Manufactures singled out iron manufacturers as 'entitled to pre-eminent rank in protection'. In 1814, in the war with Britain, American works made about 80 per cent of the bar iron the country consumed.

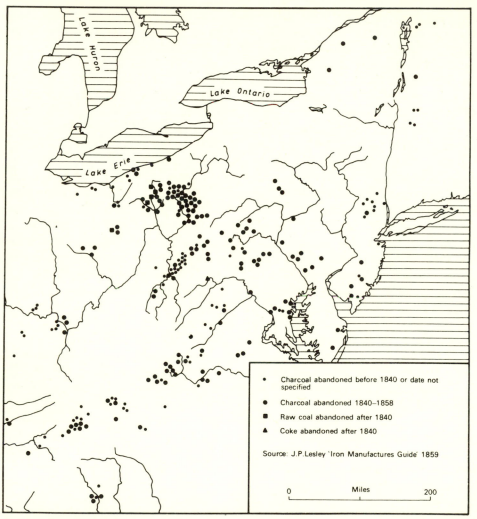

1. Abandoned blast furnaces 1859

Thereafter British iron again flooded in, and by 1817 imports supplied 58 per cent of consumption; home production had fallen to 38 per cent of the level of three years before. The tariff levels of 1816 and 1818 failed to check the flood and they were increased by the Acts of 1824 and 1828. Tariff protection was gradually reduced in the 1830s, but with general economic growth substantial increases in production occurred. The Tariff Act of 1842 reintroduced high levels of protection. In 1846 duties were lowered and by 1850, with falling prices abroad, the *ad valorem* duty ceased to give much shelter and the industry experienced a severe crisis, pig iron production in 1850 being less than 74 per cent of the 1846 level.

The effects of protection on the growth and character of the industry have been viewed from very different standpoints. Even with a high tariff American ironworks were unable to supply all the needs of the home markets, failed to adopt most British innovations with alacrity, and continued to operate small units of production. It could of course be argued that this was because protection was insufficient. There were difficulties in this case, however. For instance, from 1849 to 1854, while imports of pig and rolled iron rose to very high levels, home production stagnated. American Iron and Steel Association statistics suggest that the 1854 output was only 7,000 tons or 1·13 per cent more than in 1849. Yet in this period production of anthracite iron, though concentrated in areas exposed to foreign competition, went up, according to A.I.S.A. figures, by 192,000 tons or 167 per cent. New methods and progress here were not incompatable with keen foreign competition and relatively low tariff protection. Modernization may be speeded if competition is moderate but impossible if it is extreme; it was charcoal ironworks which found it impossibly keen. On the other hand Free Traders went to the other extreme by claiming that protection removed incentive to adopt better processes or the use of mineral fuel in smelting.

Some found the explanation of this slow growth of American iron production in high labour costs. 'If we had more of poverty, more of misery, and something of servitude, if we had an ignorant, idle, starving population, we might set up for ironmakers against the world' was how Daniel Webster put it to Congress in 1824.[5] The obvious remedy, which was eventually to render the American iron and steel industry highly competitive, was to substitute machinery for men. This had already been done in Britain and failed to proceed rapidly in the United States only because of circumstances which limited the scale economies which alone justified high capitalization. Questioned about the effect of recent coal and iron ore discoveries in the United States in 1840, Sir John Josiah Guest of the great ironworks at Dowlais, South Wales, a large supplier of rails to American lines, repeated the high-wage argument but edged a little nearer to a fuller view. 'I have conversed with persons who know the United States very well, and the general opinion is that it will be some years before they will do much with that ore to affect this country, in consequence of the high rate of wages and the coal not being adapted to make iron.'[6] Use of anthracite in smelting had begun but the limitation in its availability ensured that it alone could not revolutionize the iron trade. Nine years earlier than Guest, Gallatin, in a memorial to the Free Trade Convention, had pointed the way of advance for the American iron trade: 'A happy application of anthracite coal to the manufacture of iron, the discovery of new beds of bituminous coal, the erection of iron works in the vicinity of the most easterly beds now existing, and the improved means of transport, which may bring this at a reasonable rate to the seaboard, may hereafter enable the American ironmaster to compete with foreign rolled iron in the Atlantic districts.'[7] The key to the whole development, which would render the high wages of little account, justify the installation of machinery and bring scale economies undreamed of in Britain, which would open up distant ore fields, coal deposits, and markets, and in the light of which the tariff question was later to be seen as an irrelevance, was the organization and conquest of space, the development of efficient, long-distance bulk transport.

In the Tariff debates of 1824 Daniel Webster had pointed out that sometimes freight on Swedish iron to Philadelphia was as low as $8 a ton, which, as he put it, was equal to the cost of 50 miles land transport in the United States, 'Stockholm, therefore, for the purpose of this argument may be considered as within 50 miles of Philadelphia.' The example was colourful but inexact, for conditions of transport varied widely in different parts of the country, and even before the War of 1812 in areas near the east coast or eastern rivers road building was already widening the radius of circulation. Haulage costs by all-land routes from Philadelphia to Pittsburgh were then as much as $125 a ton.[8] By mid-1818, when hammered bar iron was $90 to $100 a ton in American sea ports, the price was $190 to $200 in Pittsburgh. Even after the opening of the Erie Canal, $40 a ton was the ruling figure for carriage between the Atlantic seaboard and western Pennsylvania.[9] Distance protected the interior producer but the same isolation forced small-scale operations on him. In the 1820s and still more the 1830s transport development began to lay the foundations for different conditions of manufacture.

Canal-building followed road-building and was even more important in changing space relationships and the economic significance of distance. For instance, in 1824 the New Jersey legislature approved the construction of the Morris Canal from the Delaware near its confluence with the Lehigh through the hills of the northern part of the State and on to Passaic. Its promoters were disappointed in their hope that Paterson might become the 'Birmingham of America' but the canal revived the decayed ore and iron districts of the State and provided a new source of minerals for the ironworks of eastern Pennsylvania.[10] By 1830 the U.S.A. had 1,343 miles of canal and navigations with another 1,828 miles under construction. The respective railroad figures were only 44 and 422 miles.[11] In the 1830s canals were being built in the Lake Erie to Ohio River belt, but the east was already well provided with them and had almost all the railroads built before 1840. Here, too, was the most concentrated market for iron, and mineral fuel and rich deposits of iron ore were in closest proximity. Six years after its introduction in Scotland the hot blast was used in a charcoal furnace at Oxford, New Jersey. This was the technical prerequisite for the successful use of anthracite fuel in iron smelting, and in 1839 Pioneer blast furnace, at Pottsville in the Schuylkill valley, was the first successful anthracite iron plant.

Over the next few years the anthracite coalfields and the lower sections of the valleys of the rivers draining them were the setting for the first great boom in American iron production. By 1854 the British Commissioner to the New York Exhibition recognized that Catasauqua, Allentown, Phillipsburg, and other anthracite furnace plants represented a new dimension in the American iron trade. 'These are all large concerns, and are constructed upon a scale and a system quite equal to the best in this country.'[12] Ores and fuel were now carried over longer distances, regional as well as local markets were supplied, and in this setting there developed the first indications of the production psychology that was to animate the whole American industry by the time the anthracite iron district had become a backwater of the metallurgical world. As J.P. Lesley put it when considering the success of the Hokendauqua furnaces in 1859, 'It is evidently no game of chance, but a trial of practical wisdom based on experience and insured by the improvement of all the means at the disposal of man. . . .' Puny though they were as compared with the industrial agglomerations along the Monongahela, crowded into the Cuyahoga flats, or sprawling out into the dunes and scrublands of Calumet half a century later, the anthracite iron works were the prototypes of the new American iron industry. In the 1850s most districts still provided little indication that a threshold was being crossed.[13]

IRON DISTRICTS IN THE 1850s. I. EAST AND SOUTH

Leading characteristics of the mid-nineteenth century American iron trade were its wide scatter, rapidly changing pattern, and lack of integration. Except for Rhode Island, Delaware, and Mississippi, every state east of the Mississippi had at least one furnace plant at the end of the fifties and there were numerous and equally widespread abandoned furnaces. (Fig. 2) Some catalan forges or bloomaries were still making crude wrought iron direct from ore scattered over a hearth of charcoal. Their extremely high consumption of charcoal and the need to minimize the amount of slag that subsequently had to be hammered out of the iron which they made, concentrated them in remoter parts where fuel was abundant, ores were rich, and where their low productivity was unimportant. Some were in high inaccessible locations exposed to suitably strong winds for the blast. The old plant at Mount Riga on the Massachussetts/New York boundary, for example, was 1,000 feet above its iron mines. There were clusters of bloomeries in interior North Carolina, in central South Carolina, in the northern Adirondacks and in a belt west of Lake Champlain. Another important group survived in the uplands of northern New Jersey. In each, as economic growth occurred, bloomaries were replaced by furnaces. Blooms

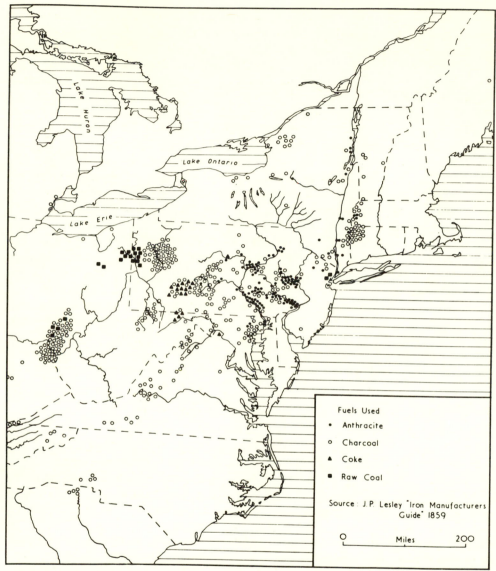

2. Blast furnaces 1859

from the Lake Champlain group were marketed in the big rolling mill centre
of Troy and in the lower Hudson Valley.[14]

Integration between blast furnaces and rolling mills was markedly
lacking.[15] Most of the biggest mill centres, including south-eastern New
England, Troy, Pittsburgh, and Wheeling had no neighbouring blast furnace
plants (Fig. 3). There were many reasons for this. Firstly, process integration
was much less important than later, for mill puddling furnaces invariably

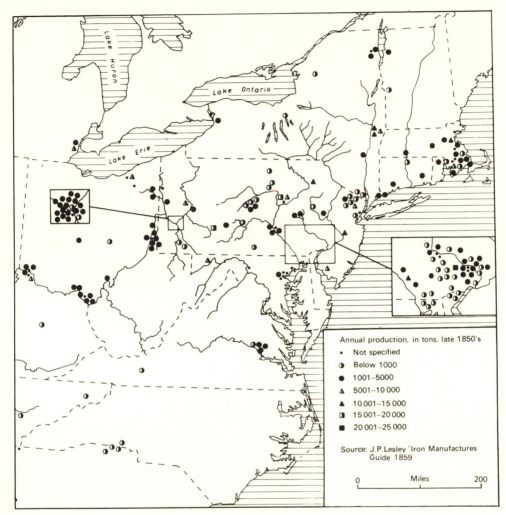

3. Rolling mills of the eastern United States 1859

worked on cold iron. Secondly, the mills used forge blooms and bars as well as puddled iron, that is both blooms and bars hammered from bloomary iron as well as those made from furnace pig iron. For the forge producing blooms and bars water power was an important localizing agent, whereas the puddled iron works and mills had high fuel consumption and were located near the coalfields, in great consumption centres or at key points in the transport network leading to major markets. Finally, many blast furnaces served local rural demands directly with simple castings for hardware, implements, or machinery. The more accessible of these would also deliver some iron to the bigger mill centres. In 1839 pig iron from the furnaces near Lancaster, Pennsylvania, was valued at $30 a ton, furnace castings $50 a ton, and bar iron, made up from their pig, $65 a ton.[16]

Under the 1832 tariff British railroad iron flooded into the United States but heavy duties on pig iron remained; in the next eight years pig output went up by an amount estimated variously at between 43 and 73 per cent.[17] Demand fell in the early 1840s but then the substantial duties placed on railroad iron by the 1842 tariff stimulated the growth of a domestic heavy rail industry, and pig output revived. New charcoal furnaces were built, nine in 1843, 23 in 1844 and 79 in the next two years. In the following two years, however, with output at a record 0·8 million tons, far beyond the dreams of ironmakers a decade before, construction of new charcoal furnaces fell to 34 and 28.[18] Anthracite iron was becoming highly competitive, although between 1844 and 1849 the increase from anthracite furnaces was only 50,000 tons, while even in the depressed conditions of 1849 national iron output was some 250,000 tons greater.[19] Then the rate of increase in production from mineral-using furnaces accelerated. A.I.S.A. figures indicate that whereas 1854 production of iron was almost identical with that of 1849, anthracite iron production increased by 185 per cent and its share of the total from 16·6 to 46·8 per cent.[20] Analysis of the American iron industry in the east in the 1850s is largely concerned with the process of adjustment to the new technology of smelting with mineral fuel. Through the new improved means of transport the influence of anthracite iron spread gradually outwards.

THE ANTHRACITE IRON DISTRICT

In 1840, in order to increase outlets for its coal, the Lehigh Coal and Navigation Company followed the Pioneer furnace into anthracite iron manufacture. By 1842 twelve anthracite iron furnaces were working in Pennsylvania. In 1847 there were forty-one and by 1856 ninety-three. The Lehigh, Schuylkill, and Susquehanna valleys remained the centres of the trade, the works concentrated not on the coalfield but in the middle and lower river courses, because the balance of raw material consumption was heavily in favour of ore rather than coal location. At the end of the 1840s, for every ton of iron the two Phoenixville furnaces consumed 1·14 tons of limestone, 2·62 tons of ore, and 1·86 tons of anthracite. In 1864 the ratio at the Bethlehem Iron Works was 1·21, 2·24, and 1·91.[21] The ore resources of the anthracite fields proved disappointing. Little was found and even that usually occurred in thin beds, was enclosed in a hard matrix, and was highly siliceous. The silica content was estimated in the 1850s to average 28 per cent, as compared with 18 per cent in the ores of the Pennsylvanian bituminous coalfield, necessitating higher consumption of flux. At Scranton, ball ore from the anthracite coal measures carried only a quarter of a mile to the furnaces cost $3·50 a ton, whereas fossiliferous ore delivered from Clinton County New York was only $3 a ton. Nearby 'mountain' ore was also used. Limestone consumption of two tons for every ton of iron indicated one advantage of coal measure ore, for the limestone, brought

60—70 miles from below Berwick on the north branch of the Susquehanna, cost $1·50 a ton at the furnaces. At Pottsville excellent iron was made from 26 per cent Fe coal measure ore, but the cost of raising was $1·50 a ton, about $6 per ton of iron, and by the late fifties working there had ceased in favour of richer ores from the east.[22] Even so fuel for puddling and mill work was cheaper in the coalfield.[23]

The lower valleys of the rivers draining the anthracite fields were attractive to the ironmaster partly because there were deposits of iron ore, established charcoal works, and markets, both in local mills and foundries and in Philadelphia, the chief iron market of the nation. Anthracite was brought down river, canal, or railroad to the works. In this area and beyond it too, distribution of some bedded ore bodies closely affected the location of the iron industry. An outstanding example over a wide area was the so-called Fossil ore, whose outcrop followed a sinuous course in the valleys of both the north and the south branches of the Susquehanna and the Juniata. Anthracite furnaces were built along or near its outcrop, as near Montour, Williamsport, at the mouth of Bald Eagle Creek, and in Mifflin County in the Juniata basin.[24] However, the most important supplies of ore came from the brown hematites of south-eastern Pennsylvania and the magnetites of New Jersey and of the Cornwall ore banks in the Lebanon Valley. The furnaces around the Cornwall ore workings were the most important exception to the river localization of furnaces in the late fifties; this area was well served by turnpike roads, the Union canal, and the Lebanon Valley Railroad.

By 1850 magnetite from New Jersey could be delivered on the Lehigh or the Schuylkill for about $3 a ton, the former, being closer, having a slight advantage. Yet although excellent iron could be made from magnetite ore alone, this practice damaged the furnaces. Consequently, each Lehigh furnace mixed with it a more than equal tonnage of hematite ores, from south or south-west of Bethlehem and Allentown. Furnace damage was reduced but production costs were raised and the iron was poorer. This was indeed the familiar problem associated with a new technique, delaying its impact on older practice. It helped preserve a market for the finer quality iron of the charcoal furnaces. Fairbairn in 1850 blamed the persistence of old ideas—the unsuitable form of the furnaces, inadequate ore preparation and the overuse of flux. There was '. . . no change or shadow of alteration from that everlasting "ton of limestone" which appears to be thrown into every iron furnace in every place and region in the United States'.[25] Limestone consumption in fact varied widely from furnace to furnace and possibly Fairbairn's diagnosis was wrong. In any case the anthracite iron companies overcame their difficulties sufficiently to use large tonnages of New Jersey magnetite.[26] By 1864 Lackawanna ironworks in Scranton, the chief single exception to the general rule of location away from the anthracite mines, raised 58,000 tons of ore at Mount Hope mines near

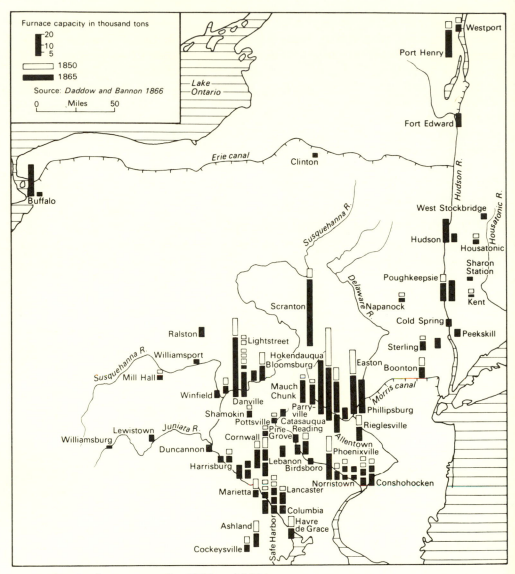

4. Anthracite furnace capacity 1850 and 1864

Morristown, New Jersey, an amount more than equal to its output of 28,000 tons of pig iron.[27]

Within the coal valleys of eastern Pennsylvania, the works of the Lehigh early became and remained pre-eminent, their share of combined output going up from 37·2 per cent in 1849 to 38·7 per cent in 1864 and 42·1 per cent in 1871.[28] (Fig. 4) Lesley listed Lehigh furnaces which made almost 84,000 tons of iron in 1856/7 and he omitted the output of one or two others. One early disadvantage there was the lack of finishing capacity. There were then only two rolling mills in the valley, at South Easton and at Weissport, and together they produced no more than 1,544 tons of rolled iron. The remaining iron had to be shipped to Schuylkill or Susquehanna mills or to still more distant markets. In 1855, however, the Lehigh Valley Railroad provided improved access, and from 1856 to 1865 new furnaces of 60,000 tons capacity were built in the valley, but only 22,000 tons of new plant on the Schuylkill. There followed a substantial growth of Lehigh valley finishing trades. In 1860 the Allentown Rolling Mills were built, in 1864 the Catasauqua Manufacturing Company established mills in Catasauqua and Ferndale, and in 1863 a new furnace, puddled iron works, and rail mill began work at Bethlehem.

COMPETITION FROM THE ANTHRACITE FURNACES

From an early stage the effect of this new, highly productive industry began to spread outwards, although the early figures are inexact (Table 1).

Table 1. *Productivity and capital in new Pennsylvanian anthracite and charcoal iron capacity 1842—1846*

Number of new Furnaces		Annual productive capacity (tons)	Annual productive capacity per furnace (tons)	Average capital investment per furnace	Average capital investment per ton of annual Product
Charcoal	67	75,200	1,112	$52,746	$47
Anthracite	36	103,000	2,861	$71,527	$25

Source: Extracts from the Report of a Committee to the Iron and Coal Association of the State of Pennsylvania 1846, quoted in *J.F.I.,* 3rd Series, vol.XII, 1846, pp.124 ff.

By the mid-fifties charcoal iron still retained its quality advantage but productivity favoured anthracite much more than in the early or mid-forties (Table 2). Many of the charcoal furnaces were west of the Alleghenies and still largely protected from competition, but the eastern industry was severely affected. National pig production in 1860 was 26 per cent greater than in 1849; output of Pennsylvania anthracite furnaces increased by 315,000 tons or 266 per cent. In 1849 seventy Pennsylvanian charcoal furnaces east of the Allegheny Front made 56,000 tons of iron. Seven new charcoal furnaces were built before 1860 but thirty-two older ones failed and the survivors made only 36,000 tons.[29]

Table 2.   *Estimated anthracite and charcoal iron production*
in Pennsylvania 1842 and 1856
(thousand tons)

	1842			1856		
	Number of furnaces	*Output*	*Output per furnace*	*Number of furnaces*	*Output*	*Output per furnace*
Anthracite	12	15	1·25	93	307	3·30
Charcoal	210	98	0·47	143	96	0·67

Based on J.P. Lesley, *The Iron Manufacturer's Guide*, 1859, p.759 and C.D. King *75 Years of Progress in Iron and Steel*, 1948.

Though the growing anthracite iron output was concentrated in the middle course of the main rivers draining the coalfields, furnace construction spread outwards. Before anthracite furnaces were built to the westward along the southern branches of the Susquehanna and Juniata the new technique was already an important disruptive influence in the Hudson Valley and western New England. Here it offered ironmasters not only high productivity but also a solution to the increasing difficulty of securing good, low-cost charcoal. Woodland was being cleared for farmland, domestic and

Table 3.   *Pig iron, fuel, and iron ore costs 1856–1858 at*
Brandon, Vermont, and at the Duncannon furnaces
near Harrisburg, Pennsylvania
($ per ton)

	Brandon (charcoal iron)	*Duncannon (anthracite iron)*
Total production cost	$19·64	17·61
Cost of ore	c$ 5·0	7·64
Cost of fuel	?	3·95

Based on Lesley, op. cit, pp.539–40 and *B.A.I.S.A.* 28 June 1871, p.343.

factory uses and now railway locomotives provided an additional, large demand for fuel. In the early 1850s near the western frontiers of settlement, around Iron Mountain, Missouri, charcoal for a ton of iron cost about $3·85, and a few years later in western Tennessee $4·40. In parts of western New England it was then as much as $10·50, though sometimes, notably with Salisbury cold blast iron, this cost could be covered in the price of a high-quality product (Table 3).[30] Improved access to Vermont by the Connecticut Western Railroad enabled Salisbury smelters to bring cordwood from a distance, but even here charcoal was more expensive than in the west, and fuel costs overall were higher than in anthracite operations. In 1850 one small anthracite ironworks in the lower Hudson and two between it and the Housatonic were reckoned to have a capacity of 13,000 tons of iron; by 1865 there were ten works with a capacity of 95,500 tons.

Along the North Branch Canal, the Chemung Canal in New York State, and the railroads, coal went to new works at Clinton, south-west of Utica, and to two works at Buffalo.[31] Anthracite smelting appeared early amidst the charcoal furnaces and the surviving bloomaries around Lake Champlain. A rationally located plant was at the outfall of the Morris Canal at Phillipsburg, New Jersey. Cooper and Hewitt had built rolling mills in Trenton in 1845 and in 1847 erected the first Phillipsburg furnace. Magnetite ore came by canal from Andover mine and coal from the Lehigh valley just across the Delaware from the ironworks. At Boonton, at the other end of the Morris Canal and only 18 miles by turnpike from Newark, was another, smaller works. In spite of an important two-way traffic in ore and coal and the carriage of anthracite iron to New Jersey mills to supplement local blooms no other anthracite works were built along the canal. To the south there were two furnaces at Havre de Grace, where the Philadelphia and Baltimore Railroad crossed the Susquehanna, with others at Baltimore Harbour and at two points some fifteen miles out on the North Central Railroad.

These outlying plants were usually linked statistically as the 'Eastern Group'. There were two exceptions—the Maryland furnaces, whose capacity, totalling 25,000 tons in 1865, was listed with that of the Lower Susquehanna group in Pennsylvania, and Phillipsburg, which was included in the Lehigh Valley statistics. The Eastern Group had rather special operating conditions. Some of the furnaces had the advantage of big local markets for iron, as with Boonton or with Buffalo. The Lake Champlain and the New Jersey furnaces were close to high-grade magnetite ore, and Buffalo could employ the new ores from the upper Great Lakes. All were remote from fuel supplies. Some owners reacted positively to this challenge by economizing on fuel. In 1864 the average consumption of coal per ton of anthracite iron was 2·03 tons, but some Hudson Valley furnaces had already reduced consumption to under 1·5 tons.[32] Remoteness and lack of ownership of coalmines, however, penalized these works at times of coal shortage. Thus at the high levels of iron production in 1864, while the anthracite ironworks as a whole worked at 71·3 per cent of rated capacity and Lehigh valley works at 80·1 per cent the Eastern Group operating rate was only 61·7 per cent.[33]

In 1865 coal production fell sharply in the Wyoming division of the field—that is in the Upper Susquehanna Valley iron district. Local iron production fell too but more distant consumers suffered even more. The autumn saw 77 per cent of the furnaces in the Lehigh, Schuylkill, and lower Susquehanna at work but only 16 out of 29 in the upper Susquehanna and Juniata basins and 10 out of the 31 outside Pennsylvania. Coal then cost $12 a ton delivered on the Hudson, whereas the average price for the three months October–December for Schuylkill White Ash Lump coal in Philadelphia was $8·99, and for the whole year the Duncannon ironworks paid only $9·66 for coal per ton of iron.[34]

Within the area of the coalfields and the rivers draining them anthracite iron soon became predominant, although even in the seventies some small charcoal ironworks were still operated for a few months each year by owners who combined iron-making with farming. In the zone beyond, anthracite iron made a severe impact. In these instances the effectiveness of the new competition was not directly related to production levels or to the tonnages of anthracite iron brought in but to the new price structure introduced by anthracite iron. The price of the marginal supply ruined the business of charcoal ironworks. The year 1873 was a record one for American pig iron production at 2·8 million net tons, 224 per cent more than in 1856. In New England anthracite output was in both years insignificant, ranging between 4,500 and 5,500 tons, but even so charcoal iron output increased only from 28,000 to 46,000 tons.[35] By the 1850s the once important industry of the cedar swamps and pine barrens of south-eastern New Jersey had been overcome by competition from anthracite iron, and only at Millville in Cumberland County was a furnace still at work. At least four other locations had been abandoned since 1840. Some idle furnace plants in this area had foundries still expanding, using anthracite pig.[36] Southwards the new technology had severe effects in Virginia. By 1840 there was already a railroad running the length of the eastern lowlands and in the forties a link was pushed out from this westwards to the middle parts of the Valley of Virginia. By Lesley's tour of 1858 only one of the sixteen furnaces east of the Blue Ridge, which had formerly smelted brown hematites, was in blast; half the others had been abandoned since 1840. Between 1850 and 1860 railroad mileage in Virginia increased from 384 to 1,379, with new links from the coast into the Shenandoah lowlands and a line through the whole of that valley. Rich ore beds outcropped along the eastern edge of the low ground whose rich farmlands in turn provided substantial demand for iron. Richmond was a major mill and foundry centre and market for Great Valley iron. Valley works, however, were troubled first by timber shortages, then by anthracite iron competition in their rather distant coastal markets, and finally by direct competition. Only some twenty-one of the furnaces were still working in 1856, and at least sixteen charcoal blast furnaces were abandoned between 1840 and 1858. The situation is illustrated by the fact that whereas rolled iron output increased by over 190 per cent in Virginia in the 1850s, pig iron production fell by almost half.[37]

Beyond Virginia the south was protected by distance. Old techniques, therefore, remained dominant. In 1858 Georgia, Alabama, and the Carolinas had fifty-six bloomaries, over a quarter of the national total, partly a reaction to the rich ores of the central Carolinas but partly also a reflection of backwardness of the region's technology. Charcoal iron firms, protected by absence of rail links with the northern network, maintained their production but remained inefficient. In 1856 the average product of an American charcoal furnace was put at 838 tons. Pennsylvania averaged only 673 tons, but south-eastern levels ranged from 498 tons in Alabama and 376 in South Carolina to 150 tons for the three furnaces in North Carolina.[38] Fuel consumption was high. A few years later Daddow and Bannan put the best yield of an improved hot blast charcoal furnace at 1 ton of pig iron to 3 cords of charcoal, but the best they knew of in the south was 1 ton to 5 cords and the worst, in Alabama, an almost unbelievable 1 ton to 30 cords, at which the bloomary was the equal of the blast furnace.[39] Small-scale operation, poor techniques and communications, and, the core of the whole complex of problems, inadequate demand frustrated all attempts to exploit the sometimes rich southern mineral endowment.[40] Later Swank recalled with wonder that at this time bar iron was still used as a local currency in Johnson and Carter counties in the hill country of Tennessee. Nearer the time Daddow and Bannan saw little evidence of improvement of iron manufacture by the use of mineral fuel: 'We notice Professor Tuomey's remarks about the *cheapness of diving* for coal in the waters of Alabama, and we have no doubt any mode of mining would be cheaper than those generally practised.'[41] While the south-east languished in isolation, west of the Allegheny Front the iron industry was responding positively to the challenge of anthracite smelting. Here there were also independent reasons for change.

IRON DISTRICTS IN THE 1850s. II. CHARCOAL AND THE ESTABLISHMENT OF COKE SMELTING IN THE WEST

By the 1840s, and still more the 1850s, economic growth was clearly moving westwards. The population of Ohio, Michigan, Indiana, Illinois, Missouri, and Iowa, only 12·5 per cent the United States total in 1850, was 25·5 per cent by 1860. In the 1850s the increase in these states was 35·3 per cent of that for the whole nation. In the early 1840s Middle Western railroad construction had scarcely begun and even in 1846 Ohio and Michigan together had only 367 miles of track. By 1850 these two, plus Indiana, Illinois, and Wisconsin, had 1,276 miles, and by 1860, 9,583 miles.

The demand situation was much more favourable than in the south. In the late 1850s Lesley listed twenty-one counties in the United States with an annual output of rolled iron of more than 10,000 tons. Twelve were eastern, one was in the south, centred on Atlanta, the other eight were west of the

Alleghenies. In 1829 and 1830 rolled iron production in Pittsburgh averaged 8,000 tons. By 1856/7 three Pittsburgh works alone, Sligo, Etna, and Juniata mills, which had made 8,000 tons of rolled iron in 1841, had increased their output to 20,000 tons. In addition 12 more rolling mills were built immediately adjacent to Pittsburgh. By 1858 the total annual make in Allegheny County was about 90,000 tons, almost three times that of the next county, Rennselaer County, New York, centred on Troy.[42]

Growth on this scale, combined with bulk transport of raw materials by rail, should have called into being large mineral-fuel-using blast furnaces like those in eastern Pennsylvania. For many years this did not happen. This provides a puzzle, significant in all the major mill centres of the west but for several reasons especially critical in relation to Pittsburgh. Firstly, Pittsburgh was the outstanding western mill centre, having over a third of all the production of rolled iron west of the Alleghenies. Secondly, Connellsville, later the greatest coking coalfield in the United States, was less than sixty miles away, nearer than the anthracite mines were to the mill centres of the lower Schuylkill or lower Susquehanna. Thirdly, Pittsburgh was already a great railway centre, the true 'gateway to the west', as Bishop still called it in the mid-sixties, so that marketing the increased output that coke pig iron might have supported presented no especial difficulty. Meanwhile, delivery of iron for the rolling mills from a host of small furnaces was costly. Finally, coke smelting was already well known long before the hot blast made the use of anthracite feasible, and had been dominant in England for three-quarters of a century.

The reasons for not using mineral fuel have long puzzled students of the American iron industry, and the various views have been re-examined by Temin.[43] He shows earlier writers attributing the delay to abundance of charcoal, ignorance on the part of iron buyers or of ironmakers, lack of suitable coal or the protection to old technologies provided by the 1842 tariff. Many years ago L.C. Hunter suggested that the concentration of western demand on bar iron suitable for easy working by the blacksmith and on high-quality iron, and especially of low sulphur content, was a dominating influence, for Connellsville coke sulphur content was double that of average anthracite. Temin follows his survey of previous views with his own analysis, in which he inevitably has to struggle with inadequate cost data. This inadequacy demands caution in forming firm conclusions. For instance, Temin suggests fuel cost per ton of pig iron in anthracite furnaces of $7 a ton when the iron was selling in Philadelphia for an average of $28 a ton. This fuel figure squares fairly well with that for the Thomas Iron Company as indicated by nine semi-annual figures for 1855-9, but over the same years the figure for Duncannon Ironworks above Harrisburg was $3·95 and for the whole of the 1850s averaged $3·72.[44] Figures for charcoal furnaces seem more doubtful. Overman's average of 180 bushels per ton of iron is accepted, giving a fuel cost of $9, at 5 cents a bushel, but as Daddow and Bannan showed, whereas a good hot blast furnace used only 120 bushels, cold blast operations needed as much as 200.[45] As Temin indicates, in 1849 four fifths of Pennsylvanian furnaces west of the Allegheny front used a cold blast, so that for them, at the same price per bushel, fuel would cost $1 a ton more.[46] Equally important, costs for charcoal varied widely, according to natural endowment and state of economic development. In the early fifties this range of costs was probably greater than random variations from average fuel costs in the anthracite furnaces. In any case these qualifications widen still further the differential between the cost of iron smelted with mineral fuel and that smelted with charcoal. In the early 1850s the two operated in markets largely insulated from each other. Even so, Temin proves conclusively that a large western coke furnace could have made iron at overall costs per ton well below those of charcoal furnaces.[47] There was a lack of large furnaces in the west with accompanying hot blast and strong blowing power, and over large areas poor transport facilities inhibited both the introduction of large-scale operations and the carriage of mineral fuel over long distances. As in the east the key to the breakdown of these conditions lay in improvement in communications, which widened the market, and changed the possibilities of material assembly, and, therefore, the practicable furnace technology. Growing western mobility in the context of rapidly growing demand for iron gave rise to much more concentration of rolled iron production. Secondly, it exposed the inadequacy of charcoal fuel and revealed a possible way out of the difficulty. This point merits further consideration. Attention will be confined to the supply of iron to the mills and foundries of Pittsburgh.

Pittsburgh first rolled iron in 1812 and by 1829 eight mills in or near the town worked up about 6,200 tons of blooms and puddled some 1,500 tons of pig a year.[48] The first western ironworks had been built in 1790 on Jacob's Creek, a tributary of the Youghiogheny River, and Fayette County nearby was the first boom area for furnace construction in western Pennsylvania. Between 1794 and 1815 at least thirteen furnaces were built there, where wooded hillsides, streamside sites, and river or river valley transport to market provided attractive conditions. In addition there were three rolling and slitting mills by 1810, with Brownsville on the Monongahela an important centre. Iron was shipped down the Ohio, and with a linking trade in sugar from New Orleans to Baltimore the circular traffic was completed at that time by caravans of Conestoga wagons bringing provisions across the Appalachian trails. By 1810 11 of the 18 furnaces in western Pennsylvania were in Fayette County, 3 in neighbouring Westmoreland County, 2 in Allegheny County, and one each in Beaver and Butler Counties.[49] Yet, despite large growth of demand from Pittsburgh works, only another 6 furnaces were built in Fayette County, and by the late fifties 15 of the 19 stacks there had been abandoned, all but one before 1840.[50] This occurred while the output of Pittsburgh mills grew from 9,300

28

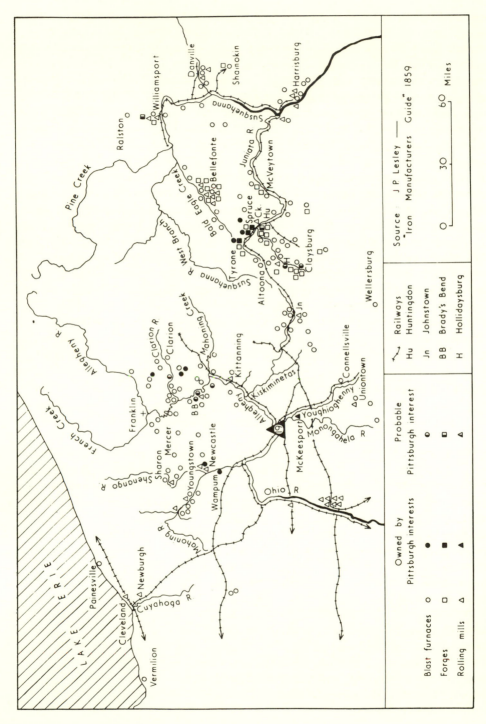

5. Pittsburgh and the iron regions of western Pennsylvania and eastern Ohio 1858

tons in 1829 to 17,700 tons in 1841,[51] and before there was any competition from mineral fuel. Virtual elimination of this ironmaking district resulted from rising costs of furnace materials, especially depletion of timber resources. (Fig. 5)

Another major iron supply source for the Pittsburgh mills was the Juniata Valley of west-central Pennsylvania, whence came some of the families who developed finishing trades at the head of the Ohio. Examples are Lyon Shorb and Company or the still more famous Schoenberger family.[52] Juniata Valley charcoal furnaces worked scattered ore deposits and exploited the huge timber resources of the ridge and valley section of the Appalachians, but even there charcoal supplies began to tighten. Rockhill furnace near Orbisonia, which ran regularly for thirteen years, making 800 tons of pig a year, would in that time consume the wood from about 200 square miles. Some furnaces had to use alternative fuels for part of each year, and of other idle furnaces, Lesley's account mentioned 'blown out for want of charcoal' or 'waiting for a second growth of timber'. By the mid-1830s, with ore still abundant, some ironworks on the Little Juniata had to haul their charcoal 10 or 12 miles 'with great expense and vexation' and were intending to try coke when rail communication made it available.[53] By the late fifties two works in this area used anthracite and there were several coke furnaces.

Table 4.   *Estimated production of pig iron, blooms, and rolling*
mill products by districts in
western Pennsylvania, mid-1850s
(tons)

	Pig iron	Blooms	Rolled iron
Juniata Valley[1]	36,788	11,532	1,660
Bald Eagle Creek[2]	8,446+	2,626	4,204
Allegheny Valley[3]	51,503	—	10,083
Monongahela-Youghiogheny[4]	4,641	—	600
North-West[5]	16,063	—	—
Cambria County	27,685	—	17,808
Allegheny County	—	—	89,779

[1] Blair, Huntingdon, Mifflin, and Bedford Counties.
[2] Centre and Clinton Counties.
[3] Venango, Clarion, Armstrong, and Indiana Counties.
[4] Westmoreland, Fayette, and Somerset Counties.
[5] Butler, Mercer, and Lawrence Counties.
Based on Lesley, op. cit.

These furnaces either supplied iron to puddled ironworks in the Pittsburgh area or hammered it into blooms in local forges and then carried it over the Alleghenies to the headstreams of the Ohio. The Juniata remained an important source until well after the Civil War, but long before that it could not keep pace with growing demands. (Table 4) The partial 1842 returns of Pennsylvania furnaces suggest a Huntingdon county capacity of at least 11,750 tons, and by 1845 three new furnaces are recorded as making 3,750 tons, although this account probably exaggerates production. Twelve years later the output of the county was little over 8,000 tons.[54] On the other hand Blair County had become a much more important source of supply. By this time the whole output of the Juniata Valley and of Bald Eagle Creek beyond their own rolling mills' requirements amounted to less than half the needs of Pittsburgh. Over the preceding quarter-century a third area of supply had developed in the basin of the Allegheny River and in the counties straddling the Pennsylvania/Ohio line.

In this area charcoal furnaces smelted coal-measure ore. Between 1841 and 1846 the number of furnaces in Clarion, Venango, Armstrong, and Mercer Counties increased from seventeen to forty-seven and estimated iron capacity from 14,000 to 48,000 tons, with an additional 5,000 tons at the coke-fired furnaces of the Great Western Iron Company at Brady's Bend.[55]

Table 5.   *Pig iron capacity or production in the counties of north-west Pennsylvania 1841, 1846, 1855–1857*
(tons)

| | Capacity | | Production |
	1841	1846	1855–7
Clarion	7,600	27,800	17,651
Venango	6,400	21,600	6,782
Mercer	—	10,000	5,279
Armstrong	—	3,000	24,568
Total	14,000	62,400	54,580

Based on *J.F.I.,* 3rd Series, vol. xii, 1846, and Lesley op. cit.

Apart from this works the area had almost no finished iron production, the product going mainly to Pittsburgh. Lyon Shorb and Company originated in the Juniata Valley, but, this source of supply to their Pittsburgh mills proving inadequate, they built one furnace in Clarion County in 1836 and two more in 1845.[56] Buchanan furnace built in 1844 sent all its output 100 miles by water to Pittsburgh.[57]

Yet by the fifties output in north-western Pennsylvania in turn was declining, the major exception being Brady's Bend. By 1867 almost all the works in Clarion County had been abandoned, partly because of exhaustion of timber (consumption being at the high figure of 175 to 225 bushels per ton of iron) partly because of increasing hauls on ore, but apparently mainly because of rising extraction costs as drift mines worked their way inwards from the surface, while iron prices were falling (Table 5).[58]

By the fifties charcoal iron production was declining throughout Pennsylvania. In the west and the neighbouring parts of Ohio and Virginia it fell by almost 12,000 tons from 1854 to 1856, and extensions of coal and coke smelting failed to compensate for the declining overall output there, increasing only from 103,000 to 107,000 tons, a total inadequate to supply the mills and foundries of Pittsburgh alone. Over a longer period, centred on the mid-fifties, the decline in charcoal iron production was considerably greater in western Pennsylvania than in the east. (Table 6) The A.I.S.A. explained this by the clearing of woodland for agriculture, the shortage and consequently high price of labour in the north-west of the state after the oil discoveries of 1859, and the competition of iron made with coal and coke.

Table 6.   *Pennsylvanian charcoal iron production east and west of the Alleghenies 1849, 1864 (output in tons)*

| | 1849 | | 1864 | |
	No. of Stacks	Output	No. of Stacks	Output
Eastern Pennsylvania	79	55,617	54	42,953
Western Pennsylvania	85	55,494	9	8,071

Source: A.I.S.A. *Report,* 13 Dec. 1865

This hardly accounts for a decline which began before the second happened and before the third became important. Between 1854 and 1856 coal and coke pig iron production in western Pennsylvania and the neighbouring parts of Ohio and Virginia fell by 3,000 tons, one-quarter of the decline in charcoal iron. Cambria Iron Works at Johnstown, which increased output by almost 18,000 tons of coke iron, was the chief exception.[59] Yet between 1856 and 1858, as Lesley showed, the average Philadelphia price of anthracite iron fell from $25 to $21 and the decline in the average price per ton of charcoal iron was from $30 to $24. In 1852 the Pennsylvania Railroad was opened to Pittsburgh and in 1853 regular quotation of anthracite iron began in the Pittsburgh market. As Temin has shown, even though this supply was marginal in tonnage it had a considerable effect on pig iron prices in the west. For western ironmasters, facing the squeeze of declining charcoal iron prices and rising production costs, the solution lay in mineral fuel, especially as the new emphasis on rolling mill puddling operations rather than direct sales to local blacksmiths made practicable the larger furnaces and necessary ancilliaries and rendered slightly less pure iron acceptable. Before mineral smelting could succeed in the west, however, there were hurdles to be surmounted.

It was early realized that use of coke or uncoked coal presented opportunities for lower costs. In 1849 a report of the Secretary of the Treasury suggested that overall production costs of anthracite iron in eastern Pennsylvania were $15·30 as against Welsh costs of $17·06 a ton, but quoted costs of coke iron in Wales of $14·71 and of raw coal pig in Scotland of as little as $9·75.[60] Welsh and Scottish labour rates were lower than American and in both these countries the ore was mostly still from the coal measures. The position of Pennsylvania anthracite iron, however, as indicated by these figures, suggested that coke iron could prove even more successful and that raw coal might also be important.

Overman in his standard treatise *The Manufacture of Iron,* 1850, is most interesting on the technical and economic context of raw coal furnaces. 'Some experiments have been made in Clarion County, Pennsylvania and some in Ohio with raw coal, which, we understand, have succeeded exceedingly well, but the demand for pig metal is very limited in the western markets, and hence the small difference in the price offered is not a sufficient inducement to substitute it for the charcoal iron at present in general use for foundry purposes.'[61] Yet in the fifties mill iron became more important, demand grew so much in the west that by 1859 Lesley estimated that two-fifths of United States rolled iron was made there, and, as shown above, charcoal iron supplies became tighter as costs rose and prices fell.

In the 1850s the raw coal iron production began in certain areas of the west. Four furnaces were built in the Hanging Rock region on the Ohio and in the Hocking Valley and two at Massillon on the Ohio Canal, but the centre was the Shenango Valley of western Pennsylvania and the neighbouring Mahoning Valley in Ohio, where thirteen raw coal furnaces were active at the time of Lesley's visit. So the Valleys, as the district containing these two rivers was called, in the mid-fifties rolling less than half the iron they made, became important suppliers of iron to Pittsburgh.[62] Raw coal smelting operations, however, were by no means an unqualified success. Productivity was higher than in charcoal furnaces but lower than with anthracite, and output failed to respond to the growth in demand in the mid-fifties. By 1858 the valleys already had four abandoned coal furnaces and, significantly, ten of the seventeen active or standing furnaces had been built in the years 1845 to 1848. There was to be further growth, but experience already suggested that major expansion of low-cost iron supplies would not be found through using new coal. It was at this point that, belatedly, coke iron became significant.

Overman was puzzled that coke furnaces could not prosper west of the Alleghenies.[63] There had been many attempts, but the two biggest works, Brady's Bend and Mount Savage, Maryland, were idle when he wrote. Coke smelting had been tried in Armstrong County in 1819, and for a month in 1835 Mary Ann Furnace in Huntingdon County used coke made from Broad Top coal, but sulphur content was too high.[64] Success was achieved at Lonaconing furnace in the Frostburg basin, Maryland, in 1837 and in the same area in 1840 at two bigger furnaces of the Mount Savage Iron Company. There were catastrophic failures, some of them considered in detail by Temin.[65] Bad trade or bad management was to blame in some cases, unsuitable coal or bad coking methods in others, but inappropriate technique and equipment seem also to have contributed, and Fulton later suggested that the prime cause was lack of blowing power in furnaces built for other fuels.[66] More fundamental were problems derived from the overall structure of the Appalachian coalfield. Westwards, towards the coalfields of the central lowlands, the volatile content of the coal increased, fixed carbon content fell, and sulphur content of the coals was too high for a fully satisfactory coke. Eastwards, towards the Allegheny Front, fixed carbon content was so high and volatile content so low that coals were too 'dry' for easy coke manufacture.[67] In this eastern section, however, at Johnstown, the first notable breakthrough was achieved. This illustrated both the causes and problems of advance in coke smelting.

In 1850 the Pennsylvania Railroad reached Johnstown, a small community with a long but unremarkable history of ironmaking. The Cambria Iron Company, in 1852, operated four charcoal furnaces in the locality but intended to make iron rails. In 1853 work began on four coke furnaces but completion was delayed. In summer 1854, when John Fritz arrived from the east as manager, he found the coke ovens and the pig iron poor.[68] After failure and reconstruction Johnstown made 130,000 tons of rails in 1856. Here was the bulk demand for a standardized pig iron which justified the use of coke in furnaces. Eventually Johnstown coke made from local fuel proved deficient. This suggests that concentrated demand for standardized iron was the key to successful coke smelting, with all that implied for high levels of furnace output.

Rail manufacture was unimportant in Pittsburgh, whose firms were favourably located for western trade but had a wider product range than Johnstown's. In 1856 the largest Pittsburgh mill, Schoenbergers, had an output under 10,000 tons of rolled iron and the average for the nineteen mills was under 5,500 tons. (Fig.6) Moreover some of these firms had considerable investments in charcoal or in raw coal furnaces. The charcoal iron trade, however, was shrinking and costs were rising, while incursions of small quantities of anthracite iron into the Pittsburgh market showed much more convincingly than trade records the low prices that mineral smelting made possible. In 1859 Graff Bennett and Company built the first coke furnace of the area at their Clinton rolling mills, a plant of about average size in the Birmingham district of Pittsburgh, opposite the Point. Significantly, unlike most of its neighbours, the Clinton rolling mills already used some coke iron in its operations.[59] The coke for the new furnace came from Connellsville, Fayette County. Causes for the delayed opening of this area, which in under forty years became the biggest coke field in the world, need examination.

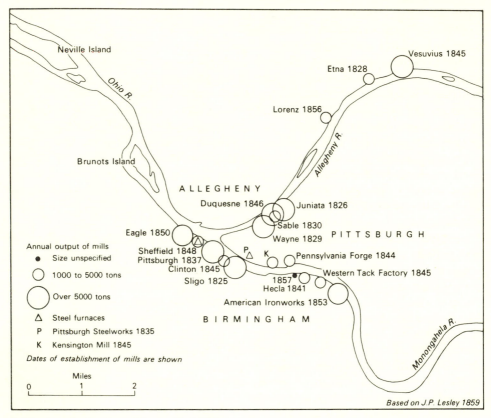

6. Pittsburgh district rolling mills 1858

Growth in demand had already for twenty years provided heavy iron consumption in the Pittsburgh area, but the division of that demand among numerous producers, the absence of rails, and the wide range of finished products, some, like sheet iron, spring or plough steel, demanding pig iron of high quality, discouraged the use of coke in ironworks. In addition to this deficiency on the demand side, from the point of view of supply various explanations for the late development of the Connellsville coke district have appeared, the chief being geographical ignorance, inadequacy of access and unsuitable techniques of both coke and iron manufacture. In anticipation of analysis one may suggest that the case for geographical ignorance is unsupportable but that inadequate techniques delayed the development of the industry until bulk movement of coke to Pittsburgh first became feasible in the 1850s.

The finest coking coals in America lay in a narrow belt some fifty miles long by no more that three miles wide along the western side of Chestnut Ridge. The core of this belt was near Connellsville on the Youghiogheny River. The area has undergone deep and regular denudation so that the upper valleys of the tributaries of the meandering main rivers 'cover the land with a meshwork of hill and valley that bids defiance to precise description'.[70] The Connellsville coal bed is geologically a continuation of the famous Pittsburgh seam, but with quality differences rendering it much better for coking. In this dissected country its regular eight to nine foot seam frequently outcropped in steep escarpments, offering easy working in primitive or 'country-bank' mines.[71] Along the Youghiogheny erosion has removed the seam from an area extending from near Sewickly Creek to within four or five miles of Connellsville. In this belt the Alliance ironworks, the first furnace plant west of the Alleghenies, was built on Jacob's Creek in 1790. Most of the numerous charcoal furnaces built in Fayette County had failed by the 1850s; but quite early sporadic attempts were made to use coke in the furnaces.

Beehive coke seems to have been made in 1817 at Plumsock ironworks near upper Middletown on Redstone Creek, ten miles south-east of Brownsville, but for use in the refinery, not the blast furnace. In 1837 coke iron was made for a time at Fairchance Furnace, seven miles south of Uniontown where the edge of the Connellsville coal abuts on Chestnut Ridge—according to Swank the first use of Connellsville coke.[72] Provance McCormick of Connellsville had two ovens in 1841; in 1850 the whole district still had only four. Technical deficiences partly account for this dismal record. In small ironworks equipped to use charcoal, with inadequate blowing power and with cold blast, failure was not surprising, but coking methods, too, were inadequate. In 1850 Overman, pessimistic about coke smelting, had some peculiarly perverse views on technology. He suggested that oven coke was unsuitable for blast furnaces.[73] Twenty years later, when there were already over 500 ovens in the Connellsville area, the visiting German, Klupfel, accustomed to the need to tackle inferior coals at home, argued that 'there was not, in the whole country, a rational coking plant'.[74] This was possibly of central importance in delaying the large-scale use of coke fuel. In the immediate vicinity of Pittsburgh advanced European coking technology might have made use of Pittsburgh Bed coal; the better coking coal of Connellsville, exploitable by inferior coking methods, was not yet accessible.

Coke was produced in Pittsburgh as early as 1813 and in 1833 there were coke ovens at the base of Coal Hill. In 1859, shortly after their first success with Connellsville coke in the Clinton furnace, Graff Bennett and Company experimented with Pittsburgh Bed coke. The result was unsatisfactory and they reverted to Connellsville coke. Klupfel later reckoned that a 'rational' coking plant could make a satisfactory fuel in Pittsburgh from the small coal being thrown away into the river, but by then coke made from the purest coal could be sold there for as little as $3 a ton. The time for such a development was past.[76]

Within the Connellsville region difficulties of access had been one factor inhibiting the adoption of better iron-smelting techniques and preventing supply to the expanding mills of Pittsburgh. Most important, location of the early known chief reserves of coking coal and easy access routes did not coincide. The Pittsburgh Bed was surveyed by the first Geological Survey of Pennsylvania in 1836-40, and its detailed form was outlined by Rogers in his Final Report of 1858. Its distribution clearly resulted from the dissection of a tract of folded country with axes parallelling Chestnut Ridge. The bed outcrops in a line five miles from the crest of Chestnut Ridge and has the same north-east—south-west strike. To the north-east it extends towards the Kiskiminetas, a tributary of the Conemaugh, and to the south-west across the Monongahela. North of a line from Dawson Station on the Youghiogheny, four miles below Connellsville, to a point on Redstone Creek down river from Uniontown and on to the Monongahela erosion has removed an upfold of the Pittsburgh Bed and exposed the barren measures beneath. Further towards Pittsburgh synclinal structures cause a reappearance of the bed but its thickness is less and as the volatile content has increased to about 40 per cent from the 30 per cent level around Connellsville the coal was classified as gas coal rather than as a coking coal.[77] (Fig.7)

This southern, favoured part of the coalfield was crossed from Fort Cumberland to Uniontown by Braddock's Road. After the War of 1812, the National Road ran from Uniontown to Brownsville near the mouth of Redstone Creek and so within the second basin of the Pittsburgh Bed. At this time forgings and iron castings were hauled between fifteen and thirty miles from Fayette County ironworks for floating down river on flat-boats from Brownsville to Ohio basin markets, but no coal or coke from the interior was used.[78] This situation is understandable: the coal bed was unknown, techniques of coke use were beyond the small, scattered western ironworks, and their product from charcoal smelting was sufficient for both local needs and Pittsburgh mills and foundries.

In the thirties and forties, numerous small mines were opened in the hills near the Monongahela, delivering to the river bank by tramway. The construction of dams and locks in 1843 and 1844 made the Monongahela suitable for slackwater navigation to Brownsville and stimulated this coal trade, but the best coking coal lay farther up river and inland away to the south-east, so that delivery to the Monongahela still required wagon haul and transhipment. By 1850 the Youghiogheny, much less suitable for

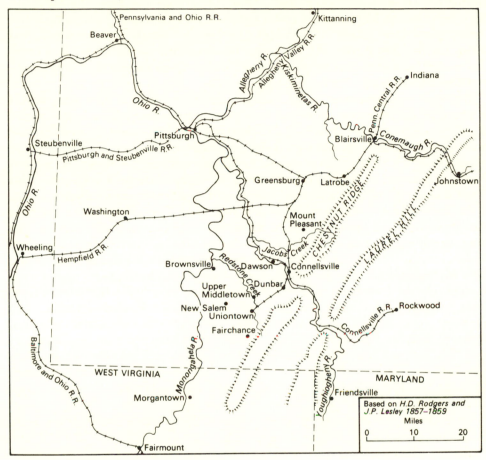

7. South western Pennsylvania 1858

navigation in its natural state, had also been improved, but navigation remained difficult and little development of commerce occurred.[79] Some coke was already being taken to Pittsburgh, by wagon. Development awaited the construction of railways into the most favoured part of the coalfield, that centred on Connellsville. The arrival of the Pennsylvania Railroad in Pittsburgh in 1852 was critical, though its effects were indirect. It opened new markets and initiated economic delivery of iron ore to Pittsburgh, for instance from Huntingdon County, or, much nearer, from Cambria County mines. In 1855 the Pittsburgh and Connellsville Railroad, following the Youghiogheny, struck the main coal basin five miles west of Connellsville.

Branches to Uniontown and Mount Pleasant opened the south-western and north-western sections of the basin. A little later the Pennsylvania Railroad built its south-west extension from Greensburg to Connellsville and on almost to Uniontown. The Monongahela above McKeesport still had no fringing railroad and was crossed by only one, the Hempfield Railroad, from Greensburg to Wheeling.[80] With these improved marketing and raw material assembly opportunities coke smelting grew rapidly in the Pittsburgh area, and with it went the growth of the coke district. Connellsville had four ovens in 1850, seventy in 1860; in that year thirty new ovens were built. Output of bituminous coal and coke furnaces was only 10 per cent of the national iron production in 1859. In 1864 pig iron production from all furnaces of the country was 263,000 tons more; 112,000 tons of this increase was from coal and coke furnaces. Equally significantly, when national output dropped by 18 per cent in 1865, the decline in anthracite iron was 30 per cent and in coke and coal iron only 10·2 per cent. A slight increase in charcoal iron production no doubt cheered many an isolated ironmaster but proved of little significance in the long run. The west now had a large market which was spreading and growing. Rail communications were welding the whole into an economic organization in which large-scale operations were desirable. Within two decades coke became the dominant fuel in iron smelting in the United States, and new supplies of rich ore from the Upper Great Lakes completed the revolution in the western raw material supply situation. Geographically the most notable result of these developments was to accelerate the westward movement of the iron trade.

NOTES

[1] M. Keir, 'Economic factors in the iron industry', *I.A.*, Apr. and May 1918; A. Nevins, *Abram S. Hewitt. With some account of Peter Cooper*, 1935, p.102.

[2] F.W. Taussig, *The Tariff History of the United States*, 1931, p.10. L.C. Hunter quoted P. Temin, *Iron and Steel in Nineteenth Century America*, 1964, p.85.

[3] Z. Allen, *Science of Mechanics*, 1829 quoted J. Swank, *Iron in All Ages*, 1892, p.539.

[4] Taussig, op. cit., and E. Stanwood, *Tariff Controversies in the Nineteenth Century*, 1903, provide general surveys, Temin, op. cit., Ch.1, a modern commentary.

[5] Taussig, op. cit., p.56.

[6] *Minutes of Evidence Select Committee on Import Duties*, British Parliamentary Papers, 1840. Evidence of Sir J.J. Guest, 13 July 1840.

[7] Quoted Taussig, op. cit., p.128.

[8] E.R. Johnson, T.W. Van Metre, G.G. Huebner, and D.S. Hanchet. *History of Domestic and Foreign Commerce of the U.S.A.*, 1915, p.210.

[9] J.L. Bishop, *A History of American Manufactures 1608–1860*, vol.2, p.242.

[10] C. Goodrich, (ed.), *Canals and American Economic Development*, 1961, p.126; Bishop, op. cit., vol.2, p.295.

[11] Bishop, op. cit., vol.2, p.345.

[12] Professor Wilson, *Report on the New York Exhibition*, British Parliamentary Papers, 1854.

[13] J.P. Lesley, *The Iron Manufacturer's Guide*, 1859, pp.8–9.

[14] Bauerman, *Metallurgy*, 1874, pp.267–8. S.H. Daddow and B. Bannan, *Coal, Iron, and Oil*, 1866, p.617. Lesley, op. cit., pp.414, 448–52, 700, 759, 760. *Economic Geology*, 2, 1907, p.154.

[15] H. Fairbairn, 'On the State of the Iron Manufacture in the United States', *Mechanics Magazine*, 26 Jan., 1850, p.68.

[16] *Tariff Proceedings and Documents 1839–1857*, Senate Documents, 1911, Part 3, pp.371–2.

[17] Temin, op. cit., p.264, Table C.1.

[18] Taussig, op. cit., p.132.

[19] Temin, op. cit., p.264.

[20] A.I.S.A. *Report*, 13 Dec. 1865.

[21] F. Overman, *The Manufacture of Iron*, 1850, p.182; Daddow and Bannan, op. cit., p.685.

[22] H.D. Rogers, *The Geology of Pennsylvania*, vol.2, part I, 1858, p.359; Daddow and Bannan, op. cit., pp.564, 785−6; Lesley, op. cit., pp.685, 686.

[23] Fairbairn, in op. cit., pp.66−7.

[24] Lesley, op. cit., (map of the anthracite furnaces).

[25] H. Fairbairn, 'On Smelting Magnetic Iron Ore', *J.F.I.*, 3rd series, 19, 1850, pp.125−9.

[26] Daddow and Bannan, op. cit., 687−93.

[27] Daddow and Bannan, op. cit., pp.540−2, 685−91.

[28] A.I.S.A. *Report*, 13 Dec. 1865. *J.I.S.I.*, 1872, p.364.

[29] A.I.S.A. *Report*, 13 Dec. 1865.

[30] Lesley, op. cit., pp.474, 605.

[31] See, 'The Lackawanna and Wyoming Coal Region', *J.F.I.*, 3rd series, 16, 1848.

[32] A.I.S.A. *Report*, 13 Dec. 1865, and Daddow and Bannan, op. cit.

[33] *A.I.S.A.* Production figures from 1864, capacity for 1865.

[34] A.I.S.A. *Report*, 13 Dec. 1865. *A.I.S.A.*, 1884, pp.60−1; *B.A.I.S.A.* 28 June 1871, p.343.

[35] Lesley, op. cit., p.759 and *A.I.S.A.*

[36] Bishop, op. cit., vol. i, p.550; Lesley, op. cit., p.62.

[37] Lesley, op. cit., p.749; K. Bruce, *Virginia Iron Manufacture in the Slave Era*, 1931, pp.275−7, 321, 322.

[38] Lesley, op. cit., p.759.

[39] Daddow and Bannan, op. cit., pp.614, 619.

[40] See J.C. Trautwine, 'Cast Iron Rails for the Hiwasee Railroad (Eastern Tennessee), *J.F.I.* 3rd series, 3, 1842, pp.21−2; Lesley, op. cit., p.403.

[41] J.M. Swank, *History of the Manufacture of Iron in All Ages*, 1892, pp.298−9; Daddow and Bannan, op. cit., p.356.

[42] Report of a Commission of Enquiry of 1831, quoted H. Scrivenor, *History of the Iron Trade*, 1841, p.379, *J.F.I.*, 3rd series, 12, 1846, p.126; Lesley, op. cit., *passim*.

[43] Temin, op. cit., pp.51−7, 77−80.

[44] Temin, op. cit., pp.63−5; R.P. Rothwell (ed.), *The Mineral Industry*, 1892, p.288; *B.A.I.S.A.*, 28 June 1871, p.343.

[45] Temin, op. cit., p.64; Daddow and Bannan, op. cit., p.619.

[46] Temin, op. cit., 60.

[47] Temin, op. cit., pp.64 and especially p.71.

[48] W.P. Shinn, 'Pittsburgh: its resources and surroundings', *T.A.I.M.E.*, 8, 1879−80, pp.15−16, and Bishop, op. cit., p.337.

[49] S.J. Buck and E.H. Buck, *The Planting of Civilisation in Western Pennsylvania*, 1939, p.305; E.C. Pechin, 'The Minerals of South Western Pennsylvania, *T.A.I.M.E.*, 3, 1874−5, p.400.

[50] Lesley, op. cit., pp.86−88.

[51] H. Scrivenor, op. cit., p.379, *J.F.I.*, 3rd series, 12, 1846, p.126.

[52] Lesley, op. cit., *passim.*

[53] H. Phillips, 'On the bituminous coalfield of Pennsylvania', *Reports of the British Association*, 1837, p.96.

[54] *J.F.I.*, 3rd series, 12, 1846, and Lesley, op. cit.

[55] *J.F.I.*, 3rd series, 12, 1846.

[56] Lesley, op. cit.

[57] *Tariff Proceedings and Documents 1839−1857*, Senate Documents 1911, Part 3, p.1815.

[58] H.M. Chance, *The Geology of Clarion County*, 1880, Ch.IX: 'The Charcoal Iron Furnaces'; *2nd Geological Survey of Pennsylvania;* Lesley, op. cit., pp.95, 100, 101.

[59] Lesley, op. cit., pp.752−753.

[60] *Report of U.S. Secretary of the Treasury 1849.*

[61] Overman, op. cit., p.174.

[62] Estimate based on Lesley's account.

[63] Overman, op. cit., p.174.

[64] J. Fulton, 'Methods of coking', in F. Platt, Special Report on the Coke Manufacture, *Second Geological Survey of Pennsylvania*, 1876, pp.120−121, 133.

[65]Lesley, op. cit., p.93. Temin, op. cit., pp.70—5.

[66]J. Fulton, *Coke*, 1905, pp.132—3.

[67]Ibid., p.24 and Chapter 10.

[68]Temin, op. cit., p.110; *Iron*, 5 July 1889, pp.8—9; J. Fritz, *Autobiography*, 1912, pp.91—2.

[69]Lesley, op. cit., pp.247—51.

[70]H.D. Rogers, *The Geology of Pennsylvania*, vol. 2, part 1, 1858, p.499.

[71]E.C. Pechin, op. cit., p.406; J. Fulton, 'Coal Mining in the Connellsville Coke Region of Pennsylvania', *T.A.I.M.E.*, 13, 1884—5, p.333.

[72]Fulton, *Coke*, p.131; *Connellsville Tribune*, 16 June 1888, quoted *I.A.*, 21 June 1888, p.1013; J.M. Swank 'The Iron and Steel Industries of Pennsylvania', *Ann. Report of the Secretary of Internal Affairs and Commonwealth of Pennsylvania, 1881—1882*, 1882. part III, p.32.

[73]*I.A.*, 21 June 1888, p.1013; F. Overman, op. cit., p.119; see also pp.174, 175.

[74]G. Klupfel, quoted. *B.A.I.S.A.*, 28 June 1871, p.341.

[75]J. Fulton, *T.A.I.M.E.*, 13, p.334.

[76]*B.A.I.S.A.*, 28 June 1871, p.341.

[77]*Second Geological Survey of Pennsylvania*, The Coke Manufacture of the Youghiogheny River Valley, 1876.

[78]E.C. Pechin, op. cit., p.400.

[79]N.B. Craig, *History of Pittsburgh* (1917), p.277; recollections of G.H. Anderson,, Secretary of the Pittsburgh Chamber of Commerce, in *Report of the Industrial Commission on Transportation*, IX, 1901, p.646.

[80]H.D. Rogers, op. cit., vol.2, part 2; Lesley, op. cit.

CHAPTER 3

The Westward Movement of the Iron Trade

WESTERN ADVANTAGES

Already in the 1850s certain trades were growing more rapidly in the west, and following the Civil War this trend became much more noticeable. New activities shunned the east, and existing firms in some cases moved their whole operations from one to the other. While new growth points emerged and flourished on the one hand, on the other whole metal-working regions seemed on the edge of extinction. In the seventies, however, the change was less cataclysmic than later (Table 7).

Table 7.　　　*Rolled iron and steel production by states*
1875, 1880, 1890
(percentage of national total)

	1875	*1880*	*1890*
New England	8·3	6·4	3·0
New York	8·9	6·3	1·9
Pennsylvania	39·0	44·2	58·5
Ohio	13·1	13·2	12·4
Illinois Indiana	8·0	8·0	14·1

Note: 1875, 1880 rolled iron only.
Based on *A.I.S.A.*

Reasons for the shift westwards are to be found in the rapid economic growth there (Table 8), and the development of raw material supplies scarcely if at all touched when Lesley's survey was completed in 1859. The key to the whole was the development of the western railroads linking the plants to distant raw materials, opening regional rather than local markets, and themselves constituting by far the biggest single consumer of iron. Older plant and, equally important, less expansive habits of thought also penalized the eastern works.

Eastern minerals throughout this period were generally worked on a small scale and often inefficiently. At the beginning this was little disadvantage for the same conditions characterized the western workings, but in the seventies and still more in the eighties western mining and systems of mineral transport were rationalized. Bell described the New Jersey ore districts in the mid-1870s, many of the 100 or so magnetite ore mines being 'of an extremely insignificant nature', and two years later Richard Akerman condemned the commercial policy adopted in the Cornwall ore operations,

Table 8.   *Value of manufacturing production*
              north-eastern and north-central states 1860—1900
                            (million $)

	North-east	North-central
1860	1,214	225
1870	2,675	803
1880	3,186	1,200
1890	4,896	2,518
1900	6,498	3,506

North-east: New England, New York, New Jersey, Pennsylvania.
North-central: Indiana, Illinois, Michigan, Wisconsin, Minnesota, Iowa, Missouri, North Dakota, South Dakota, Nebraska, Kansas.
Note: Much Pennsylvanian growth was west of the Alleghenies.
Based on Bureau of the Census, *Census of Manufactures.*

the biggest of the Mid-Atlantic district. The ore could be cheaply worked, but 'the whole deposit also belongs to a single family, which is said to be uncommonly conservative, and to prefer a high price with a limited sale to the plan otherwise common in America of working on as large a scale as possible.' The 1885 output was reckoned only one-quarter of the ore banks' capacity. Even so, in respect of ore supply the position of the east was little worse than that of the west until the late 1880s, as Holley's discussion of ore supply for Bethlehem and Johnstown in 1877 and 1878 makes clear. In 1880 ore output per man year in the Lake Superior district was 310 tons. In 1885 in the Cornwall ore banks it was 254·1 tons.[1] With fuel the advantage of the west was greater even earlier.

DEVELOPING PATTERNS OF WESTERN MINERAL SUPPLY

In mid-century most American ironworks obtained their raw materials locally. Lesley's account is helpful here though the evidence is only partial. Until that time the areas west of the Appalachians had shorter mineral hauls than those in the east. A report from the New York Exhibition observed that whereas a calculation of assembly costs was necessary to decide location of plants east of the Appalachians where coal and ore were not found together, to the west of the Appalachians this was not necessary. There, and discounting the charcoal furnaces, it was observed that coal measures contained both clayband and, sometimes, blackband ore. Development of the use of a specialized and localized fuel, Connellsville coke, and of Lake Superior ore along with bulk lake shipment facilities upset this situation completely, so that, by the nineties, it was the east, with local hematites and anthracite which, for a time, had shorter hauls than the west. Even there the need to bring in outside supplies of Lake Superior or foreign ore, and increased use of Connellsville coke, first in mixtures with anthracite and then alone, softened the contrast. By 1897 Birkibine summed up the revolution which had occurred: 'At the present time the ore supply is a local consideration for but a small proportion of American blast furnaces, and 700 miles or more separate a greater number from the mines producing the ore in 1897 than was supplied in 1857 with ore carried 50 miles.'[2] In 1860 shipments of Lake Superior ore were still a little under 4 per cent of the national production, but by 1879 Michigan produced 22·9 per cent of the U.S. total, and by the end of the 1880s had eclipsed Pennsylvania, for so long the leading producer. With the westward movement of mineral production went that of demand—the U.S. centre of population was on the west Virginia side of the Ohio near Parkersburg in 1850 but by 1890 in the extreme south-east of Indiana. As settlement moved west so too, though confined to a very large extent within the north-east, did manufacturing and the consumption of iron and steel. By 1892 when Mesabi shipped its first ore the balance of advantage in mineral assembly was shifting not only from east to west of the Alleghenies, but from the Ohio valley towards the shores of the Great Lakes.

THE DEVELOPMENT OF ORE MINING IN THE LAKE RANGES

Writing early in the twentieth century H.R. Mussey distinguished three major phases in the development of the Lake Superior ore districts.[3] Until the depression of 1873, ore prices were high, mining was conducted in many small units and was labour-intensive. Prices fell after that and cost reduction became essential. At the same time with the exhaustion of surface workings mining conditions became more difficult. New mines were opened, but each operation had to be bigger and more capital-intensive. The 1893 panic brought prices to a new low, and at the same time Mesabi introduced another cheap producer. Big steel companies had already begun to acquire mine interests, but the process was now accelerated, and the full integration of mining and transportation was carried to a high degree of excellence. In 'the period of exploitation', as Mussey termed the first division, there was little sign of all this.

Although white men are said to have known of Lake Superior ore resources as early as 1830, the Teal Lake ore body along the northern edge of Marquette range was not proved until 1844. In the following year the Jackson Mining Company obtained control of Jackson Mountain. In 1850–2 experimental shipments were made to furnaces in the Shenango Valley, but as late as 1854 Marquette shipped only 3,000 tons. The high cost of mining and delivery checked more rapid development. Their cause in turn was a difficult environment and remoteness; as late as 1865 R.H. Lambourn, Secretary of the A.I.S.A., speaking of charcoal iron manufacture on the upper lakes, stressed that '. . . The difficulties to be encountered in building large structures, erecting new machinery, and collecting necessary labour in a distant and hyperborean region are numerous and serious'.[4] The opening of Marquette almost exactly coincided with development at Pilot Knob and Iron Mountain in Missouri, and, for a time, it seemed that the latter would be more important. They were more accessible, St. Louis was a nearby market for the charcoal furnaces which were built near the ore fields in the late forties, and water transport was available via the Missouri, the Mississippi and its tributaries. By 1860 Pilot Knob was connected by rail to St. Louis, already an important railway junction, whereas there were then no railroad connections within hundreds of miles of Marquette. In 1851—2, John Fritz of Safe Harbour Works, eastern Pennsylvania, journeyed to Marquette to look at the new orefields, calling at iron works in Sharon and Newcastle on the way, presumably to see how they managed with the new ore. Having seen the mines, he determined to draw eastern interest to them, but the iron business was then in a bad way and he was unable to arouse its interest.[5] In 1845, when the Jackson Company acquired its ore properties, the aggregate tonnage of vessels on the Great Lakes above Niagara had been 76,000 tons. At the time of Fritz's visit there were only two vessels on Lake Superior and they were cut off completely from the other lakes on the United States side. The first shipments of ore to Newcastle had to be hauled round the falls of Sault St. Marie on a strip railroad one and a quarter miles long.[6] The Soo Locks, opened in 1855, were limited to vessels of 11.·5 feet draft but as the typical ore boat of the time was only a ninety foot schooner they were adequate enough to remove the bottle neck.[7] The canal was improved by the government in the seventies. Specialized ore boats gradually developed from general cargo vessels and carrying capacity was increased. The first schooner through the Soo in 1855 carried 300 tons of ore, but by 1867 loads of around 550 tons were usual and by 1872 1,500 tons.

The difficult country between the mines and the lakes was wooded with no natural trackways. The Marquette Iron Company had tried to supply the forge which it operated at Marquette from 1849 to 1853 with ore from Cleveland mine, fifteen miles away, by the use of twenty or more double teams of mules. In winter ore was carried by sleigh. A plank road was built in 1853 and in the years 1855—7 carriage was by means of a primitive railroad, wood covered with iron strips providing the road bed. Wagons carrying three to four tons of ore, each pulled by teams of mules, made one round trip a day. There were many derailments. Locomotives replaced the mule teams in 1857.[8] By 1860 the cost of freight to the ore dock had been cut to $1·09 a ton as compared with $3 a ton in 1855. At the same time handling facilities at Marquette were improved. Until 1857 the ore docks were platforms built on wooden piles twelve feet above water level. When the railroad was completed in 1857, the first pocket storage docks were also built. Subsequently, as lake vessels increased in height, the ore docks were raised.

From 1866 to 1873 ore prices at Lake Erie ports ranged from $8 to $12 but the average price at the mine was $3·50, the rail freight to Marquette was $1, and the lake freight $3 to $5. In 1865 Escanaba joined Marquette as a shipping port after the Chigago and North Western Railroad had opened its Peninsular line from Negaunee and built ore docks on Lake Michigan costing $200,000.[9] There were other causes for the trend towards higher capitalization which this line exemplified.

In the early days ore was quarried from shallow openings in the hillsides. There was little need for drainage, but, as deeper cuts were made, by the early sixties it was necessary to provide for the removal of water. Until nitroglycerine and power drills were introduced on a small scale in the late seventies, sledge hammers, picks and the explosive force of black powder were the only tools. By 1897 drift driving cost $16 a foot but at Cleveland mine in 1871 the cost had been as much as $100.[10]

A sharp fall in output followed the 1873 panic. In 1875 shipments were 271,000 tons or 21 per cent lower than in 1873 and the number of mines at work fell from forty to twenty-nine. 1873 output levels were not again reached until 1879. (Table 9)

Table 9.   *Marquette Range Mines and average shipments*
1856 to 1873

Date	No. of mines	Shipments per mine (tons)
1856	2	18,171
1858	3	5,292
1863	3	67,685
1865	7	33,732
1870	27	30,776
1872	29	31,065
1873	40	29,061

Based on J.M. Swank, *Iron in All Ages*, 1892, and H.R. Mussey, *Combinations in the Mining industry*, 1905.

Gogebic, Menominee, Vermilion had each been known to contain iron ore since 1848/9; exploration began in 1860, 1866, and 1876 respectively. The first Menominee shaft had been sunk in 1872 and Gogebic's first in 1873, but because of depression shipments did not begin until 1877 and 1884. Meanwhile Marquette production costs were rising as underground working gradually superseded open pit operations. On the Gogebic range the multiplication of units was repeated, fifty companies opening it up most wastefully. Vermilion development was more rational. The Minnesota Iron Company monopolised its development. In 1882, two years before shipments began, it built the Duluth and Iron Range Railroad across the sixty-eight miles of wasteland separating Tower from the port of Two Harbours, which it equipped with large ore docks. The company also organized the Minnesota Steamship Company with six large vessels to free itself from the uncertainties of freight rate fluctuations, and formed trading links with major iron companies. Two agreements were made for the dispatch of ore east of Buffalo, but by far the most important development occurred in 1887 when interests connected with Illinois Steel Company obtained control. The large concern, integrated from mine to furnace over distances of up to 500 miles or more, had appeared for the first time, with cost reductions in every section of the business. By 1887 ore on rail cars at Vermilion mines cost $2 a ton, the rail rate to the lake was $1—though this was by now higher than in the short haul from the Marquette mines—and the contract lake freight to Cleveland was $2. In the same year Marquette ore cost $5·10 on Lake Superior and the average price for Bessemer ore from all lake ranges at Lake Erie ports in 1887 was $7·25. Operators on other ranges now had to rationalize as well; in the eighties average mine output on Marquette increased from 39,000 to 90,000 tons. Vermilion development was the model for the development of Mesabi, and for the bigger integrated ore, transport, and smelting groups of the nineties.

THE COKE TRADE FROM 1860

Civil War demand led to a great growth in iron-making in the Pittsburgh area, and by 1870 Connellsville had 300 ovens. It was, however, the boom of the early seventies which marked the real breakthrough. By 1874 there were eleven furnaces in Allegheny County, two years later 3,000 ovens were at work in Connellsville. By this time production and marketing were being more effectively organized. Henry Clay Frick had begun making coke in the Mount Pleasant area at the end of the sixties and Frick and Company with 300 acres of coal land and 50 ovens failed in the 1873 panic, but were saved by a $20,000 loan from the Mellon interests. By ploughing back profits, linking with competitors and later by importing relatively easily managed eastern European labour, Frick gradually extended his control. Oven construction continued even during the depression, and in 1875 output of coke and raw coal pig iron passed that of anthracite iron. By spring 1879 there were reckoned to be 3,668 ovens in the Connellsville district, and another 550 or so in and about Pittsburgh itself or along the Pennsylvania, Pittsburgh, Chicago, and St. Louis Railroads. They were capable of producing about 5,500 tons of coke daily.[11] In the trade revival of the next four years national pig production went up two-thirds, and in Allegheny County more than doubled, and by 1882 there were 8,400 ovens in Connellsville. From then to 1887, another 800 or so were built each year.[12] By then coke works were strung out through the coalfield from Latrobe to Morgantown in west Virginia, but the name Connellsville was extended to the whole district. In 1890 this small area produced half the nation's coke.

For long, coking practice was primitive but the coal was so good and cheap that costs were low. In 1875 the visiting English ironmaster and authority on iron manufacture, Lowthian Bell, found that 32 to 40 cwts. of coal were used per ton of coke 'and yet the coke itself was being put into the wagons at the pits at 5s. a ton, or not more than half the price which Durham coke commands at the present time'.[13] Later, the most rationally constructed of the oven plants curved along the contours of the hillsides to reduce haulage costs. Unionization of the foreign labour force was almost impossible because of the variety of language and the brief seasonal stay of some of the European workers. Waste of by-products, the pollution of the air by smoke and gas, were universal features at a time when abundance gave no incentive for conservation. The coking coal district was now transformed into a rapidly expanding economy with isolated settlements heavily dependent on mines and coke ovens.

Two aspects of Connellsville organization deserve special attention—its links with other districts and its own lack of economic diversification. As coke proved its value, it pushed out other fuels. By 1890 it was already a more important contributor than anthracite to eastern Pennsylvanian iron production. By carefully covering the wagons with tarpaulin it was eventually possible to reduce damage from wetting to a very low level but abrasion reduced the quality of coke so that nearness to Connellsville remained for technical as well as freight cost reasons an important Pittsburgh advantage. Most companies bought coke in the open market, but as with iron ore, though even earlier because the mineral was more accessible, the bigger, more far-sighted companies saw the advantages of controlling their own supplies.

In 1875, when smelting Lake Superior iron to make Bessemer pig, Cambria Iron Company found that its own local coal, though it made a purer coke than that from Connellsville, was neither hard enough nor so suitable in cellular structure. It therefore began to buy from Connellsville, and in 1880 built its Morrell coke ovens at the mouth of its own mine a few miles south of Connellsville. In 1881 Carnegie Brothers decided to obtain control of coking coal. Frick now controlled almost 1,000 ovens, and in and after 1882 Carnegie bought its way into this company, eventually becoming the main shareholder. In return Frick gained a key position in the steel organization. By 1887 Frick controlled 5,000 acres coal land and produced 6,000 tons coke a day, equal to over a third of the total shipments of the following year.

Control of coke supplies removed one of the cost uncertainties of the pig iron business. In times of bad trade ovens could be closed, and men laid off with no obligation to the firm. June 1888 was a lull in a generally good year for pig iron. Eighteen of a total of seventy-five coke plants were wholly idle and sixteen partially idle. Almost a quarter of the ovens were out of production. In times of boom price rises could be contained by the company which controlled its supplies. 1890 was a record iron year and Connellsville coke shipments were 6·46 million net tons or 30·5 per cent above the level of 1888 but the price was 63 per cent higher, $1·94 as compared with $1·19 per net ton.[14]

Coke-making remained a coalfield activity. In 1880 45·8 per cent of United States coke was made in Fayette County, 27·4 per cent in Westmoreland County. Allegheny County made only 3·5 per cent. In 1890 Jones and Laughlin began a new policy when they acquired the Vesta Mines near the Monongahela and barged coal down to ovens on the north side of the river in the Hazelwood district of Pittsburgh, only a little way above their blast furnaces. In the following year they bought their first tow boat to transfer coal down the river. By this development they anticipated by 20—25 years the general trend of coal working westwards towards the navigable Monongahela and away from the unusable Youghiogheny. By this means delivery costs were reduced below those paid by other Pittsburgh iron-makers, including the Carnegie plants whose coke was made on the coalfield and railed to the furnaces. By 1899 the Pittsburgh area made 0·6 million tons of coke or only 6·2 per cent of the Connellsville total. 166,582 carloads of coke were shipped from the coking coalfield into Pittsburgh.[15]

Yet in spite of Connellsville's key position in the iron industry, it had few metal plants of its own. There were blast furnaces at Scottdale, Dunbar, Uniontown, and Fairchance by 1890 and in 1886/7 a two-converter Bessemer plant was built at Uniontown. By 1890 this was in the 50,000—100,000 tons a year capacity range, but by 1904 had been abandoned. In 1904, in the north of the district, in and near Scottdale, the Steel Corporation operated three rolling mills. One by one these plants failed. The general absence of metal industries is at first sight illogical, for, shipping to a very large extent by rail, hundreds of wagons a day had to return empty at a time when specialization of wagon function had not generally gone so far as to prohibit coal, coke, or iron ore occupying the same wagon. Pittsburgh and the Ohio river towns, Cleveland and other lake Erie ore ports usurped the apparently logical position of the Upper Lakes iron ranges and the coke district as locations for iron manufacture. The Upper Lakes region was deficient in markets, but once in railway wagons, ore could travel without break to Connellsville or Uniontown, as well as to Pittsburgh or Youngstown.

However, Pittsburgh had long been established as the chief foundry and mill centre in the United States when smelting with coke began. Rolling mill firms, therefore, chose to build their own ironworks locally rather than on the cokefield. Moreover, even in this time of high fuel use per ton of iron, the structure of rail freight rates made an ore port or intermediate location a cheaper assembly point for ore and coke per ton of pig iron than the cokefield. (Table 10) Progressive fuel economy reduced still more the

Table 10.   *Rail freight charges on ore and coke per ton of*
pig iron—April 1887
(dollars)

	Ore	Coke	Total
Cleveland	—	2·44	2·44
Youngstown	1·53	1·83	3·36
Dunbar or Scottdale	3·60	—	3·60

Note: Assumes 55 per cent Fe ore, Fuel use of 1·85 tons coal equivalent and a 60 per cent coal/coke yield.
Based on *I.A.* 14 Apr. 21 Apr. 1887.

attractiveness of the coalfield even before by-product coking was introduced on a large scale. It has been estimated that 2·1 tons of coal equivalent were used for every ton of pig iron in 1879, 1·85 tons in 1889 and 1·72 tons in 1899. At the same time as there was a direct decline in blast furnace fuel consumption, the advance of steelmaking at the expense of puddled iron decreased the tonnage of coal employed in finishing trades and so still more lessened the attraction of a purely coalfield location.[16]

OTHER IMPEDIMENTS OF THE EAST

Apart from its undoubtedly inferior mineral endowment the East was penalized in its competition with rising western centres of iron manufacture by a variety of other factors.

Labour, it is true, cost more in the west and was less tractable there, but the loyal Pennsylvania Dutch furnace labour of which Fritz thought so highly may have been a disadvantage in the long run by delaying mechanization. Puddling and skilled mill labour rates in Pittsburgh were 30 to 50 per cent higher than those in the east in the mid-seventies, but with cheaper fuel it was reckoned that Pittsburgh had a cost advantage on merchant iron of $6·18 a ton over eastern works by 1877, and within five years was said to be underselling all rivals.[17] Eastern entrepreneurs generally seem to have been less enterprising, less ready to innovate. This may be seen in railroad policy. In 1875 Bell found that assembly costs on Lake ore at Pittsburgh were 25 shillings ($6·00) a ton of iron whereas New Jersey ore for a ton of iron cost 10 shillings ($2·40) on the Lehigh. Connellsville coke at 4½ shillings ($1·08) per ton of iron in Pittsburgh compared very favourably with 6 shillings ($1·44) for anthracite on the Lehigh.[18] There were numerous complaints about high cost ore in the west but even more general agreement that eastern freight rates were too high.

Aided by superior fuel ironmasters west of the Alleghenies soon became pioneers of new technology. In 1859 Lesley had extolled the Thomas furnaces on the Lehigh as the largest and most productive in the country, but the initiative in the construction of big furnaces moved westwards with surprising speed. The early Pittsburgh furnaces were forty-five feet high, but by the late sixties, while eastern furnaces were still usually built of stone, western furnaces, sheathed in boiler plate, crept upwards and assumed a more 'modern' appearance. Eastern works employed blast temperatures of 620°F when western furnaces were operating at 1,000° to 1,200°F. As early as 1873 *Iron Age*, describing the new seventy-five foot Lucy furnace on the Allegheny, wrote: 'To one accustomed to the methods of blast-furnace construction as practised east of the Allegheny Mountains, the Lucy Furnace possesses much interest, It may be said to embody the best features of western practice, both in construction and management, and will well repay a visit from any Eastern ironmaster who may find himself in Pittsburgh, either on business or pleasure.' Two years later, Bell wrote of his visit to American ironmaking centres: 'Of course the chief subject for consideration is the question of fuel consumption, and here I am bound to say, as a rule, the Lehigh masters are perhaps a little behind the age.' However, as Bell was accustomed to furnace practice with the excellent coke of County Durham he may well have been a biased interpreter of eastern anthracite iron practice.[20] Yet for the large outputs which the Civil War or the boom of the early seventies required, coke was a superior fuel to any other. By 1880 the average daily capacity of coke and raw coal furnaces was 38 tons as compared with 29 for anthracite or anthracite and coke and 12 tons for charcoal furnaces. When trade was bad the high output of coke furnaces helped to reduce costs and extend their market, so that between 1873 and 1876, when pig output generally fell 750,000 tons, coke and coal furnaces increased their share. In the next decade came the greatest breakthrough of all, when Kennedy and Jones more than doubled the output of the Edgar Thomson furnaces by use of even higher temperatures and a larger blast volume without altering the plant. In that decade the average capacity of charcoal furnaces doubled, that of anthracite and mixed anthracite and coke furnaces went up by 89 per cent, but coke and coal furnace average capacity increased by 165 per cent. Coke had become the dominant fuel and with that the primacy of the west seemed assured.

WESTERN MOVEMENT, TRADES AND FIRMS

The anthracite iron industry growth in the forties and fifties checked the western movement for a while, but with large-scale coal and coke smelting it was resumed. Already by 1856, Abram Hewitt was predicting that a century hence, half the world's supply of 100 million tons of iron and steel a year would come from the area between central Pennsylvania and the Missouri.[21]

Some firms moved west or built new plants which became growth points with more vitality than the parent plants. Dewes Wood from the Schuylkill set up plant near Pittsburgh in 1861; by 1900 this had grown into major operations at McKeesport and at Wellsville on the Ohio. A Unionsville, Connecticut, nut and bolt firm built a branch in Cleveland in the seventies which soon became the firm's headquarters, later the foundation for the integrated works of the Upson Nut Company and eventually one of the units of the Republic Steel Corporation. In 1878 the important group of merchant mills making bars, wire products, tyres, springs and plough steel operated by the D.G. Gautier Company of Jersey City was moved to Johnstown. Later examples of the same trend were western developments by Wickwire Spencer and Washburn and Moen, both Massachusetts wire firms. The rail trade, considered more fully later, was an outstanding case of westward movement. In 1856 New England, New York, and New Jersey produced 31·4 per cent of the nation's rails; twenty years later only 9·4 per cent.

In spite of the drift westwards, eastern production as a whole continued to grow, and certain centres were very successful. As many as twenty new anthracite furnaces were put between 1868 and 1870 and even in the seventies seventy-five new ones were built.[22] In Bessemer steel the east had a good start with two of the first three works. Troy entered the Bessemer trade in 1865 and two years later the Pennsylvania Railroad chose a point east of Harrisburg, in the central section of its system, to manufacture its own steel rails. At Lewistown the Freedom Iron and Steel Works Bessemer project lasted only one year, but Bethlehem and Lackawanna were other early major eastern steel rail plants. By August 1875 there were thirteen open hearth steelworks in the country; eight of them east of the mountains. Perhaps it would be wise to regard some of this development as innovation forced upon an ill-favoured area but in some respects the east was well endowed in the early steel age.

Cornwall provided Bessemer grade ores, and for a number of years Mediterranean ores, shipped as ballast, could supplement these as cheaply as western works could obtain Lake Superior ore from still independent and often inefficient iron ore concerns. In 1879 the United States imported 0·28 million tons of ore, mostly from Europe, and in 1881 0·78 million. Some of this was taken right through to Pittsburgh but bit by bit in the eighties Lake ore began to come east in greater tonnages, first to Rochester or Syracuse and then right over into the main iron districts. By 1890 0·68 million tons of Lake ore came east of the Alleghenies.

Good management, a product for which a theoretically bad location was not particularly important, or a happy relationship with transport enabled some eastern centres to fare better than others. Pennsylvania steel at Steelton was an outstanding example, and Johnstown also had close relationships with the Pennsylvania Railroad. Johnstown indeed was in a bridge position between east and west with ore and coal suitable for puddled iron within its own yards, with better access than eastern works to high-quality coke and Lake ore, and with shorter rail hauls to western markets or to shipping points on the Ohio. In the long run this intermediate position, neither east nor west, was to prove its undoing.

THE IRON AND STEEL INDUSTRY OF THE OHIO AND MISSISSIPPI VALLEYS

The Ohio River was at its peak importance for freight movement in the 1860s. The big event of the day in Pittsburgh was the arrival of the steamboat. Manufactured iron, though dwarfed by the tonnages of Pittsburgh coal, went down river to western and southern markets; ores from the Cumberland basin and from Missouri came upstream.[23] When railroads were built to parallel the river, they caused a fall in the traffic, but were forced by the existence of the alternative route to quote low rates to waterside plants.

Iron prospects seemed very bright for a time in the Hanging Rock district of the Scioto Valley of Ohio and Adams and Greenup County, Kentucky. (Table 11) In addition to the markets of Cincinnati there were local mills and foundries, but after the sixties, when large profits were made, prospects dimmed as iron prices fell and as suitable charcoal timber became scarce.[24]

Table 11.   *Production and capacity of furnaces in the*
Hanging Rock district of Ohio 1872, 1880, 1884
(thousand net tons)

1872 Production		1880 Production		1884 Capacity[1]			
Charcoal	*Coke*	*Charcoal*	*Coke*	*Charcoal*		*Coke*	
				at work	*idle*	*at work*	*idle*
87	23	65	60	43	85	62	69

[1] A fifty week year is assumed though this overstates the importance of charcoal iron.
Based on *A.I.S.A.*

The trade survived into the twentieth century gradually converting to coke fuel. Portsmouth, the only steel plant which survived, was not then integrated with ironmaking.

The Hocking Valley, further north and away from the river, produced no iron until 1875. By the end of the following year it had four raw coal furnaces. A year later there were thirteen, the increase in the rest of the

United States in the same year being only eight furnaces. After 1880 only three more furnaces were built and by 1903 only one was at work, two were idle, and the rest had been abandoned. Raw coal gave a low productivity, the area had a long rail haul for coke, and suffered in competition with Hanging Rock from lack of water shipment for its produce and from the higher rail rates charged to landlocked centres. The only nearby mill towns were Zanesville and Columbus. (Table 12)

Table 12.   *Number and capacity of furnaces using*
mineral fuel. Hanging Rock district and
Hocking Valley 1884, 1887, 1892
(weekly capacity in tons)

		1884 July	1887 January	1892 January
Hanging Rock	Number In blast	7	11	8
	Capacity	1,250	2,464	1,833
	Idle	8	1	7
	Capacity	1,385	85	1,550
Hocking Valley	Number In blast	1	7	4
	Capacity	112	1,438	1,380
	Idle	14	8	10
	Capacity	2,490	1,027	2,513

Based on *I.A.*

In 1850 when river locations were not unreasonably regarded as better than Great Lakes ones, Horace Greeley forecast a great future for Cincinnati both as market and manufacturer: 'It requires no keenness of observation to perceive that Cincinnati is destined to become the focus and mart for the grandest circle of manufacturing thrift on this continent ... I doubt if there is another spot on the earth where food, fuel, cotton, timber, iron can all be concentrated so cheaply ...'[25] Important though its steamship works, plough, tea and sugar kettle trades, and its mills and foundries were, Cincinnati failed to realize these high destinies. Consumption of coal in puddling and rolling mill operations was high, so that the area was at a considerable disadvantage as compared with the upper Ohio. As Lake ore became more important a location twice as far from Lake Erie as Pittsburgh was another telling disadvantage.

Five rolling mills were built in and near St. Louis before 1860 and in 1863 a coke blast furnace was put up at Carondolet. Coke was made from Illinois coal but Connellsville coke was also used. In 1872 the Vulcan Rolling Mill was built to roll iron rails, and four years later a Bessemer plant was added.

Well located to supply the west St. Louis seemed to bid fair to outpace both Pittsburgh and Chicago, but quite failed to do so. The explanation is to be found both in raw material disadvantages and in human factors. Missouri ore proved much less competitive than had been anticipated in the early sixties. Every ton of coke made near the furnaces required the carriage of between 32 and 40 cwt. of Connellsville coal, and coke from Illinois coal proved incapable of providing the high-quality pig needed for the Bessemer process. If Connellsville coke was used freight costs were high—55 cents a ton more than to Chicago by 1887. As far as mill operations were concerned Vulcan's rail output was hindered by the attempt to roll steel in a mill built for iron, and into the mid-1880s the whole operation was said to have been mismanaged.[26] Compared with the hesitancy or disappointments of lower Ohio developments growth upstream was impressive.

WHEELING, THE VALLEYS, AND PITTSBURGH

With local coal, a position where the National Road crossed the Ohio, and access along that river to Mississippi basin markets, the Wheeling area developed a rolling mill business similar to, though much smaller than that of Pittsburgh. The first mill was established in 1832 by the Schoenbergers to work up Pittsburgh iron and to ship it on as plates and nails. Entrepreneurs from this mill built a second plant in 1847, and in 1849 and 1851 in the same manner a third and a fourth were established. The area played a small and not very successful role in the iron rail business in the fifties.

The first blast furnace was built at Martins Ferry in 1857, two years before the first furnace in Pittsburgh. From local minerals it soon turned to use Connellsville coke shipped by water. The second furnace in 1866 used Connellsville coke and Lake ore. By this time complex inter-plant links were beginning to emerge, for instance those between the La Belle mill and the Martins Ferry furnace, and the Jefferson works at Steubenville and the furnace at Mingo Junction. By 1862 the Wheeling neighbourhood had seven large rolling mills, three nail works, and seven foundries. After the war it added new lines, notably sheet iron, but was above all prominent in the nail trade.

New England had made 25,000 tons of the national output of 81,000 tons of nails in 1856, but twenty years later, when the national total had reached 208,000 tons, its output was only 23,000 tons. By this time West Virginia produced more than twice as many kegs of nails as New England, Pennsylvania more than three times as many, with Pittsburgh as an important centre. By the mid-1880s there were 28,000 nail machines east of the Alleghenies but 38,000 to the west. (Table 13) The Wheeling area and also Central Pennsylvania were by then much more important in this trade than Allegheny County. Ohio and Marshall counties in West Virginia and Belmont and Jefferson counties in Ohio made 27·3 per cent of the national output of cut nails in 1880.[27] Steel began to replace iron in cut nails, and in the years 1884–6, to meet this change, five Bessemer plants were built in the Wheeling area. Opposing the introduction of the new material without higher rates of pay, the nailers went on strike. This failed to halt the advance of

Table 13. *Cut nail production 1856, 1880, 1884*
(thousand tons)

	1856	1880	1884
Allegheny County	14·2	20·9	22·9
Wheeling Area	6·5	73·5	99·6
Central Pennsylvania	4·6(?)	?	54·2
U.S.A.	81·5	268·5	379·1

Based on Lesley and *A.I.S.A.*

steel, but gave wire nails, made elsewhere, an especial boost, and by 1892 half the nails made in the United States were of wire. Hanging on to the old trade for too long, Wheeling failed to find a large place in wire nail manufacture. New products to occupy the capital and labour of the area were found to a large extent in iron and then steel pipe and in sheet, galvanized sheet, and, after the McKinley Tariff Act, tinplate. Wheeling thus came to have a product range not dissimilar to that of the Valleys, though as a result of backward integration rather than forward integration from iron-making as in the Valleys district. As with the Valleys district there was no important local market, there were serious site problems, the high bluffs fringing the Ohio later limiting rational plant expansion. River transport permitted a linking of the various stages of iron and steel manufacture but this was sometimes accompanied by an uneconomically wide scatter of plants and much loss of heat.

In 1804 and 1806 the first blast furnaces in Ohio were built in Poland Township and at Struthers near Youngstown. [28] They supplied simple castings and hardware to local farmers and tradesmen, but from quite an early date iron was sent from this area to foundries and rolling mills in Pittsburgh. Local mills and forges were established at Newcastle as early as 1806, at another point on Neshannock Creek four years later, and at Niles in 1809, but the district was landbound and, therefore, its market radius was limited. Coal working in the Youngstown area began in 1840 and by 1845 it had been proved that coal from Brier Hill just north of Youngstown could be used uncoked in the blast furnace. Within two years one furnace at Lowell, another on Mill Creek and two in Youngstown were using this 'block' coal, and by 1850 eleven block coal furnaces were at work in both valleys. By this time also 'black band' or carbonaceous iron ore had been discovered below the coal seams of Mineral Ridge south of Niles, and was beginning to replace bog ore in the furnace charges.

Transport conditions were improved when the Pennsylvania and Ohio Canal was completed through the Mahoning Valley in 1839, and the next year, through the Erie Extension Canal, the Shenango Valley also obtained water transport to both the Ohio and Lake Erie. In 1856 the Cleveland and Mahoning Railroad provided a further impetus. The Shenango Valley could not match this advance until 1864 when the railroad from Newcastle through Sharon to the much less important port of Erie was opened.

The canals and railroads made possible a switch to richer, distant supplies of minerals, opened new markets, and justified new plant with its higher productivity. The black band ore was siliceous and iron made from it had been unsatisfactory for many purposes. In order to improve the grade of iron a good deal of mixing of ores went on, some from Lake Champlain or elsewhere in the Lake Ontario region coming as back-haul freight in grain ships using the canals of the St. Lawrence or the Erie Canal, and Lesley listed two furnaces which smelted some Canadian magnetite ore. Missouri ore was an important source of Valley iron in the 1870s. In 1851 the Sharon Iron Company bought Jackson mine in the Marquette district and had smelted some Lake ore before the lower delivery costs resulting from the opening of the Soo canal induced other Valley furnaces to turn to its use. By 1858 Lake ore still cost three times as much as black band ore per ton in parts of eastern Ohio. Even so, it was replacing it there for it enabled furnaces to make a far better iron with higher productivity and less coal.[30]

Commenting on the expansion brought about by the Civil War the *Youngstown Telegram* observed in 1863 that the town would soon have seven blast furnaces, three rolling mills, a steelworks, and two machine shops and foundries.[31] In fact, markets for Valley iron were still to only a small extent local. Pittsburgh conditions and prices continued to determine the prosperity of the iron towns.

There were thirteen coal furnace plants in or on the margins of the Valleys district by 1859 but only four rolling mills, three of which were small, with annual capacities only in the 1 to 5,000 tons range. Pittsburgh with twenty mills and a production of about 90,000 tons of rolled iron a year was engaged in building its first blast furnace, and Valley furnaces were increasing their relative importance in supplying it with iron at the expense of works in the Allegheny or Juniata valleys. Allegheny County made 158,000 gross tons of pig iron in 1873, but forges, mills, and foundries there were estimated to need 400,000 tons. In fact some estimates suggest that 280,000 tons of pig iron were brought in by rail and 18,000 tons by river. Growth in Pittsburgh finishing justified much of the tenfold increase in Valley pig iron production from 1856 to 1872. At this time iron production there was well over half as much again as in Allegheny County, and Valley iron was, as Shinn put it, 'by far the chief supply' of outside pig iron to Pittsburgh.[32]

Pittsburgh mill capacity grew through the 1870s, and after 1875 large steel works added their requirements to those of the iron rolling firms. By now, however, extension of Pittsburgh furnace capacity was also pushing ahead rapidly. (Table 14) Big, and burning coke, these furnaces had a higher

Table 14.   *Allegheny County production of pig iron,*
rolled iron, and steel rails 1856, 1874, 1880
(percentage of U.S. total)

	1856	*1874*	*1880*
Pig iron	nil	5·3	7·0
Rolled iron	9·0[1]	24·7	21·1
Steel rails	nil	nil	N.A.

[1] Rails, nails, plate, and sheet only.
Based on Lesley, and *A.I.S.A.*

productivity than the older coal furnaces of the Valleys, and very low marketing costs. When trade was bad, as in the mid-seventies, Pittsburgh furnaces were able to perform better than more distant suppliers. Allegheny County pig iron production in 1875 was almost one-fifth, or 20,000 tons, up on 1872, but output in the Valleys was down by a similar proportion, or 54,000 tons. Subsequent recovery soon carried Valley production far beyond the level of 1872, but further growth was noticeably slower than in Pittsburgh. Valley iron was still of great importance in times of exceptional demand, but by the mid-1880s, most of the iron used in Pittsburgh was made there.

Table 15.   *Pittsburgh iron ore supply 1873*
(thousand gross tons)

Lake Superior	203
Iron Mountain—by river	88
by rail	25
Lake Champlain	3
Native ore	1
	321

Based on *A.I.S.A.*, 1874.

Following the construction of Clinton furnace, other Pittsburgh furnaces were built by rolling mill interests, as by the Laughlins in 1861 and the Schoenbergers in 1865. Superior furnace, built in 1863, was established before its finishing mill. Market attraction rather than low-cost iron production seems to have stimulated early growth: indeed, pig iron costs were probably lower both to east and west. None of the counties immediately around Pittsburgh produced any of Pennsylvania's $3·9 million worth of ore output in 1870, and Connellsville coke could not prove itself as a superior fuel until bigger furnaces, with higher temperatures and stronger blasts, were built. (Table 15) Not until 1872 with the four Isabella and Lucy furnaces, seventy-five feet high and operating under improved conditions, were the full advantages of the local fuel exploited.

Table 16.   *The iron supply for Pittsburgh 1873*
                          (thousand net tons)

Allegheny County pig iron production	159
Received from outside: Pig Iron by rail	308
Pig Iron by river	19
Blooms and scrap	13
Total	499

Based on *A.I.S.A.* 1874, pp.61–2.

In 1870 the seven Pittsburgh furnaces made 48,000 tons, by 1872 100,000 tons. At this time there were 609 puddling furnaces in the neighbourhood. In 1873 local furnaces were unable to meet as much as one-third of local demand, and as late as 1878 the area was said to bring in 275,000 tons of pig iron a year by rail or river (Table 16). The large local market

Table 17.   *Allegheny County pig iron production 1873–1885*
                      (percentage of U.S. total)

1873	5·54	1880	7·00
1874	5·34	1881	8·34
1875	5·82	1882	6·93
1876	6·14	1883	11·51
1877	6·12	1884	10·61
1878	8·40	1885	12·93
1879	8·70		

Based on *A.I.S.A.*

enabled the iron industry to perform well throughout most of the depression which began in 1873. Again in 1886, Shinn pointed out how well the area performed in times of depression, and had the 1885 figures to back his statement.[33] (Table 17) Though the local market for iron could at last be locally satisfied in years of low production, the productivity of local furnaces was now so much higher than that of most other districts and costs so much lower, that the iron industry and the puddling, steel, and mill operations depending on it remained highly competitive (Table 18, 19).

Table 18.   *Pig iron, steel, and rolled iron production*
Allegheny County
(thousand net tons)

	1874	1880	1885	1886
Pig iron	144	300	586	737
Steel ingots	24	222	407	620
Rolled iron	275	389	413	543

Based on *A.I.S.A.*

Table 19.   *Active furnaces and average furnace capacity of*
leading iron districts, July 1884

District	No. of active furnaces	Average weekly capacity (tons)
Lehigh Valley	25	309
Mahoning Valley	8	382
Shenango Valley	10	539
Allegheny County	10	773

Source: *I.A.*, 10 July 1884, p.17.

Traffic in coke and ore had now been efficiently organized. With a 55 per cent Fe ore, the 1887 freight per ton of pig iron from Lake vessel to Pittsburgh was $2·72 as compared with $1·52 to the Valleys. With a coke rate as low as 1·25 tons the freight to the Valleys was about $2·06 per ton of pig iron while to Pittsburgh probably below $1·30. Theoretically, therefore, Valley furnaces could make iron for perhaps 45 cents per ton less than in Pittsburgh but, as many of them had been built to use raw coal, even with coal/coke mixtures or wholly on coke their productivity was much lower. Shipping pig to Pittsburgh they had to bear freight charges of about 40 to 50 cents a ton.[34]

Pittsburgh had long been the leading centre of crucible steel manufacture, but not until 1875, when the nation made 335,000 gross tons of Bessemer steel, did it have a share in this output. This was partly due to the district's very great advantages as a wrought iron centre, for puddling was a fuel-intensive process. Coal suitable for mill and puddling work could be cheaply won and cheaply delivered even within the city. By the early 1880s some works along the river could obtain slack coal for as little as 36 cents a ton, and even away from the river good coal cost no more than $1 to $1·20 a ton to concerns mining their own. Natural gas, first used in the mid-1870s in puddling, and a few years later in mill work, still more increased the district's competitiveness.[35]

In and around Pittsburgh there were 609 puddling furnaces in 1871, seven years later Allegheny County had 790 and by 1886 1,009. By this time its iron rolling mill capacity was 600,000 tons a year. A wide variety of finished forms were rolled but rails were of very little significance. Perhaps part of the explanation is to be found in the high cost of pig iron in Pittsburgh in the 1840s and 1850s when it had no furnaces. By the 1870s also puddlers' wages were said to be a third to a half higher in Pittsburgh than in the east so that a low-value product like rails would be an unsuitable line.

The peak 'natural' advantage of the Pittsburgh area was probably reached in the last decades of the puddled iron era when coke consumption was still high in pig iron making. Late in his life Henry Bessemer recalled how when he saw the conversion of his first experimental charge of pig iron his mind grasped 'what all this meant, what a perfect revolution it threatened in every iron-making district in the world . . .'[37] Ten years after this Daddow and Bannan observed of the process '. . . The chief economy is in the fuel, which is an important fact to be considered by the inhabitants of those regions where fuel is dear or of inferior quality; and the process may lead to some most important revolutions in the iron industry of the world by transferring the business to regions hitherto unfrequented by iron manufacturers.'[37] Eventually this was to encourage the growth of iron and steel manufacture away from the Appalachian coalfield and the Ohio Valley. Encouraged by the development of Lake traffic and the extension of rail transport and local metal-using industries the nucleus for this growth already existed.

LAKE ERIE AND CHICAGO

At the time of Lesley's survey there were almost a dozen abandoned charcoal furnaces and two active plants in the Lake zone between Sandusky and Erie, and two furnaces near Detroit. Rolling mills were at work in Buffalo, Cleveland, and Detroit—five mills in all with an annual mill output in the three areas respectively of 3,000, 14,400, and 10,300 tons, according to Lesley's figures. But at this stage the Erie shore could not be considered in any way a leading iron-making centre. There was prosperity, expansion, and new building in and after the Civil War, but in the seventies and eighties some of the new works closed. In Buffalo very promising early beginnings ended in long stagnation. Detroit had a few charcoal furnaces smelting Lake ore, a place in history as the first centre of American Bessemer steel manufacture, and a small rolling mill trade. Cleveland was much more important. A leading reason for the slowness of the industry to establish itself seems to have been what was later one of its chief advantages—a central position in relation to the raw materials and markets of the areas to the east of the Appalachians, in the Ohio Valley, and the Middle West. The first two of these areas had bigger mineral resources and all three had larger markets. When local charcoal supplies were exhausted iron smelting suffered from lack of fuel. The Charleston furnace on the Black River near Lorain cleared $65,000 in 1865, but after fire destroyed it in 1871 shortage of charcoal made rebuilding unjustified.[38]

The first iron mills had been built in Buffalo in 1846, largely to supply the huge hardware trade then done from the town. These mills, the Buffalo Rolling Mill and Ironworks, brought in eastern puddled iron by the Erie Canal, and worked it up into nails and bar iron. The first blast furnaces in the area were the Union Iron Works of 1861. Buffalo was still mainly a shipping point between a western raw-material-producing area, and the highly developed east which produced most of the manufactures. There was scope here for processing and for working up iron for western mines, fields, and cabins, but Chicago, nearer to these markets, or the coal-based plants of the upper Ohio were much better placed. In 1856 when mills along the Ohio made over 21,000 tons of nails, Buffalo produced only 1,400 tons. Even so there was a considerable growth during the Civil War and in the post-war years up to the panic of 1873. The Union Iron Company built two more furnaces, puddling furnaces, and mills, while single furnaces were built at Black Rock and North Tonawanda. In the early 1870s Union Iron Works could make 36,000 tons of rolled iron a year, and was for a time perhaps the leading maker of iron beams in the country, rolling the first iron plates for Great Lakes boats, some of which it built itself. Having started with ore from the Adirondacks, by this time it smelted about 50,000 tons a year, mostly from Lake Superior, its fuel being 45,000 tons of anthracite brought in by the Erie and Lehigh Valley railroads. In 1876 the company failed and the plant seems to have been idle until 1892. The North Tonawanda furnace was blown out a few years after construction, and was not at work again until 1889. Black Rock stack turned over to coke fuel in 1881, but four years later was abandoned. As late as 1888 no pig iron, finished iron, or steel was produced in the Buffalo area.[39]

Detroit smelted Lake Superior ores in the late 1850s, but the Wyandotte Bessemer operations were a brilliant flash in the pan, though they were followed by a second mill built in 1877. By 1890 the area was of little importance, having six small charcoal furnaces, three rolling mills and works making crucible steel. There were rolling mills at East Toledo, and at Lockport on the canal east of Buffalo, and apart from these the only other iron and steel plants along the whole lake shore were at Cleveland.

After the first trial shipments of Lake Superior ore came into its harbour in 1852, Cleveland dominated the Lake carrying business and became the chief transhipment centre for ore, the Cleveland–Pittsburgh railroad opening in 1853. It was the main outlet on the Lakes for the products of the mines and factories of the Ohio Valley. Local shipbuilding was a logical development. There was a large hinterland to which the town could supply manufactured goods, and as a consequence it soon had a big local market as well. By 1865 local sales of manufactured iron were over $6,000,000. At Newburgh, the Cleveland Rolling Mill Company had been established in 1857 to reroll old rails, but in the following year a puddling plant and a small blast furnace were added and in 1860 a second furnace. At first some Brier Hill coal from the Mahoning Valley was used, but by the late 1870s Connellsville coke was the only fuel. By 1868 Bessemer steel was made, and within two years wire, plate, and sheet. In the early 1870s came a rush of new construction among which was the 1874 establishment of the Otis Iron and Steel Works, the first American works built expressly to produce open hearth steel to be finished as plate and sheet.

Chicago had six iron foundries as early as 1847, in 1852 the first bridge works was established, and at about the same time locomotive manufacture was begun. For rolling mill operations it benefited from nearness to western markets, an advantage which became more obvious with the building of railroads out towards the west, but for pig iron manufacture or puddling its poor fuel supply was a problem. The new North Chicago Rolling Mill was completed in summer 1858 working on old rails. When, with growing demand, it started rolling rails from new material, it bought its iron from Ohio, Pennsylvania, to some extent from Missouri and even from Scotland, though probably in the last case especially for foundry work. In 1863 or 1864 puddling furnaces were added to the North Chicago mills and in 1864 rail capacity doubled to 200 tons a day when a second rail mill was installed. In a series of articles in 1864, urging the area's advantages for pig iron manufacture, the *Chicago Tribune* pointed out that ore could be brought down through Lake Michigan in boats which had carried grain and other foodstuffs eastwards, but for four more years all the iron used in puddling was brought in, that for a new puddling department built in 1866 coming from Pittsburgh. In December 1868 the first blast furnace plant, with the very small annual output of 12,000 tons, was established by an independent concern, the Chicago Iron Company. In the next two years the Chicago Rolling Mill Company built its own two furnaces. By this time other growth points had begun to emerge in the area—the iron rail mill of the Union Works on the South Branch of the Chicago River in 1863, the Bay View Works at Milwaukee in 1868 and the Joliet iron rail mill two years later.[40]

Already the superior marketing attractions of Chicago were being made plain. Although as late as 1873 the total rolled iron output of Illinois,

Indiana, and Wisconsin was less than one-tenth the national total the west exercised a strong attraction for the capitalist further east. In 1857 E.B. Ward, owner of the Wyandotte mill, realizing the superior marketing advantages of Chicago, built the first rolling mill on the North Chicago River: in 1865, when Wyandotte made experimental Bessemer Steel, it was North Chicago which rolled the first American steel rails from it. Henry Chisholm of the Cleveland Rolling Mill Company reorganized the old Union Rolling Mill Company's plant to produce the first Chicago Bessemer steel in 1871. The North Chicago works and Joliet began Bessemer production in 1872 and 1873 respectively.

By the decade after the Civil War it was clear that locational advantage was shifting west of the Alleghenies. In the west it was not at all clear yet where the greatest advantages would be found, whether on the great river routes or on the rapidly developing lake shores. In 1867 and 1868 Abram Hewitt, one of the most respected eastern ironmasters—and, as his choice of Phillipsburg for anthracite iron manufacture twenty years before had shown, one of the most conscious of locational considerations—speculated on the best location for rail manufacture now that his firm was examining the prospects of steel-headed rails. 'I have no doubt', he wrote, 'that if our mill were moved to Chicago, and Lake Superior iron used, we could make a success in this business.' In the spring of 1868 he went west to look for a location. 'Generally, I incline to the valley of the Ohio River somewhere between Wheeling and Cairo', he noted, but he had also considered the shores of the Great Lakes as far west as Chicago. The rail trade was to show both the strength of westward movement and the importance of imponderable factors in distorting its general progress.[41]

NOTES

[1] I.L. Bell, 'Notes of a visit to the Coal and Iron Mines and Iron Works in the United States, 1875', *Iron*, 28 July 1877, p.98; *J.I.S.I.*, 1886, 1, p.232; A.L. Holley and L. Smith, 'American Iron and Steel Works', a series in *Eng.*, 1877—1880: 'Bethlehem', 24 Aug. 1877, pp.139—140; 'Johnstown', 31 May 1879, p.422, 21 June 1878, pp.485—486, 12 July 1878, p.24.

[2] J. Birkibine, 'The Iron Ore Supply', *T.A.I.M.E.*, 27, 1897, p.519.

[3] H.R. Mussey, *Combinations in the Mining Industry. A Study of Concentration in Lake Superior Iron Ore Production*, 1905.

[4] R.H. Lambourn quoted S.H. Daddow and B. Bannan, *Coal, Iron, and Oil*, 1866, p.549.

[5] J. Fritz, *Autobiography*, 1912, pp.68, 71, 74.

[6] J.L. Ringwalt, *Development of Transportation Systems in the United States*, 1888, p.291; Fritz, op. cit., p.70; J.M. Swank, *Iron in All Ages*, 1892.

[7] Mussey, op. cit., p.82.

[8] Works projects administration, *Technology, Employment and Output per Man in Iron Mining*, 1940, provides a valuable summary of the development of the ore fields.

[9] Mussey, op. cit., pp.60—1, 69.

[10] D.H. Baron, 'The Development of Lake Superior Iron Ore', *T.A.I.M.E.*, 27, 1897, p.343.

[11] W.P. Shinn, 'Pittsburgh—Its Resources and Surroundings', *T.A.I.M.E.*, 8, 1879—80, p.24.

[12] *I.A.*, 21 June 1888, p.1013.

[13] I.L. Bell, 'Report on the Iron Manufacture of the United States and a comparison of it with that of Great Britain', *Iron*, 13 Oct. 1877, p.454.

[14] *I.A.*, 21 June 1888, p.1013.

[15] U.S. Geological Survey, Annual Reports.

[16] W. Isard, 'Some Locational Factors in the Iron and Steel Industry since the early Nineteenth Century', *Journal of Political Economy*, 56, 1948, pp.205, 207–208, 211.

[17] *American Manufacturer*, quoted in *Iron*, 23 June 1877, p.781, and *Eng.*, 8 Dec. 1882, p.552.

[18] Bell, op. cit., pp.34–8.

[19] *I.A.*, 20 Jan. 1887, p.17, 28 Mar. 1889, 4 Apr. 1889, p.515.

[20] J.H. Bridge, *The Inside History of the Carnegie Steel Company*, 1903, p.58. Bell, op. cit., p.34.

[21] 'The Statistics and Geography of the Production of Iron.' Lecture to the American Geographical and Statistical Society, quoted A. Nevins, *Abram S. Hewitt*, 1935, p.139.

[22] *Eng.*, 25 Feb. 1870, p.112. R.H. Sweetser in *I.A.*, 30 Dec. 1943, p.32.

[23] C.H. Ambler, *A History of Transportation in the Ohio Valley*, 1932; E.C. Pechin, 'The Minerals of South Western Pennsylvania', *T.A.I.M.E.*, 3, 1874–5, p.400.

[24] F.H. Rowe, *History of the Iron and Steel Industry in Scioto County, Ohio*, 1938.

[25] Quoted F.P. Weisenburger, *The History of the State of Ohio*, vol.3, 1941, p.33.

[26] *Eng.*, 18 Apr. 1879, p.315; *I.A.*, 21 Apr. 1887; *Iron*, 13 Oct. 1877, p.454; J.L. Ringwalt, op. cit., p.201.

[27] E. May, *Principio to Wheeling 1715–1945*, 1945; L. White, 'The Iron and Steel Industry of Wheeling, West Virginia'. *Economic Geography*, 8, 1932. *Eng.*, 9 Oct. 1885, p.361.

[28] A.L. Rodgers, 'The Iron and Steel Industry of the Mahoning and Shenango Valleys', unpublished Ph.D. Thesis, University of Wisconsin, 1949 and L. White, 'The Iron and Steel Industry of Youngstown, Ohio', *Dennison University Bulletin*, 1930.

[29] F.P. Weisenburger, op. cit., pp.77–9; Rodgers, op. cit., J.P. Lesley, *The Iron Manufacturer's Guide*, 1859, p.107.

[30] Lesley, op. cit., p.488.

[31] Quoted *I.A.*, 3 Oct. 1912, p.763.

[32] W.P. Shinn, 'Pittsburgh and Vicinity. A Brief Record of Seven Years' Progress', *T.A.I.M.E.*, 1885–6.

[33] Ibid., p.667.

[34] *I.A.*, 14 Apr. 1887, 21 Apr. 1887, 13 Oct. 1892, p.676.

[35] The Coal Trade of Pittsburgh, *Reports of H.M. Secretaries of Embassy and Legation Respecting Coal*, 1867; A. Carnegie, 'On Natural Gas Fuel and its Application to Manufacturing Purposes', *J.I.S.I.*, 1885. 1, pp.168–186.

[36] Sir Henry Bessemer, *An Autobiography*, 1905, p.153.

[37] Daddow and Bannan, op. cit., p.648.

[38] R.B. Frost, 'Lorain, Ohio', *Ohio Journal of Science*, 35, 3 May 1935, p.165.

[39] See the comprehensive survey, E.F. Entwisle 'The Iron and Steel Industry of the Niagara Frontier', *I. & S. Eng.*, Jan. 1945, pp.93–104. On Union Iron Works see A.L. Holley and L. Smith., *Eng.*, 22 June 1877, pp.473–4.

[40] B.L. Pierce, *A History of Chicago*, vol.2, *1848–1871*, 1940.

[41] Nevins, op. cit., pp.243, 256.

CHAPTER 4

Iron Production beyond the Manufacturing Belt

Outside the evolving manufacturing belt of the north-east the iron industry made much slower and uncertain progress. In the old north-west or Upper Lakes district and in the west there were raw material supply difficulties, a critical deficiency of fuel in the former and problems with both fuel and iron ore in the latter. The South, particularly the areas around Chattanooga and the new town of Birmingham, was well endowed with both coal and ore. Their quality, it is true, was inferior to that of Connellsville coal or Lake ore but a major compensating advantage was their close proximity. All three peripheral regions suffered especially from marketing problems. Consumption was small and scattered, the density of demand was low. Similarly each region was remote from other major markets. As a result scales of operation remained generally small and techniques lagged behind those of the north-east. The peculiar circumstances of manufacture in these outlying iron districts caused a high mortality of projects or operations and gave them a bad reputation. Finally by the later years of the century bigger producers from the manufacturing belt were invading, and in some cases dominating their home markets.

THE SOUTHERN IRON INDUSTRY TO THE 1890s

Long before the Civil War the markets for iron in the South were supplied to a large extent from outside. Castings, nails, ploughs, engines, sugar mill equipment, stoves, and a whole range of implements were already flooding south along the rivers from Pittsburgh, Wheeling, Cincinnati, and Louisville in the 1830s and 1840s. There was some return trade, as with pig iron and blooms from works along the lower Cumberland or Tennessee or finished iron from Richmond to New England, where it had a high reputation, but generally the flow was southwards. Dependence on outside finished products and subsidiary sales of semi-finished material to be worked up somewhere else was a pattern which was to persist. Iron districts away from coast or from navigable rivers were protected from outside competition but in turn had to depend on local outlets. These were slight.

By the mid nineteenth century the population of the south was smaller than that of the north, growing more slowly, and much poorer. There was little manufacturing and little demand for iron. In 1850 the southern states had only 242,000 cotton spindles out of the national 3,500,000 and the value of farm implements and machinery per acre was only 48 per cent of that of the northern states. There was a small market for bar and foundry iron in the southern ports for the building and repair of steamships, but the railroad market was a poor one in this section, and this in particular adversely affected the whole competitive position of southern works. By 1860 a close mesh of railroads had been laid down between the Northern Appalachians and the Mississippi, but southern construction lagged. Georgia and the Carolinas had 1,180 miles of track in 1850 compared with 1,145 miles in Ohio, Indiana, and Michigan. By 1860 the figures were 3,280 and 5,223 miles respectively. There was even prejudice in the South against southern railroad iron. Tredegar works in Richmond began to roll light rails in 1837 but not until the Gate City mill was built at Atlanta in the late 1850s did the section make heavy rails. Most of the iron for the 3,000 miles of track built in Florida and Alabama in the late sixties and early seventies was British.[1] The South was a major market for nails and La Belle ironworks in Richmond had a high reputation for this product, but it was the only nail factory in the whole section and Lesley estimated that in 1856, when national nail production was 81,460 tons, only 1,075 tons of this came from south of the Ohio and Potomac. Two Wheeling mills strategically placed for the southern market made 6,465 tons.[2]

By the late 1850s the bloomaries and blast furnaces of the South were widely scattered, supplying local customers with unsophisticated products such as direct castings and bloomary bars. Pennsylvanian anthracite iron had already played havoc with this trade in the more accessible areas. The rolling mills of the South were few and strongly localized. South of Maryland and of the mills along the Ohio and near the mouth of the Cumberland and Tennessee—all only marginally southern, though of great importance in southern markets—only two counties at that time rolled over 1,000 tons of iron a year, Henrico County, Virginia, whose centre was Richmond, and Fulton County, Georgia, whose mill was in Atlanta. The Civil War devastated this under-developed industry.

CIVIL WAR AND THE EARLY RECONSTRUCTION PERIOD

Iron production was feverishly extended as blockade cut off foreign supplies. Not surprisingly, many war projects were economically indefensible. Before 1861 Alabama had 9 primitive blast furnaces, 17 forges and 1 rolling mill, but 12 new furnaces were built with Confederate money, and altogether during the war the state operated 16 furnaces and 6 rolling mills.[3] Production rose to 30,000 tons of pig iron and 10,000 tons of bar iron a year, but by the end of the war each furnace and mill had been destroyed. A number of Shenandoah Valley ironworks were wrecked, a not surprising fate in view of the routeway nature of the area. At other times their long supply lines to Richmond were cut. Dependent on local consumption, southern iron works were crippled for a number of years after the war by the distress of the section's farm economy, and much of the modest progress in railroad building made in the fifties had been lost by the need to pull up existing track for re-laying in more critical war zones, or by

Union destruction. However, by 1872/3, with the north booming, southern states had begun to pull out of stagnation. They were already showing gains over iron production levels of the late 1850s. But these gains were relatively small, and some parts still lagged far behind. In the boom year 1873, Alabama, which had rolled up to 10,000 tons iron a decade earlier, turned out only 500 tons (Table 20).

Table 20.   *Production of pig iron and rolled iron by states*
1856 and 1872/1873
(thousand tons)

| | 1856 | | 1872/3[1] | |
	Pig Iron	Rolled Iron	Pig Iron	Rolled Iron
Pennsylvania	449	241	1,401	788
New England				
New Jersey	142	154	436	382
New York				
Ohio	87	31	400	248
Maryland	31	15	63	58
West Virginia	1	20	21	52
Virginia	6	6	21	13
Kentucky	71	19	110	54
Tennessee				
Alabama	1	—	12	0.5
North Carolina				
South Carolina	5	2	4	11
Georgia				

[1] 1872 for pig iron, 1873 for rolled iron.
Based on Lesley, op. cit. and *A.I.S.A.*

Lack of capital for reconstruction and operation of the works, shortage of skilled labour, and a general pessimism about the future of the section retarded recovery. By July 1873 James Noble from Rome, in the formerly promising iron district of north west Georgia, noted how few furnaces there were in his own state or Alabama, except those owned almost wholly by capitalists from other parts of the country.[4] South Carolina provided an extreme case, where loss of the capital represented by the slaves brought the complete ruin of a significant iron industry. The Secretary of the A.I.S.A. remarked that it was a 'singular fact' that even in the boom of 1872 and 1873 not one of the eight charcoal furnaces in the state was at work.[5] There was no capital for essential repairs and, in fact, these furnaces never worked again. The rolled iron trade was no better. Spartanburg and Gaffney in the

Piedmont section of the state had two fair-sized mills built in 1835 and 1840. The larger plant had been continuously busy until 1865, drawing on the mineral resources of 11,000 acres of land to supply four furnaces, rolling mill, nailworks, forge, and foundry. By the early seventies it operated only its foundry. The other works too was idle through the boom.[6]

For a time favoured parts of the South shared to some extent in northern prosperity by choosing to emphasize the least capital-intensive trades, shipping large tonnages of iron ore north to the Ohio river and into Indiana between 1872 and 1875.[7] When, with the business revival of 1879, iron-making at last began a very marked advance, it was to a large extent with northern, or even foreign, capital and entrepreneurial skill. At the same time, the geography of iron production was changing. For a number of years Virginia remained the leading state, but there was a move southwards first to Tennessee and then to Alabama. The Piedmont was displaced by the Great Valley (Table 21).

Table 21.   *Southern rolled iron and steel production*
1873, 1880, 1893
(thousand net tons)

	1873	1880	1893
Virginia	12·8	37·7	37·8
Georgia	10·6	1·5	—
Tennessee	16·5	25·4	9·4
Alabama	0·5	6·6	27·0

Based on *A.I.S.A.*

CHATTANOOGA AND BIRMINGHAM AS SOUTHERN IRON CENTRES

Atlanta had been the only important railroad junction of the South in 1850, but construction in the next decade made Chattanooga of at least equal importance. By this time also the advantages of its rich 'dye stone' ores and of the Coosa coalfield on the edge of the city had been recognized. As early as spring 1860 John Fritz had examined coal and ore properties near Chattanooga.[8] By 1873 Swank, Secretary of the A.I.S.A., was hailing it as the future leading iron city of the South.[9] Union soldiers who had camped nearby during the war noted its undeveloped mineral resources, and from their initiative stemmed some early postwar growth. J.T. Wilder, in 1867, brought capital from his home state of Ohio and from Indiana, to found the Roane Iron Company, built a coke furnace near Nashville, and in 1871 resumed rail-making in the idle Chattanooga mill. In 1874 the Chattanooga Iron Company blew in a blast furnace, and in 1877 the Vulcan ironworks was reopened with a capital of $400,000 to make bars, rails, nails, and spikes. In the same year the Tennessee Iron and Steel Works made bars and T-rails, and Roane Iron Company was building a small open hearth plant in the town.[10] A little below Chattanooga, the development of the Sequatchee Valley was begun by an 1876 contract between the Nashville, Chattanooga and St. Louis Railroad and an English group headed by John Bowron from the Cleveland iron district. This Southern States Coal Iron and Land Company undertook to open collieries in the valley and to build blast furnaces where it opened into the Tennessee, at which point South Pittsburg was laid out in 1876. For every $1,000 spent on railroad extensions, Southern States agreed to invest $2,000 in plant, and to spend at least $300,000 in five years.[11] In the mid-eighties furnaces were built further down the Tennessee at Decatur, Sheffield, and Florence. Forty miles northwest of Chattanooga, the Tennessee Coal and Railroad Company built coke ovens at its Tracy City coalmines, and in 1880 put up a blast furnace at Cowan on the mainline of the Nashville, Chattanooga and St. Louis railroad. As late as 1886, when Roane invested $1,000,000 in Bessemer plant, the Philadelphia correspondent of *Engineering* still referred to Chattanooga as 'the centre of iron-making in the South'.[12] Within the next few years Birmingham, a cornfield when Chattanooga was already developing rapidly, became the focus of Southern growth.[13]

Reconnaissance work for a railroad to link north and south Alabama through the central mineral districts began in 1858. Jones Valley, lying along the western edge of Red Mountain, was at that time 'one vast garden as far as the eye could reach'.[14] In the Civil War the first three blast furnaces were built near the Red Mountain ore body. Later, when the Alabama South and North Railroad was connected through the Nashville and Decatur into the Louisville and Nashville system, the basis for much more substantial growth was laid. Birmingham was established at a junction point in the new railroad system. At first consideration its attractions to the iron manufacturer were inferior to Chattanooga's, an established, bigger rail focus, with lower freight rates to the north by rail and in addition able to ship by the Tennessee. The mineral endowment of the Chattanooga district was inferior. By 1887 1·5 tons of coke per ton of iron was usual around Birmingham, but Chattanooga coal was poorer and the coke rate was about 2 tons. Coke at furnaces there cost $2·50 as compared with $2 a ton, around Birmingham.[15] Around Chattanooga the Clinton ore beds were thinner and there were not the same superlative assembly conditions as at Birmingham. There in Jones Valley, the absence of some of the sandstones and slates between the red iron ore and the limestone, and of other measures between the ore and coal meant that iron-making raw materials were available in close proximity. The Louisville and Nashville owned 500,000 acres of mineral land in central Alabama by 1876, and built a large mileage of tracks linking iron towns, mines, and ore workings. In the fifteen months to March 1887, eleven blast furnaces were built or contracted for in Tennessee, and fifteen in Alabama, of which twelve were in the tract between Birmingham and the new town of Bessemer. In 1880, Alabama made 70,000 tons of iron, only 6,000 tons more than Tennessee, but by 1890, at 831,000 tons, its output was just over three times as great. The Louisville and Nashville had by this time thirty-two furnaces on its lines.[16] A key point in development occurred in 1886, when the Tennessee Coal Iron and Railroad Company bought its way into the Birmingham area. Coke smelting now became established and grew rapidly. (Table 22) Even though coal was costly to work, needed more preparation before coking and gave a rather poorer coke than in the north, it proved possible to make foundry iron at highly competitive prices. (Fig. 8)

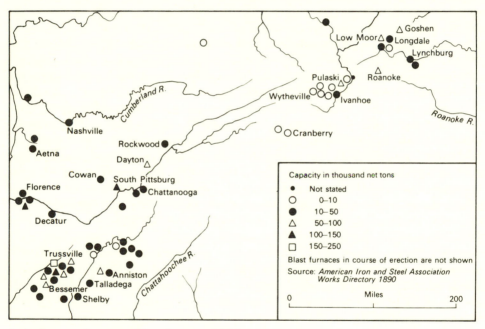

8. Active south eastern blast furnaces 1890

Table 22. *Coke production in Alabama and Tennessee 1880–1895*
(thousand short tons)

	1880	1885	1890	1895
Alabama	61	301	1,073	1,444
Tennessee	131	219	349	397

Source: U.S. Geological Survey.

COSTS OF PRODUCTION, IRON MARKETS, AND COMPETITION

As early as 1873 a Select Committee was assured that with the red ore 'all you have to do is to blast it, throw it into the valleys and knock it to pieces' and '. . . whenever we get our means of communication completed . . . the mines and the coal will take care of themselves'.[17] In fact, the mineral endowment of the South was of indifferent quality, the area's only outstanding advantage being the proximity of its coal and ore as compared with the almost 1,000 miles which separated the far better minerals of Lake Superior and of the northern Appalachian coalfield. Bewitched by prospects of cheap labour and low cost mineral assembly, outsiders parted with their capital far too readily. De Bardeleben, one of the entrepreneurs, typified the ambitions of the new southern ironmasters hoping to build up great new iron centres. '. . . there's nothing like taking a wild bit of land, all rock and woods—ground not fit to feed a goat on—and turning it into a settlement of men and women; making payrolls; bringing the railroads in; starting things going . . .'[18] The town of Bessemer was an outstanding example of the transformation that was achieved in a few cases. Until 1887 it was a forest-covered tract, but in 1891 had 4,544 people and was in direct connection with seven railroads.[19] Many entrepreneurs chose badly. Even when they succeeded there was dispute as to the quality of their plants. Occasionally, commentators reckoned the equipment of southern works was good. In 1888 *Iron Age* noted 'one of the most striking features of the southern iron industry is the very excellent character of the appliances used.'[20] The concensus of opinion was undoubtedly that southern equipment was indifferent or bad. In 1890 visiting members of the Iron and Steel Institute reckoned coalmining operations inefficient, and William Colquhoun of Cardiff described the furnaces as variable in quality, with some very poor examples. 'Some of the furnaces erected in the South are a miracle of frailty . . . furnaces have been known to topple over in a high wind, or to have been driven hard for several weeks without metal appearing through the tap-hole.'[21] A little later his opinion was echoed by Haller.[22] Looking back to the boom years of 1878—82, Axel Sahlin described the connection between southern promotion, the inflow of northern capital, poor plant, and the disappointed hopes which turned boom towns into depressed ones. 'Estimates were procured from contractors for furnace plants of the flimsiest description. On the basis of these, glowing prospectuses were drawn up proving that small investments would reap golden harvests. The moderate sums considered sufficient were subscribed by northern capitalists, who were attracted by the hope of profit, but as a rule knew nothing of the business they were embarking in.' In the event '. . . one plant after another was put out of blast, towns were abandoned, trees grew up between the rails of the mineral railways, decay and hopelessness succeeded the feeling of extreme buoyancy.'[23] These conditions were commonest in the small towns and with the less notable companies, but even Birmingham and the Tennessee Coal,

Iron and Railroad Company were not wholly free of them, the latter sometimes ending one day's business without knowing where the money for the
next day's operations would come from.[24] Carnegie, Hewitt, and other
northern industrialists visited the South in 1889. They were impressed by its
resources and potential competitiveness, but as Carnegie wrote shortly
afterwards: 'Still it is always to be remembered that success is largely
dependent upon management; poor management in the South, as in any
other place in the world, will ruin the best enterprise.'[25] To a member of
Carnegie Steel who once suggested the purchase of T.C.I. Carnegie rejoined
that he would never buy such a 'football' proposition. Southern management, interpreting the term in a wide sense, could certainly provide him with
examples enough of bad management, and southern labour then, and long
afterwards, was inefficient and in real terms much more costly than pay
packet figures suggest.[26]

The greatest weakness of the iron industry of the South remained lack of
local consumption. In 1877 the British trade journal *Iron* asked 'Is it only a
wild flight of the imagination to suppose that there may some day be a
Birmingham in Alabama which may rival Birmingham in England?' Thirteen
years later the President of the Alabama National Bank, welcoming the Iron
and Steel Institute to Birmingham, was reckoning that by 1900 the place
would have grown so much that a much larger pattern would be needed and
the name would be changed to London.[27] In fact the name was particularly
inapt, for Birmingham, England was above all a finishing centre, working up
wrought iron and, later, steel made elsewhere. Samuel Noble, founder of
Anniston, painted the remarkable contrast in 1883, though perhaps in
exaggerated form. 'The great trouble is, we have no home market beyond the
demand created by the iron furnaces themselves. The whole state of
Alabama cannot take the product of a single blast furnace for a month. We
depend entirely on the North and great West to keep our furnaces going
. . .'[28] In fact, in the eighties southern outlets did grow in importance. Some
Hudson Valley stove makers built southern plants to get cheap iron, by 1888
the U.S. Rolling Stock Company of New York had built an Anniston plant
making car wheels, axles, and castings. The cast iron pipe trade was growing
at the expense of old centres in eastern Pennsylvania. Nevertheless, by far
the greatest bulk of southern iron had to be sold outside the region—
estimates in the 1890s ranging up to well over 90 per cent of the output.[29]

The first outlets were in the Ohio Valley and the Middle West, but in the
1884/5 depression came the first big invasion of north-eastern markets, and
by November 1884 southern iron was said to have a 50 cents to $1 a ton
advantage in New England over iron from the Lehigh Valley. By early spring
1886 two weeks of contracts for southern iron delivered in Philadelphia,
Chicago, Cincinnati, Louisville, and some western points amounted to
40,000 to 50,000 tons. This inflow eliminated the last elements of the
eastern Pennsylvania charcoal iron industry.[30] Yet although it could wreak

havoc in the north, the position of southern iron there was insecure. Markets for forge iron were shrinking rapidly as steel replaced finished iron; moreover, Pennsylvanian ironworks reacted to the challenge by cutting prices, and while the Louisville and Nashville did all it could to extend the southern marketing area, northern railroads which made the final delivery tried to protect their local works. In 1888 Lehigh Valley ironworks received freight concessions to allow them to meet the low prices of southern pig, and in 1890 the Pennsylvania Railroad raised rates on southern iron by 60 cents a ton. Southern lines countered by absorbing it in their own charges. By the late nineties low costs at southern furnaces were largely cancelled out by freight charges ranging between $1·75 and $2·75 a ton for delivery to the Ohio and $3·25 and $3·75 to the north-east.[31] Especially in times of recession distant marketing was an uncertain business as was shown by the panic which began in 1893. 1894 pig iron production in the country as a whole was 72·6 per cent of the high 1892 figure; Pennsylvania production was 80·3 per cent. Tennessee managed only 70·6 per cent and Alabama 64·7 per cent. After this, with Mesabi ore and rationalization, costs at northern works west of the Appalachians fell rapidly. Southern firms now began to export iron on a considerable scale. Over the years 1896 to 1900 the Sloss-Sheffield Company exported 196,000 tons of pig iron, and in 1900 Alabama's foreign sales were 84 per cent the U.S. total. Severe competition was feared by some foreigners. J.S. Jeans of the British Iron Trade Association remarked of his American journeys in 1901, 'time did not allow me to visit Chicago. I had to choose between Chicago and Birmingham and I deemed the latter the more important centre from the point of view of future competition.'[32] Foreign markets were, however, notoriously uncertain, and Alabama and its southern neighbours were not the only low-cost producers in the field. Over the years 1902 to 1907 the average cost of Alabama iron delivered New York was $17·24 a ton. Virginia Iron, Coal and Coke Company pig iron could be delivered there 1907/8 for $18·28. It was claimed that English iron could be delivered, duty paid and with profit, for $18.[33] The only satisfactory long-term solution for the southern iron trade was to go into steelmaking. This raised new problems.

THE UPPER LAKES

The founders of the city of Superior are said to have thought of it as a rival for Chicago. Later both it and Duluth were spoken of as potential Pittsburghs of the west, and in 1892 it was predicted that Duluth would be the greatest steel centre in the country within twenty years.[34] The projections were all sadly awry, ambitions remained unfulfilled, for in each case there was a strange lack of appreciation of the economics of iron and steel-making, and the intrinsic unsuitability of the upper lakes district.

Iron markets in the Upper Lakes district were negligible even by southern standards. The ore mines, or those of the copper country, took some, the

lumber camps a little more, but not until the railways were pushed
westwards did an apparent natural market area appear. Writing in 1886
Carnegie described this new wider market in tones which show how easy
others must have found it to be overcome with the fever of speculation in
northern iron. 'Look at the great North-West. Scarcely a decade has passed
since it was represented as a barren, icy plain, wild, inhospitable and scarcely
habitable. The railway has changed it as by a wizard's touch. Minnesota has
more than a million inhabitants. The population of Dakota has quadrupled
in five years, and is now half a million. Towns are springing up with magical
rapidity. Its wheat crop last year was thirty million bushels—twice as great as
the whole crop of Egypt.'[35] The wheatfields required implements, the
pasturelands were beginning to use large tonnages of wire fencing, the
railroad track and rolling stock needs were immense and the towns
demanded a host of other iron products, but what was easily overlooked was
that Chicago was already well established in the iron trade, its mills enjoyed
a gigantic local market, were almost as near to western consumers as the
head of Lake Superior, and were joined to them by far better rail links. On
this rock, and on that of high fuel costs, one north western plant project
after another foundered.

1880—1900. FRUSTRATED HOPES

There were twenty-four furnaces in the Lake Superior ore district by 1876
but only one rolling mill. In the following year the Escanaba blast furnace
was removed to become the first Edgar Thomson furnace. Two years later
the rolling mill built at Jackson, Michigan, only seven years before was
removed to Springfield, Illinois.[36] The eighties seemed to promise better
things. Ore shipments grew rapidly. Menominee was a wilderness until 1877
but in 1880 shipped 524,000 tons, and in 1881 1,100,000. Between 1875
and 1885 deliveries from all the lake ranges went up from 890,000 to
2,400,000 tons.[37] As the traffic grew a return cargo became more and more
desirable, and with the encouragement of ballast rates the coal traffic up the
lakes, long expected, at last increased. By 1886 the Marquette to Ohio port
rate on ore was $2·15—though it fell to $1 or $1·15 in the following years.
Birkibine put the Cleveland to Duluth freight for coke at $1·25 with another
60 cents added for dock charges and unloading. Allowing for loss and waste,
Connellsville coke could be delivered in Duluth for $5·60 a net ton, only 30
cents more than its cost delivered by rail in Chicago. If coke was made at
Duluth instead it seemed possible to cut costs to $5·25, allowing for over-
heads on the ovens.[38] By 1891 return freight was so much in demand that
the usual Erie port to Duluth coal rate was 25 cents a ton and in the summer
as low as 10 cents.[39] It was tempting to conclude, not only that pig iron
could be made more cheaply at an upper lakes port, but that a business there
could be made highly successful. So by 1888, when the average price of
Bessemer pig iron at Pittsburgh was $17·38, it was reckoned that Duluth

could make iron for $14·80 a ton. There was talk of a $2,000,000, 150,000 tons a year integrated works to make rails, using Vermilion and Gogebic ore and coking coal from the Reynoldsville basin, Pennsylvania. In the same year the Lehigh Coal and Iron Company built the first upper lake coke works at West Superior, a battery of 60,000 tons capacity. In 1890 Marquette Board of Trade made a study of the prospects of coke iron manufacture.[40]

Achievement fell far short of hopes.. A Duluth coke furnace built in 1889 stood idle to at least the autumn of 1891. There was one active mill on the lake in 1890, but two others in upper Michigan had been recently abandoned, and the only other one in the whole north-west, at St. Paul, had been idle four years.[41] The West Superior Iron and Steel Company embarked upon large-scale developments from its foundry and machine shop plant in 1889. It began a large coke furnace, which it never finished, installed two Bessemer converters which started work in January 1892, and rolled bars, plates, and structurals. The 1893 depression dragged the plant down almost at once. It was unable to make a full recovery and by 1896 was in receivers' hands. In 1899 the Duluth Manufacturing Company, the Wisconsin Steel Company, and the Ironton Structural Steel Company, all of whose works were idle, were merged as the Lake Superior Steel Company, but its attempt to revive production in the Upper Lakes was unsuccessful. Within the next three years the two Duluth mills were abandoned, the West Superior works remained closed—apart from its cast iron pipe plant—while to the south-west Republic Iron and Steel Company, rationalizing, had abandoned the steelworks and made idle the tie and shape mill at Columbia Heights, Minneapolis. Another ambitious project in 1903 to make iron, steel, and a variety of other products in a new works at Sault Ste Marie proved yet another flash in the pan. North of a line from Saginaw Bay to Minneapolis nine charcoal ironworks survived—out of a considerably larger number in the early nineties—but there were no active coke stacks, steelworks, or mills.[42]

It was easy to find scapegoats for this failure, though these in turn usually depended on deeper-rooted, inherent regional disadvantages. In the early 1890s, for instance, Rockefeller who had interests in the West Superior Company was shocked to find that it had not been established purely as a business proposition but partly to promote the importance of the town of Superior. A few years later Birkibine referred to shortage of capital and preoccupation with land speculation.[43] If trade prospects had been sounder these side benefits would have been less attractive and there would have been no shortage of capital. In view of the events of the next few years it is well to survey the basic conditions of production and trade in the upper lakes district.

In 1898 Birkibine submitted a substantial report in which he reckoned that Marquette should be able to produce coke pig iron in a similar furnace for 24 cents a ton less than Buffalo or Cleveland. In 1898 ore freights down the lake were 50·8 cents a ton, the back-haul rate on coal or coke 23·4 cents.

Assembly costs on ore and coal per ton of pig iron were $3·03 for Chicago, $2·72 for Pittsburgh and $1·86 for Marquette. None the less, Birkibine's overall conclusions were restrained, for the area was seen to have no great cost advantages over other districts.[44] There is reason to suppose that even this opinion was too sanguine. Scale and regularity of operations was of major importance in affecting process costs, and here an isolated area, still very sparsely populated and with very few foundries or metal fabricating plants, was at a disadvantage. In the nineties the plants proposed or built were Bessemer ones but there were no large local scrap arisings able to support a big open hearth steel plant. Duluth did become a collection point for western scrap but Chicago was even better placed in this respect, and additionally had huge local supplies. The north-western steel market was growing but was still very small. It was facile to provide impressive figures of deliveries of steel from other districts, and from these to conclude that there was an ample market for an integrated works, forgetting that a host of different categories were needed, implying either a very large investment in different types of mill, none of which would be fully stretched for most of the time, or frequent changes of rolls if different sections were to be produced on the same mill. The twentieth century brought no radical change in these conditions. (Fig. 9)

THE MINNESOTA IRON ORE TAX AND THE DULUTH STEEL PROJECT

By the early twentieth century the ironworks of the western half of the manufacturing belt were almost wholly dependent on Lake Superior ore, and increasingly on that from Mesabi. Ore production by this time was dominated by the United States Steel Corporation. Increasing dependence on Minnesota ore, the evidence that, in spite of this, Pittsburgh, Chicago, Lorain, and now Gary received the lion's share of plant expenditure, and growing irritation with Pittsburgh Plus prices for steel purchases, gave Minnesota both means and incentive to influence the Steel Corporation's development plans. (Table 23) In April 1907 United States Steel announced plans for a Duluth integrated plant to make rails, structurals, and merchant steel.[45] During the next two years test pits were sunk there, the site was graded, and some construction work was undertaken on the railroad to connect the site to the main lines into Duluth. This leisurely progress was in sharp contrast with the urgency with which construction was pushed ahead at Gary.

By 1908 the Minnesota Tax Commission estimated that the Steel Corporation controlled 76·6 per cent of the state's ore or 913 million tons. In 1909 a tonnage tax of about 5 cents a ton was suggested in a bill before the Minnesota legislature, which, on 1909 shipments from Menominee and Mesabi, would have cost U.S. Steel almost $1 million. Among the arguments used against the bill, it was said that promise to build the Duluth works had been given on the express understanding that no tonnage tax would be

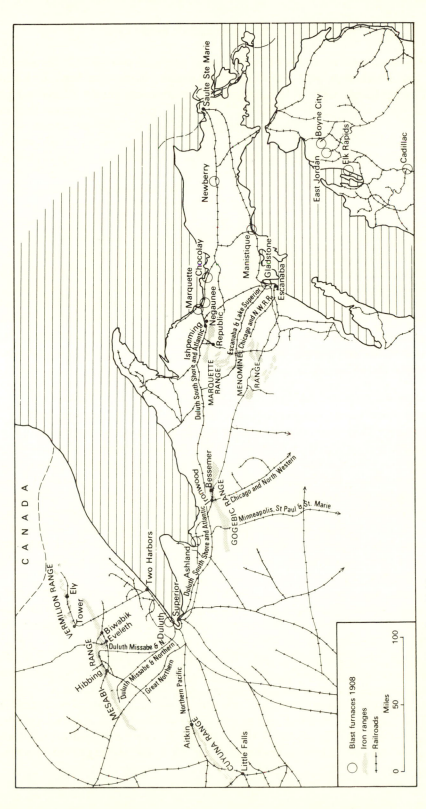

9. The upper Lakes district 1908–1910

imposed. Delay in construction may have reflected a Steel Corporation attitude of 'wait and see', though on the other hand it may have been no more than the result of preoccupation with work at Gary. Passed by the

Table 23.   *Production of Lake Superior and Minnesota*
iron ore by the United States Steel Corporation
1901 and 1906
(million tons)

	Lake Superior	of which Minnesota
1901	13·02	8·21
1906	20·64	14·06

Based on Bureau of Corporations *Report on the Steel Industry*, 1, 1911, p.270.

legislature the bill was vetoed by the Governor, but, as one commentator neatly put it, 'as the governorship is not a life job it looked well to the Steel Corporation to go ahead with the plant.'[46] Speaking to Duluth Commercial Club in September 1911, James Farrell, President of United States Steel, assured his hearers that they intended to build a works 'complete in every respect for the manufacture of every class of steel product for which a market can be found.'[47] The last phrase was to prove of great significance. Even after this, construction was slow and production did not begin until the spring of 1915, eight years after the first announcement—compared with four years for the much bigger Gary project.

In the next few years further light was shed on the decision to build at Duluth. In the Pittsburgh Plus hearings E.H. Gary revealed that Duluth interests had objected to the new pricing system, and pressed the Corporation to build there. Other concerns could not have afforded to do so.[48] To the Duluth Chamber of Commerce in 1918, he had been even more frank. He analysed why the bulk of expansion had been at Gary.

> Why did the Steel Corporation build a plant on a sandy desert along the southern shores of Lake Michigan? Because of a love for Indiana? Oh no, none of us have any particular interest in that state. It was purely a business proposition. We would much rather have spent the money expended in Gary right here in Duluth. *Our friends are here. But the proposition would have been a failure from a business point of view.* There fuel was easily obtainable, as were other things which are necessary to the making of steel. There was a market. For the manufacture of pig iron Duluth is well situated, perhaps nearly as well as almost any other city. But Birmingham can manufacture pig iron $3·05 more cheaply a ton than can Duluth. As to steel products, Duluth is behind Gary by 38 per cent, Pittsburgh by 13 per cent.[49]

The Duluth works of Minnesota Steel Company, as the new subsidiary was called, was built on a two-mile frontage of the St. Louis River—a choice of site which burdened it with the need to tranship its coal from the lake docks and rail it to the works.[50] This site was 2,000 acres, and only one-fifth of this was occupied by the plant, a situation which encouraged expectations that, as at Gary, other Steel Corporation companies would build finishing mills.[51] Initial annual rolling capacity was 350,000 tons of rails, shapes, and merchant products. Operations did not prove very profitable, other United State Steel subsidiaries did not occupy the rest of the tract surrounding the works, nor did local steel-using industies boom. As early as 1917 it was being urged that Duluth should be made a steel basing point so that lower steel prices could stimulate local consumption.

From the start the Duluth works operated well below capacity. Able to make 350,000 tons of finished products it shipped only 89,000 tons in 1919, and in 1920, nationally a very good year for the industry, 108,000 tons. 1921 was a year of sharp depression which cut American output of rolled steel to 45·6 per cent of the 1920 level, but Carnegie Steel shipments were almost 50 per cent and Illinois Steel managed a 57·4 per cent performance. Minnesota Steel was closed for most of the year and shipped only 6,000 tons. In 1920 just over one-fifth of the Duluth plant's steel was consumed in Minnesota, and other north-central states, apparently in its market area, took little. Illinois and Indiana received the largest share, 50,244 tons, but Illinois Steel Company supplied 115,721 tons of finished steel to the state of Minnesota, 8,000 tons more than the local plant shipped altogether.[53] In the light of these figures it is difficult to avoid the conclusion that its newest integrated steel works was an embarrassment to United States Steel even though under Pittsburgh Plus pricing, and with rail freight rates advancing rapidly, there was every incentive to supply local orders from the nearest mill so long as production costs were low. By 1918 rail freight from Pittsburgh, to Minnesota Steel a phantom freight, or bonus, was $6·58 a ton and in 1920 $13·20. Yet Carnegie Steel supplied Minnesota customers with 1,391 tons of steel in 1921 and Illinois Steel shipped 91,834 tons. There could scarcely be a more conclusive evidence of the dangers facing a smallish plant, located apparently in accordance with a minimum assembly cost principle for the raw materials of iron-making, but ignoring everything else.

The small Upper Lake iron and steel industry shrank in the late 1920s. U.S. Steel abandoned its Milwaukee blast furnaces in 1928, concentrating production at Chicago and Gary. Furnaces at Ashland closed, and in 1934 Youngstown Sheet and Tube Company dismantled the 275,000 ton Mayville ironworks, Wisconsin. There were suggestions of a Ford steelworks for St. Paul but this came to nothing.[54] In the twenties Gary ingot capacity was increased by 1·4 million tons or 45·7 per cent but Duluth remained unchanged at 540,000 tons. 1929 was a high point for steel nationally, production being 81·4 per cent of rated capacity, but Minnesota managed only 71 per cent. In 1932 output was 19·6 per cent of the capacity figure of

1930 as compared with 21·4 per cent in Minnesota, but, whereas 1933 was a year of sharp improvement nationally, Minnesota output dropped to less than half the level of 1932.

In the thirties Duluth was placed in the control of U.S. Steel's subsidiary American Steel and Wire. One of its blast furnaces and five open hearth furnaces were abandoned in 1935, it being explained that they were both obsolete and uneconomically located—a strange admission for a plant just twenty years old. 1938 steel capacity was 300,000 tons, 44·4 per cent less than in 1930. By this time the structural mill had ceased work and rolling capacity was in bars and wire rods. When, in June 1938, the Steel Corporation abolished most of the inter-basing point price differentials, it retained those for Worcester, Massachusetts, and Duluth, observing that production costs were high, partly due to the need for frequent roll changing, and that distribution costs were pushed up by the scattered nature of the market.[55]

THE WEST TO THE FIRST WORLD WAR

FURNACES AND MILLS TO 1880

The Far West was a considerable consumer of iron and later of steel long before it was an important producer. Railroad demand soon exceeded that from farms, mines, and towns. Between 1871 and 1880 alone 2,439 miles of track were built in the Pacific states, a total equal to a consumption of almost 600,000 tons of rails. California, the only western producer, turned out a little over 43,000 tons in this period.[56] By 1874/5 the Pacific coast was said to import 250,000 to 300,000 tons of iron in all forms a year. This figure was inflated by the inclusion of highly manufactured articles, but by this time iron finishing trades, bringing in most of their material, were developing. San Francisco in 1877 had 47 foundries, one rolling mill, and a wire and wire rope works.[57]

Marketing, and the process of scrap collection, together helped to explain the location of mills at great railroad junction points. Iron scrap was usually not in a form to be easily used again without puddling, but worn rails from western tracks provided an important exception. Main-line rails could be rerolled into lighter sections or slit and rolled into bars. A number of the mills in this business were small-capacity plants brought from the east where they were no longer competitive. The Decatur, Illinois, rail mill, built in 1870, was moved 300 miles westwards to the neighbourhood of Kansas City five years later. A few years afterwards a mill from Dunleith in the extreme north-west of Illinois was moved to Omaha. In 1877 the mill from Danville on the Susquehanna was transferred to Pueblo to reroll rails, only to be moved on to Denver in the following year.[58] Between 1873 and 1885 other iron rail rerolling plants were built at Burlington on the Missouri, at Topeka, at Houston, at Laramie, and in San Francisco. They had short lives and were often idle. On the other hand they were cheaply built, had no large investment tied up in primary iron-making capacity, and their greatest strength was remoteness from the bigger eastern mills. This enabled them to operate successfully in spite of higher unit costs. Eventually, this advantage was eroded by improved rail service and decreasing eastern production costs. On the Pacific, European and east coast works could offer keen competition by virtue of the ballast rates in returning grain boats.

MINERAL RESOURCES AND BLAST FURNACE OPERATIONS IN THE WEST

There were high-grade iron ore deposits in the west, and, in some parts, thick woodlands ideal for charcoal. Frequently these two materials were far apart, and still more often away from any main line of movement to market. Coals suitable for coking were usually remote from ore or markets. In 1854 magnetities of over 64 per cent iron content and 45 per cent hematites were found in the foothills of the Sierra Nevada, forty-two miles north-east of Sacramento. Fifteen years later a futile attempt was made to smelt the ore. In 1874/5 there was another fiasco, this time with ore from Shasta County smelted on a site provided at Sacramento by that city and developed with New York capital. In the 1850s Utah began an iron-making career, to be chequered with failures for seventy years. At Iron City in the iron ore country of the south-west of the state, the Great Western Iron Manufacturing Company built a charcoal furnace which lingered on for a few years of the 1870s, and at the end of that decade occurred a failure at Ogden.

The Oswego Iron Company first made iron on the Williamette River in 1867, and opened an iron ore and timber estate served by a narrow gauge railroad. By 1890, it was processing half its iron in a pipe factory, the only one west of St. Louis. Even so the blast furnace was idle throughout 1886 and 1887, and after blowing out in 1894 was not at work again until at least 1904.[59] East of the continental divide hopes similarly flickered and died. In the early sixties the Union Pacific Railroad had proved ore and coal deposits near Fort Laramie and Cheyenne and built a small blast furnace at Langford. Only remoteness could justify the high operating costs—ore delivered for $9 per ton of iron, the 250–300 bushels of charcoal costing $25 to $30, and pig production costs overall anticipated as $45 a ton.[60] A year later the plant was abandoned, costs of transport by ox-team having proved ruinous.

Lack of good coking coal reasonably near iron ore and iron markets was a critical impediment to larger, more efficient operations. Washington coal, the only possible source in the coast states, had a high ash and sulphur content. Utah coal was better, but the field was away from the main railroads, ore was 200 miles to the south-west, and although the state had a nodal position in relation to the main coastal centres, it was far away and its own population was small. Coking coal was found in southern Colorado and northern New Mexico, and in 1880 this gave rise to the first big coke iron plant, which for sixty years was to be the main western primary metallurgical centre.

South Pueblo, on the Denver and Rio Grande Railroad, was chosen by the Colorado Coal and Iron Company as the location for the new works. The earliest coal properties were to the south-east in Las Animas County, but others were opened 140 miles to the west around Crested Butte. Iron ore was brought from properties to the west and north-west and up to 70 miles or so away. Railway nodality made mineral assembly relatively cheap and conferred marketing advantages. Pueblo made nails, spikes, bolts, and bars, but rails soon became its chief product. It had a hard time in this trade. Rail capacity was 54,000 tons by 1890, but in the following year only 34,000 tons of rails were rolled, though in 1893 output went up to almost 49,000 tons.

Both raw materials and marketing problems were involved in Pueblo's uncertain growth. Local iron ores seem to have been poor, so that it had to begin to draw from further afield. By 1900 it was still using ore of 50 per cent iron content from Orient, Colorado, to the west of the plant, its old source, but also smelted ore from Hartville, Wyoming, 350 miles away, and ore of 61–2 per cent iron railed 600 miles from Fierro in south-western New Mexico.[61] By this time its hard coke came mainly from 90 miles south of the works and steam coal from little more than half as far away. In the early eighties the Great Plains were in their heyday as open range cattle country, and soon passed into ranch and farm expansion. However, Pueblo's markets were not only dispersed but also subject to wide variations in western prosperity both in the farm economy and in mine output. It also became clear that the distance of over 700 miles which separated Pueblo from St. Louis and 1,000 from the mills of Chicago gave less protection than might at first be expected. Railroad expansion in some of the states immediately bounding Colorado was very small in the 1890s; track mileage in Nebraska, Utah, and Wyoming went up only from 7,675 to 8,662 miles. In Kansas there was a huge growth, from 4,227 to 9,000 miles over the years 1884 to 1900, but much of this area was almost as accessible to Chicago mills. A position in the centre of the western roads mileage between the Mississippi and the Pacific was, however, a considerable advantage, and early in 1893 the company had contracts reckoned equal to about 40 per cent of this area's rail consumption, orders totalling 60,000 tons for three of the leading western rail roads, the Santa Fe, the Denver and Rio Grande, and the Union Pacific.[62] However, freight rate policies ensured that Pueblo often had no advantage over Chicago in west-coast markets. (Table 24)

In 1895 the Interstate Commerce Commission ordered that the tariff on iron and steel from Pueblo to San Francisco should at no time be more than 75 per cent of the rate from Chicago. However, by 1898, railroads had again raised their charges to a level comparable to the Chicago rate, and in 1900 refused an appeal by Colorado Fuel and Iron against this. C.F.I. gained some advantages through the connection which it had with the Gould railroad interests—in 1905, to coerce United States Steel to make the promised connection with the Wabash in Pittsburgh, Gould lowered freights on C.F.I. rails and wire which extended Pueblo's market sphere at the expense of that company's Chicago works.[63] In Denver C.F.I. reaped a handsome phantom

Table 24.   *Freight rates on steel rails to San Francisco 1895*
($ per 100 lbs.)

From:	$
Pueblo	1·60
Chicago	0·60
Pittsburgh via Chicago	0·72
Pittsburgh via New York	0·71

Source: Interstate Commerce Commission, *Report for 1895,* 1896, p.41.

Table 25.   *Iron and steel plants west of the Rockies*
1890, 1896, 1901, 1904

	1890	1896	1901	1904
Blast Furnaces				
Utah	—	Ogden (P)	?	Ogden (P)
Oregon	Oswego	Oswego*	Oswego*	Oswego*
Washington	Irondale	Irondale*	Irondale	Irondale
Steelworks				
California	San Francisco	San Francisco	San Francisco	San Francisco (3)
Washington	—	—	Irondale (p)	—
	—	—	—	West Seattle (P)
Oregon	—	—	—	Portland
Rolling Mills				
Wyoming	1	1	1	1
California	4	5	3	5 and 1 (P)
Washington	—	2	2	1 and 2 (P)
Oregon	—	1	1	2

*Idle. By 1904 Oswego had been idle for ten years.
(P) building or projected.
Based on A.I.S.I. *Works Directories.*

freight by making its prices just competitive with the delivered price of Chicago steel. In retaliation, local consumers continued to buy large tonnages of eastern material.

Important extensions were made at Pueblo at the end of the nineties, and by 1904 its rail capacity was almost six times the 1890 level and it ranked as one of the ten biggest in the country. After this the plant grew more slowly though it remained a major railmaker for western roads.

As compared with Pueblo a west-coast integrated works would have had costlier material assembly, suffered from exposure to foreign and eastern competition, and lacked the advantages of a central position in the track system of the western railroads. Steel consumption of the western interior was possibly then at its maximum in relation to the country's total consumption. The situation was complicated in various ways, as with the freight rate irregularities discussed above, or Harriman's decision in 1907 to buy rails from T.C.I. even though Pueblo was in the centre of the vast western system which he controlled. Growth in the rest of the west was very slow and uncertain, but in the long run Pueblo was to prove a location inferior to the Pacific coast. (Table 25).

UTAH AND THE PACIFIC COAST IN THE EARLY TWENTIETH CENTURY

Utah attracted a good deal of general development survey work. Joseph Sellwood of Pickands Mather made a study of Utah minerals soon after 1900. He estimated that there were at least 500 million tons of iron ore in the south-western part of the state. Before 1909 U.S. Steel engineers concluded that 'when the time comes that Salt Lake City can distribute large quantities of that material' it would be a suitable point for iron and steel making.[64] For the time being the Steel Corporation did not feel the proposals justified acquisition of Utah ore, for, to defray the big assembly and marketing bill, the plant would have to be big, and western consumption prospects were reckoned inadequate. A few years later failure of small steel-making operations at Midvale, Utah, just south of Salt Lake City, seemed to endorse the Steel Corporation's decision. After this Utah merely provided pig iron to be used in west-coast melting shops and foundries. Elsewhere neither raw material supply nor marketing was even as favourable as in Utah, and all the other plans, dreams, achievements, and failures were coastal.

After early disappointments in smelting, growth of Pacific coast steel-making remained dependent on high scrap charges. Local scrap and the focusing of the railways which tapped country scrap helped to reinforce the strong market attractions of the three main population groupings on the coast, but the division of this market between them, its still modest size, and raw material supply difficulties caused the failure of one plan after another for big operations. Alternatives to coking coal were tried. Technically the electric blast furnace of the Nobel Electric Steel Company in Shasta County was a success. It made excellent foundry iron but eastern iron was cheaper, and in the First World War the company switched to the manufacture of ferro alloys. On the Hood River in Oregon black sands were smelted in electric furnaces. In California attempts were made to use fuel oil. In 1912 the California Industrial Company was planning to smelt Mexican and Southern Californian ores at Wilmington, Los Angeles, using crude oil in specially designed burners and supplying iron to open hearth furnaces, rail, and other mills.[65] The iron-making part of this project was a failure. Renamed the Southern California Iron and Steel Company, by 1915 it was using scrap and also pig iron from the Hanyang works in the Yangtse Valley.[66] Recognizing fuel as the greatest obstacle to southern Californian iron-making Hardar in 1912 described Colorado as the nearest source of coking coal. He believed coal might also come from Alaska. Utah coals he ignored, probably because of their deficiencies until the narrow slot-type oven was introduced some five years later. He had already recognized the future key role in ore supply of Eagle Mountain—'. . . it is beyond doubt, however, that, before long, blast furnace and steel plants will be established at points in southern California, as they are now being established on Puget Sound, and, in this event, the Eagle Mountain ores will probably be among the first to be utilized.'[67]

The competitive situation of coastal plants was complicated by the need to take into account foreign and east-coast sources of iron and steel supply. In 1909, at a time of new west-coast projects, the Payne Tariff Act reduced duties on most steel goods. Normal freight charges on iron and steel from Europe to Pacific ports ranged from $3·50 to $5 a ton, but ballast rates were often little more than nominal, so that European mills paying freight of perhaps $2 and customs duties of $6 had delivery charges there only half those of the overland route from the eastern states.[68] Bigger eastern firms began to improve their position in the west by developing an efficient warehousing organization enabling them to combine economies of scale with prompt service. By 1912 the Steel Corporation held 3 to 4 million dollars' worth of stock in its San Francisco warehouse, and sold 100,000 tons from it. There were always 25,000 to 30,000 tons of material on their way there, the material being transhipped and railed across the isthmus of Tehuantepec.[69] The opening of the Panama Canal threw in another complication. In 1915, in reaction, the railroads cut their transcontinental rates. Until the middle of that year there was a blanket rate from points east of the Mississippi of 80 cents per 100 lbs. or $16 per net ton, but then the rate from Chicago was reduced to $11, from Pittsburgh to $13, and from other mills east of Chicago to $15 a ton.[70] All these developments might be expected to have worked against the west, but war-time growth of demand stimulated expansion. In 1915 melting shops were built for the first time at Midvale, Utah, at Tacoma, and at Wilmington, Los Angeles. Yet in spite of market expansion, except for the last they were only short-term successes.

NOTES

[1] J.F. Stover, *The Railroads of the South 1865—1900,* 1955, pp.18—19. *Reports by H.M. Consuls on British Trade Abroad,* III, 1873, Command No. 824, p.759.

[2] J.P. Lesley, *The Iron Manufacturer's Guide,* 1859, p.758.

[3] Ethel Armes, *The Story of Coal and Iron in Alabama,* 1910, pp.157—69, 185—6 (Miss Armes lists the Civil War Works p.86); B.C. Colcard (President of Woodward Iron Company), 'The History of Pig Iron Manufacture in Alabama', *B.F.S.P.* Feb. 1951.

[4] Senate Document. *Report of the Select Committee on Transportation Routes to the Seaboard,* 1874, p.719.

[5] *B.A.I.S.A.,* 1874, p.35.

[6] *B.A.I.S.A.,* quoted *I.C.T.R.,* 16 Dec. 1874, p.778.

[7] V.S. Clark, *History of American Manufactures,* 1929, vol.2, p.63; Senate Document. *Report of the Select Committee on Transportation Routes to the Seaboard,* 1874, pp.497—8.

[8] J. Fritz, *Autobiography,* p.135.

[9] *A.I.S.A.,* 1873, p.58.

[10] *American Manufacturer* quoted *Iron,* 3 Nov. 1877, p.554.

[11] *Colliery Guardian* (London), 8 Dec. 1876, p.895. *Iron,* 6 Aug. 1880, p.119.

[12] *Eng.,* 30 July 1886, p.107.

[13] *Eng.,* 4 Mar. 1887, p.214.

[14] Armes, op. cit., p.113.

[15] *I.A.,* 24 Feb. 1887, p.27.

[16] C.V. Woodward, *Origins of the New South,* 1951, pp.126—7.

[17] Senate Documents. *Report of the Select Committee on Transportation Routes to the Seaboard,* 1874, pp.834, 836.

[18] Armes, op. cit., p.XXX.

[19] *Iron,* 31 July 1891, p.91, 7 Aug. 1891, p.114.

[20] *I.A.,* 14 June 1888, p.971.

[21] The Iron and Steel Institute, *The Iron and Steel Institute in America 1890,* 1891, p.350.

[22] C. Haller in *Stahl und Eisen* quoted *J.I.S.I.,* 1897, 2, p.393.

[23] Axel Sahlin in J.S. Jeans, *American Industrial Conditions and Competition,* 1902, p.499.

[24] *I.A.,* 22 Nov. 1906, pp.1388—90.

[25] *Manufacturer's Record,* 15 Mar. 1889, quoted *J.I.S.I.,* 1889, 1, p.208.

[26] On this question see E.O. Hopkins, President of Sloss—Sheffield, *Report of Industrial Commission,* XIII, *Trusts and Industrial Combinations,* 1901, p.508; H.H. Campbell, *The Manufacture and Properties of Iron and Steel,* 1907, p.483.

[27] *Iron,* 3 Nov. 1877, p.555; *The Iron and Steel Institute in America, 1890,* pp.462—463.

[28] Quoted Armes, op. cit., pp.301—6.

[29] *I.A.,* 24 July 1890, p.261; *Eng.,* 2 Mar. 1888, p.210; Armes, op. cit., p.315; *I.A.,* 24 Nov. 1892, p.984.

[30] Woodward, op. cit., p.127; *Eng.,* 21 Nov. 1884; *Eng.,* 23 Apr. 1886, p.409; *J.I.S.I.,* 1896. 2, p.348.

[31] Clark, op. cit., vol.2, p.287; *I.A.,* 17 July 1890, p.100, 24 July 1890, p.261, 6 Jan. 1898, p.19; *Report of Industrial Commission,* XIII, p.516.

[32] Jeans, op. cit., p.1.

[33] *Tariff Hearings,* 1908—1909, p.1426.

[34] *Iron,* 2 Dec. 1892, p.500.

[35] A. Carnegie, *Triumphant Democracy,* 1886, p.27.

[36] *A.I.S.A.,* 1876, p.145; *A.I.S.A.,* 1878, p.14.

[37] J. Birkibine, 'The Resources of the Lake Superior District', *T.A.I.M.E.,* 16, 1887—8, pp.172, 175.

[38] Birkibine, op. cit., and *T.A.I.M.E.,* 27, 1898.

[39] *I.A.,* 17 Dec. 1891, p.1977.

[40] *J.I.S.I.,* 1888, 2, p.378; U.S. Geological Survey, *Annual Report 1895—1896, Metallic Products and Coal,* p.66; *I.A.,* 3 May 1888; *J.I.S.I.,* 1891, 1, p.302; *I.A.,* 4 Aug. 1898, p.10.

[41] *I.A.,* 15 Oct. 1891, p.642. A.I.S.A., *Works Directory,* 1890.

[42] A.I.S.A., *Works Directory,* 1896; *A.I.S.A.,* 1891, p.41; J.I.S.I., 1892, 1, p.431.

[43] A. Nevins, *A Study in Power,* Vol.2, 1953, p.206, *I.A.,* 4 Aug. 1898, p.14.

[44] The Report was Summarized in *I.A.,* 4 Aug. 1898, pp.9—14.

[45] *I.A.,* 4 Apr. 1907, p.1078.

[46] *I.C.T.R.,* 1 July 1910, p.12.
[47] *I.A.,* 21 Sept. 1911, p.626.
[48] *I.A.,* 23 Nov. 1922, p.1367.
[49] *I.A.,* 6 Mar. 1919, p.611 (my italics); *I.A.,* 9 Oct. 1924, p.918.
[50] L. White and G. Primmer, 'The Iron and Steel Industry of Duluth: A Study in Locational Maladjustment', *Geographical Review,* 27, 1937.
[51] *I.A.,* 30 Nov. 1911, p.1198.
[52] *I.A.,* 23 Nov. 1922, p.1367; *I.A.,* 12 June 1924, p.1726.
[53] *I.A.,* 19 Feb. 1925, pp.545, 547.
[54] *I.A.,* 1 Jan. 1925.
[55] *I.A.,* 7 Apr. 1938, p.70B; A.I.S.I., *Works Directories;* Temporary National Economic Committee (T.N.E.C.), *Investigation of Monopoly Power,* part 19, 1939–1940, pp.10542, 10554.
[56] *A.I.S.A.,* and Ringwalt, *Development of Transportation Systems in the United States,* 1888, p.201.
[57] *J.I.S.I.,* 1875, 1, p.333. But see *J.I.S.I.,* 1882, 2, pp.726–7; *Iron,* 18 Mar. 1876, p.363.
[58] Swank, *Iron in All Ages,* 1892, pp.343, 344.
[59] *J.I.S.I.,* 1882, 2, p.726–7; *I.A.,* 10 July 1890, p.58.
[60] Secretary of the Treasury, *Statistics of the Foreign and Domestic Commerce of the United States,* 1864, p.206. Swank, op. cit., p.343.
[61] H.H. Campbell, op. cit., 1907, p.492.
[62] *I.A.,* 2 Mar. 1893, p.484.
[63] *I.A.,* 27 Apr. 1905, p.1399.
[64] *I.A.,* 18 Apr. 1912, pp.981–2.
[65] E.C. Hardar, *Iron Ore Deposits of the Eagle Mountain, California,* U.S. Geological Survey, 1912, p.7, *I.A.,* 22 Aug. 1912, p.406.
[66] *I.A.,* 20 May 1915, p.1110.
[67] E.C. Hardar, op. cit.
[68] *I.A.,* 23 Sept. 1909, p.931.
[69] J. Farrell, *I.A.,* 22 May 1913, p.1242.
[70] *I.A.,* 4 May 1916, p.1086.

CHAPTER 5

The Rail Trade

THE ESTABLISHMENT OF THE AMERICAN RAIL TRADE

The United States had 4,185 miles of railroad before the first heavy rail was rolled in the country. The Lehigh Coal and Navigation Company made cast iron rails in 1826, but its works had a short life. Flat rails, strips of iron which were spiked down onto the ties, were also produced, but no heavy, wrought iron rails. American orders flooded into Wales, one agent alone placing orders for 40,000 tons in Britain in 1836. However, as demand increased, so too keen competition among Welsh works caused the quality of their product to decline, and in 1842, helped by a tariff of $25 a ton, the American industry was ready to begin production. Early in 1844 Mount Savage rolled the first heavy T rails in North America (Fig. 10). By this time the duty of $25 a ton was more than the cost of production in England, but the New Jersey Iron Company at Trenton let it be known that it would supply large tonnages at the import price of $55 a ton, and within less than five years expected to be able to produce for as little as $40 a ton.[1] By 1847 eight mills were rolling rails, at Mount Savage, Danville, Boston, Providence, Trenton, Phoenixville, Scranton, and, beyond the mountains, and in a works expressly built for the rail trade, the Great Western Iron Company at Brady's Bend on the Allegheny. Within three years seven other mills could roll heavy rails. Some of the locations were ill-chosen and had low survival value. Brady's Bend was an outstanding example of a poor location, and Ringwalt noted the large role of expediency in the choice of Scranton, a major factor in the rail business throughout the rest of the century. The Scranton brothers seized on the urgency with which the Erie Railroad had to complete its line to Binghampton to raise capital to build the Lackawanna rail mill.[2]

In July 1846 a new tariff act introduced an *ad valorem* duty of 30 per cent on rails. As prices were then high abroad this gave good protection for a time, but the British market was depressed in 1847, prices there fell, and the tariff became less and less protective. British iron exports more than doubled from 1845 to 1850, in 1850 the United States made only two-thirds as much iron as two years before, and in the early months of 1850 only two of the fifteen rail mills were at work, and then only because of their inland position. In 1853 only a quarter of the American consumption of 400,000 tons of rails was of home manufacture, but by 1856 things had improved and the home make was almost half the consumption.

In 1856 114,500 out of 142,000 tons of home-rolled rails came from plants east of the Alleghenies, but half of the rails imported 1850—7 were laid down in the Atlantic states and so only two of the six mills built

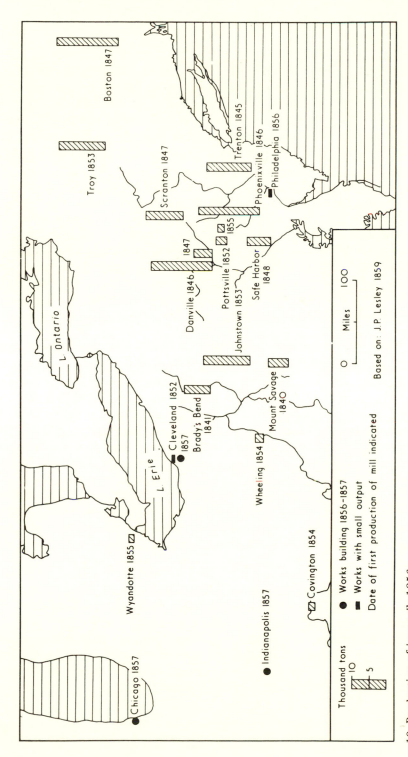

Boston 1847

Troy 1853

Scranton 1847

Trenton 1845

Phoenixville 1846

Philadelphia 1856

L. Ontario

Danville 1846 1847

1855

Pottsville 1852

Johnstown 1853

Safe Harbor
1848

Cleveland 1852

Brady's Bend
1841

1857

L. Erie

Mount Savage
1840

Wheeling 1854

Wyandotte 1855

Chicago 1857

● Indianapolis 1857

Covington 1854

Thousand tons
10
5

● Works building 1856-1857
■ Works with small output
Date of first production of mill indicated

0   Miles   100

Based on: J.P. Lesley 1859

10. Production of iron rails 1856

between 1850 and 1856, Troy and Palo Alto at Pottsville, were in the east. The others were Johnstown, Wheeling, Covington and Wyandotte (Fig. 11). Meanwhile a number of eastern mills were abandoned or changed to less competitive trades. This happened with the Providence mill, the Treaty mill in Philadelphia, Glendon in East Boston, and Avalon mill near Baltimore.[3] As Lesley's account was written new mills were being completed in Cleveland, at Newburgh, Indianapolis, and Chicago. Some of these produced little other than rerolled rails for a number of years, but they marked the firm establishment of a western rail trade. Protected from foreign competition, and with consumption shifting rapidly in their direction, they could meet the competition of bigger eastern integrated works. Growth in the south-eastern railway system was also rapid at this stage, and at Atlanta, only rivalled as a southern railroad junction by Chattanooga, the first important southern rail mill was completed in 1858 with a capacity of 1,200 tons a year, about one thirty-sixth of the annual rail need of the south-eastern group of states in the fifties for new track alone. Some of the southern supply came from the north, much from Britain.

By the late fifties, the big markets had definitely moved across the Appalachians, though the quick wearing-out of iron rails provided an eastern outlet for rails for track replacement. (Table 26) Between 1860 and 1870

Table 26.   *Railway mileage by sections 1850–1860*

	1850	*1860*
New England	2,505	3,666
Mid Atlantic[1]	2,722	6,318
South-eastern[2]	1,848	6,215
Great Lakes, Ohio River[3]	1,275	9,012

[1] New York, New Jersey, Pennsylvania, Maryland, Delaware.
[2] Virginia, North and South Carolina, Georgia, Florida, Alabama.
[3] Ohio, Indiana, Illinois, Michigan, Wisconsin.
Source: J.L. Ringwalt, *Development of Transportation Systems in the U.S.,* 1888.

track mileage went up by 834 miles in New England, 4285 miles in the middle section, and 12,476 miles in the mid-west and west. By 1864 western mills had become much more prominent than in 1856, though all five mills west of Cleveland still only produced rerolled rails. (Table 27)

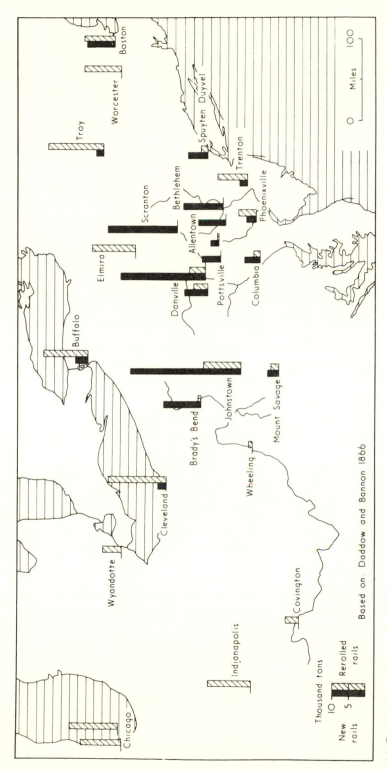

11. Production of iron rails 1864

The position of the eastern mills, and that of any other producer with high processing costs, worsened in the depression which began in September 1873, and which for the rail business deepened as Bessemer rails began to cut

Table 27.   *Iron rail production 1864*
                    (thousand tons)

	New Rails	Rerolled Rails
New England	8·7	21·5
New York and New Jersey	13·6	55·4
Eastern and Central Pennsylvania	119·5	28·5
Western Pennsylvania	10·8	0·6
Ohio	2·1	18·1
Indiana and Illinois	—	39·6
West Virginia and Kentucky	—	5·2
Michigan	—	5·6
Maryland	3·2	2·2
Total:	158·1	177·2

Source: S.H. Daddow and B. Bannan, *Coal, Iron, and Oil,* 1866.

into the market. The average price for iron rails at eastern Pennsylvanian mills was $85·12 a ton in 1872, but four years later $41·25.[4] (Table 28) Mount Savage gave up rail making at an early date, and the works was dismantled in 1875. Safe Harbor last rolled iron rails in 1861, the National Iron Company at Danville failed in 1873, and in 1877 Pottsville works turned to shape manufacture. Phoenix Iron of Phoenixville failed in the 1857 depression, and though it was still important in rails to the end of the Civil War it gradually transferred to beams. Trenton began to roll steel-headed rails in 1866 which sold at $135 a ton, when Bessemer rails were $170 and iron rails in 1868 were under $79. By the Autumn of 1869 Trenton was rolling these at the rate of 24,000 tons a year and making $10 a ton profit, but within a few years all-steel rails were so cheap that this trade was no longer tenable.[5] Other eastern firms turned over to Bessemer rail production, the higher prices for these making their locations viable again, but in the process they solved their immediate problems by laying up trouble for the highly competitive days in steel rails twenty years ahead. In the long run nothing could shelter them from the fact that both consumption and low-cost production were shifting westwards.

Table 28. *Production of rails of all types*
1856, 1864, 1871, 1876
(thousand tons)

	1856	1864	1871	1876
New England	17·9	30·3	42·2	25·7
New York	13·5	57·3	87·0	57·3
New Jersey	13·0	11·7	6·7	0·2
Pennsylvania—eastern	122·6	36·2		
central	41·5	67·6	335·6	353·9
western	20·7	55·6		
West Virginia	2·3	0·8	5·0	0·5
Kentucky	1·9	4·4	6·0	1·5
Ohio	—	20·3	75·8	100·8
Michigan	1·8	5·6	14·0	1·6
Indiana	—	12·7	12·8	29·4
Illinois	—	26·9	91·2	181·5
Maryland	7·1	5·5	44·9	18·8
Wisconsin Missouri Western States	—	—	36·9	77·9
U.S. total:	141·5	335·5	775·7[1]	879·6[2]

[1] of which 735, [2] of which 467 were iron rails.
Based on Lesley, op. cit., Daddow and Bannan, op. cit., and J. Pecher, *Coal and Iron in all Countries of the World*, 1878.

BESSEMER RAILS

In 1861 the Pennsylvania Railroad was persuaded to buy the first Bessemer rails used in the United States from Charles Cammell of Sheffield. By 1869 John Browns, also of Sheffield, had supplied 51,000 tons of steel rails, or more than all other British mills. The first American Bessemer steel was made in September 1864 at Wyandotte, Michigan, and the following May the first rails were rolled from this at the North Chicago rolling mill. When the Cambria Works at Johnstown rolled the first strictly commercial order of rails in 1867, the national output was only 2,550 tons. In spite of tariff protection these and other early works were not operated with profit.[6]

Eight of the first ten steel rail plants were at existing puddled iron and iron rail works, and when the output was small it was practicable to roll steel on the old mill. As production increased, it became clear that this was not satisfactory, for steel was much tougher. Soon the pattern of steel rail production came to differ in important respects from than of iron rails. 129,000 tons of steel rails were rolled in 1874, and one-third of these were made in Illinois. Output in 1875 was double that of the previous year, and in 1875 Pittsburgh first had a hand in the trade, the Edgar Thomson works rolling that year under 6,000 tons rails. By the late 1880s Cook County, Illinois still retained its lead over Allegheny County, Pennsylvania in rails, and sometimes also in Bessemer ingots. After 1882 the North works and Union works were joined by the big new lake-shore South Chicago works, and beyond the county line Joliet was busy in the trade. In spite of this, right through to the end of the 1880s, the biggest tonnage of rails came from eastern and central Pennsylvania, from works whose pre-eminence in iron rails had been founded on the success of anthracite iron manufacture forty years before. These eastern works were still highly competitive. This may be illustrated by the situation of depressed demand in 1884 and 1885.

With prices falling in the summer of 1884 the correspondent of *Engineering* wrote:

> It is not difficult to see where the pinching will begin, with rails quoted at $34 in Chicago and deliveries from Pennsylvania made in that city at $30·50 to $30·75, it is easy to see that the western mills must feel the pressure of competition and suffer first and longest. The Pennsylvania rail makers, after their failure to harmonise steel rail interests have resolved on another course, more effective and more in consonance with business principles, viz. to take care of themselves and let other influences take the hindmost.

By late August both Bethlehem and Scranton were offering rails from stock at $27·50.[7] In fact, the actual performances of 1884 and 1885 proved *Engineering* to be only half right in its forecasts. National output in 1884 and 1885 was 86·7 and 83·5 per cent respectively of the high level of 1883 but Illinois output was up by 25·7 and 33·5 per cent. Output at the Edgar Thomson works and in the rest of Pennsylvania fell away. Yet works east of the Allegheny Front continued to be highly successful, rolling 41 per cent of U.S. rails even in such a difficult year as 1888. A number of factors account for this continued viability of eastern mills during the period when new track construction was moving westwards.

Their market situation was by no means as bad as appeared at first sight. For southern or Gulf orders water carriage gave them important advantages, a certain amount of new building still went on in the east, and renewal orders were important. At the beginning of the eighties there were 82,000 miles of iron rail track in the country. By 1888 a quarter of this had been taken up and replaced with steel. This business gave eastern mills the higher level of operations necessary to reduce costs enough to compete in more distant markets. As steel rails lasted so much longer this advantage later declined. In Autumn 1884 it was reckoned that production costs for rails were $26·83 per ton in western Pennsylvania and $1 less in the east of that state. One factor in favour of eastern works as compared with those of Chicago, though not with Pittsburgh mills, was the possession of local fuel for the cupolas used to melt down the pig iron before conversion. Although as late as 1892 the metallurgist Howe reckoned the metal was better from cupola practice, the progressive advance of hot metal operations in the eighties eliminated the cupolas and this asset.[8] Cheaper labour was another help. In the early autumn of 1884 Pennsylvania Steel cut its wage rates by 10 per cent, Scranton and Bethlehem by 15 to 20 per cent but in January 1885, to compete, wages in the Carnegie mills had to be reduced 20 to 33 per cent.[9] Later in the same year Carnegie found a far more effective reply to the rather lower labour costs of the east when, with the introduction of new machinery at Edgar Thomson, 57 of the 69 men on the heating furnaces and 51 of the 63 on the rail mill train were dispensed with.[10] This was a clue to another factor of great importance in the competitive power of the mills—the ability or willingness to innovate. Equipment was not yet so standardized as to exclude a breakthrough in process costs which might cancel out the apparent disadvantages of location. Technical excellence, an aggressive sales policy, might still mean more than nearness to materials or markets. As Schwab put it many years later 'In those days there was a great opportunity for the chemist and the engineer to reduce costs that doesn't exist today. The most skilful men with unlimited capital couldn't reduce the price a $1 a ton.'[11] As a result the two Scranton works were successful when Joliet and still more the Vulcan works at St. Louis, each apparently very well situated, were unprofitable for many years and passed through failure and reorganization. In the late 1870s the North Chicago works was making records in spite of its cramped site and bad layout. A little later, with technical excellence this time accentuating already very great natural advantages, Potter was reaching new high levels of production at South Chicago.

The whole situation must be set in its price context. Prices for Bessemer rails in 1868 were about twice as high as those for iron rails. After that, as production costs were lowered, prices fell, but demand remained buoyant so that production went up steadily through the crisis which began in 1873, and from 1867 to 1882 each year's output was greater than the last. Even at its lowest position in the seventies, the average price for rails in 1878 was $42·25. Although competition was sometimes keen this gave room enough for a large number of mills to survive and even for some poorly located ones to do well. Early in 1886 prices were $34 to $35, in the spring of 1888 they ranged from $31 to $32 at eastern mills, to $33 to $34 at Chicago. By the beginning of 1889 the price was under $29. There was a rise at the beginning of 1890 but after that the price was not to go above $30 and at times was below $20. Lower prices, reduced demand, and consequently sharply increased competition, taken along with a number of other basic changes in the production situation, caused major alterations in the geography of rail production in the next ten years. Changes in raw material supply and in marketing were important factors, but were by no means the only ones, and the degree to which both opportunities and challenges were taken up was, as always, but perhaps never more than in this trade and at this period, affected by organizational genius, commercial policy, and plant modernization decisions.

THE RAIL POOL AND CONDITIONS OF TRADE FROM THE LATE 1880s TO 1893

The rail situation was unique within steel production and marketing. Demand was dominated by a few large concerns, the product had only one use, it could be made only by equipment which cost much more to install than that for almost any other category of steel and yet which had been greatly overbuilt. As Potter put it, 'the rail is today the cheapest finished product in the whole domain of iron and steel manufacture, and is at the same time the most difficult to make. It requires an expenditure of at least $3 million before a single rail can be economically turned out.'[12] By 1884 seventy-one mills made rails in the United States, but most were small iron rail mills, and by August 1887 the first rail pool, which included all the important works, had only fifteen members—the Carnegie allocation being divided between Carnegie Phipps and Co. and Carnegie Brothers. The pool's aim was to avoid the sort of struggle which would not only eliminate the weakest mills, but leave even the most powerful with carrying charges and running costs inadequately covered. An allotment of 800,000 tons was divided between the firms, and a further allotment of 250,000 tons was given to the Board of Control to be given 'as and to whom it may be deemed equitable in the adjustment of any differences that may arise'. (Table 29) There were said to be penalties of $1·50 to $2·50 a ton for production in excess of the allocation. Even though 1888 output was less than two-thirds that of 1887 a collapse of prices was avoided.

The four east and central Pennsylvanian members of the pool received only 34·8 per cent of the total allocation, but in fact in the year these two sections of the state made 41·5 per cent of the national output of rails, the North Chicago Company with 12·5 per cent of the allotment made 162,000 tons of rails at its North and South works, 11·6 per cent of the total. Even exluding Homestead's production, Carnegie made 148,000 tons of rails at the Edgar Thomson works, but this production was below the company's allocation.[13]

In the second half of 1893, at a time of depression, the rail pool collapsed. Allegheny County output fell slightly; for Edgar Thomson, its chief producer, the share of the national output was down from 21·4 per cent in 1892 to 20·3 per cent in 1893. Decline in eastern Pennsylvania, mainly at Bethlehem, was greater, but on the other hand Cambria and Lackawanna did better. Illinois Steel performed badly. National output of rails was down 26·7 per cent on the levels of the previous year but the decline in Illinois state was 48·3 per cent and there were times when every mill in the Chicago area was idle. Union rail mill did not work at all that year, North works was laid off during the last half, Joliet ran for only six weeks, and the South Chicago works, the obvious choice for a concentration of production,

Table 29.   *1888 Rail tonnage allotments among*
                           rail pool members
                              (percentage)

Eastern works:	Bethlehem	9·0)	
	Lackawanna	8·0)	
	Scranton	8·0)	
	Pennsylvania	9·8)	40·7
	Troy	4·5)	
	Worcester	1·4)	
Western Pennsylvania and Ohio:			
	Cambria	8·0)	
	Carnegie	13·5)	26·3
	Cleveland	4·8)	
Western:	North Chicago	12·5)	
	Joliet	8·0)	
	Springfield Iron Company	—)	33·0
	Union	8·0)	
	Western Steel Co.	4·5)	

Source: Report of the Commissioner of Corporations, *The Steel Industry,* 1911, p.69.

operated for only eight and a half months.[14]   For much of the mid-1890s Illinois Steel was in serious straits. (Fig. 12)

The rail pool was re-established in the autumn of 1893, but its course was not smooth, and in November 1894 the firms had to hold another meeting in Chicago after negotiations had got into difficulties. That year Maryland Steel sold its rail allotment for $400,000. The allotment for 1895 excluded Illinois Steel which was now prevented from joining by state law. (Table 30) Eastern

Table 30.   *Steel rail pool, 1895 allotments*
                              (percentage)

Pennsylvania and Maryland Steel	15·74
Lackawanna	15·74
Cambria	7·87
Bethlehem	7·87
Carnegie, for western works	52·78

Source: *I.A.* 11 Feb. 1897, p.18.

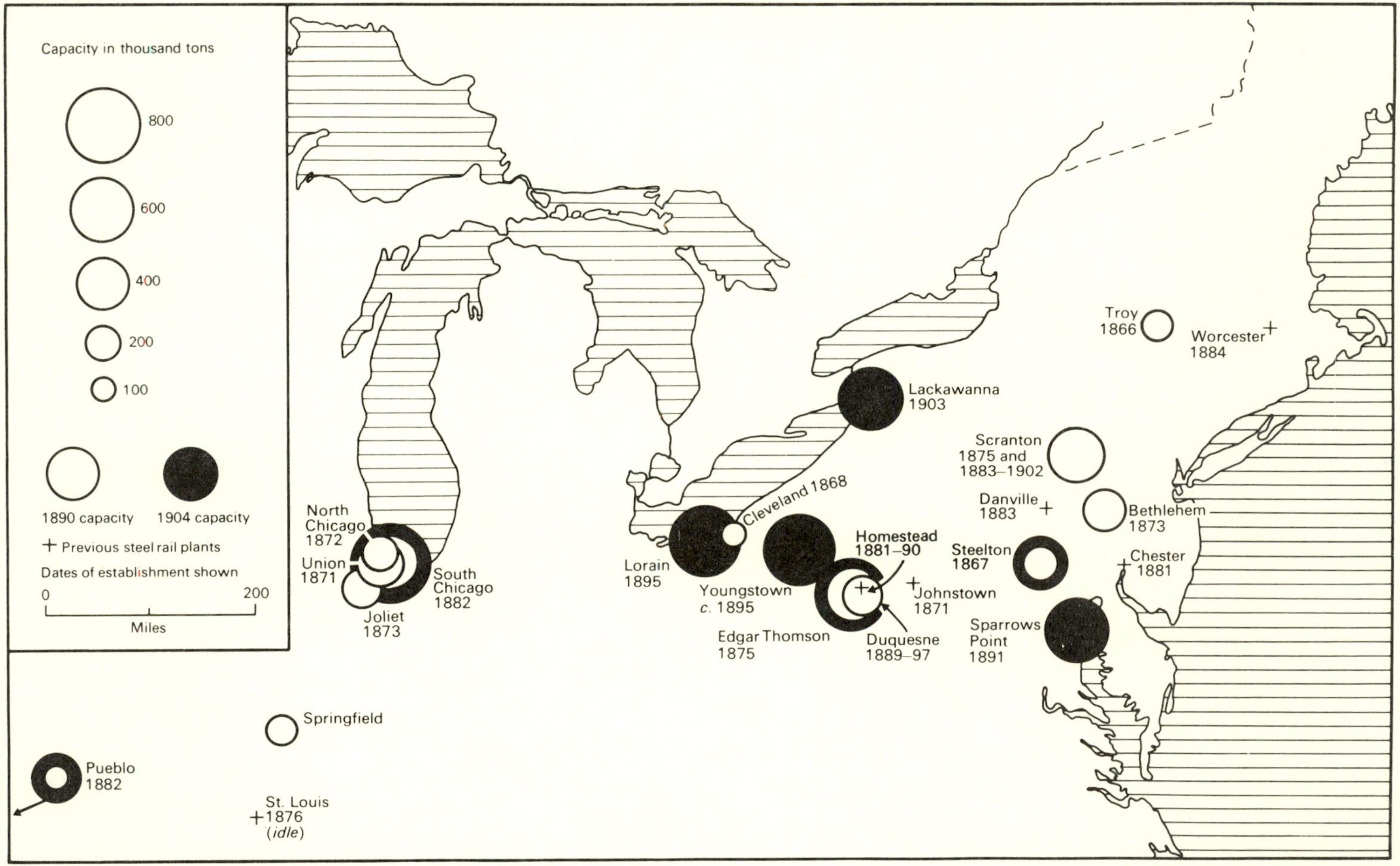

12. Steel rail mills 1890 and 1904

Capacity in thousand tons

800
600
400
200
100

1890 capacity   1904 capacity
+ Previous steel rail plants
Dates of establishment shown

0   Miles   200

Troy
1866

Worcester +
1884

Bethlehem
1873

Chester
+ 1881

Scranton
1875 and
1883–1902

Danville +
1883

Steelton
1867

Sparrows
Point
1891

Lackawanna
1903

Cleveland 1868

Homestead
1881–90

+ Johnstown
1871

Duquesne
1889–97

Youngstown
c. 1895

Edgar Thomson
1875

Lorain
1895

North
Chicago
1872

Union
1871

South
Chicago
1882

Joliet
1873

Springfield

St. Louis
+ 1876
(idle)

Pueblo
1882

works still kept a very important share. At the end of 1895 the pool members over-estimated the 1896 demand by hundreds of thousands of tons. There were by this time not only penalties but subsidies. Two Ohio mills were given subsidies for 1896, in one case amounting to $160,000.[15] In December 1896 meetings were held to arrange the 1897 allocation. (Table 31) New producers had now to be reckoned with, as at Indianapolis and the new basic Bessemer steel and rail plant at Troy. The proceedings were stormy. Although Illinois Steel was not formally a member, there seems to have been a tacit understanding between that company and Carnegie which the other concerns did not know about until the two disagreed and asked for arbitration. In return for a fancied wrong by the Carnegie Company, Illinois Steel began to slash rail prices. Angered by the belief that the price agreement was not being held to and that, as a result, it was losing business, Lackawanna withdrew from the pool at the beginning of 1897.[16] This sparked off the exceptionally sharp competition which exposed to the full the increasing weakness of the east and the outstanding quality of the Carnegie organization.

In tonnage 1897 was the best year since 1890, but prices fell, though output rose. Without the pool arrangements, for the last ten months the price was normally $18 a ton, $10 less than in 1896. Some business was done at $16 or even $14 a ton. Under these conditions none of the eastern firms proved able to win their share of the 1897 allocations.

It was at this time that Carnegie Steel carried its competition into the territory of the Illinois mills. Illinois rail production was 40·1 per cent up on 1896 but that of Allegheny County went up by 76·3 per cent. Connecting through to the lakes by the Bessemer and Lake Erie Railroad, Carnegie could deliver cheaply in the north-west when the Great Lakes navigation was open,

Table 31.   *Steel rail pool, 1897 allotments*
                              (percentage)

Pennsylvania	8·25
Maryland Steel	2·75[1]
Lackawanna	19·00
Cambria	8·25
Bethlehem	8·25
Carnegie	53·50

[1] In addition was to receive a subsidy.
Based on A. Berglund, *The United States Steel Corporation*, 1907 pp.34—5.

and in the south-west when the Ohio was high. However, it sold mostly east of Indiana. Illinois Steel at this time marketed 95 per cent or more of its product west of this line, according to a later estimate of E.H. Gary. He also reckoned that it could not compete with Carnegie more than 100 miles east of its mills, if as far as that. Some Carnegie rails were sold in Chicago for $18 a ton, below the price which Illinois Steel could meet if proper accounting practices were followed, while Carnegie claimed that it made a profit at $16 a ton. Gary reckoned that if these conditions had continued Illinois would have been driven out of business, and recalled that only very narrowly did it escape receivership, the papers indeed having been drawn.[17] After the price war, a sales agency, the Empire Rail Company of New Jersey, was proposed to handle the output of the big firms, but Carnegie opposition secured the failure of this idea. Scranton recalled 'there was a general feeling that it would be a godsend if Mr. Carnegie was out of the rail business.'[18]

1898 production was higher than in 1897 but prices remained low, in the autumn nominally as low as $17 at Pennsylvania mills, and still later, after the collapse of the common sales agency plan, there was a new struggle between the eastern, upper Ohio, and western producers. The last two divided their market area, fixing prices at $17 a ton in Pittsburgh, $18 at Chicago, $20 at Pueblo, and leaving the eastern territory open to free competition, in practice to be invaded by western mills. A year later Carnegie wrote, 'my view is that sooner or later Harrisburg, Sparrows Point and Scranton will cease to make rails like Bethlehem. The autumn of last year seemed as good a time to force them out of business as any other. It did not prove so. The boom came and cost us a great deal of money.'[19] The loss referred to was presumably in contracts entered into for future supply at the low prices of 1898. In December 1898 rail prices were $17 a ton, but by November—December 1899 they were $35. Rail war prices were very variable. J.W. Gates recalled that when he was head of Illinois Steel 'the price was $23 one day, the next day it was $18 and the next day it was $15.' According to Gayley the companies '. . . made agreements to recuperate and then there would be another fight. Agreements were necessary from time to time, or ultimately one company would have secured a monopoly of the business.'[20] In autumn 1899, when the western companies invaded the east, the mills there fought with a good deal of success to hold on to their trade. Of 125,000 to 150,000 tons of rail orders placed at this time from this section, they secured all but 15,000 tons, which was for a trunk line, and 4,000 tons for one road in New England.[21] In 1898 Illinois did even better than Allegheny County, and central Pennsylvania, dominated by Lackawanna, also performed well, but Cambria production fell, and the other producers in the eastern counties of the state were all but eliminated from the trade. By May 1899 Schwab claimed in a letter to Frick that the Carnegie Company could make rails for less than $12 a ton, which, even with Carnegie's addition of $3 for interest, depreciation, and loss through bad debts, made the Edgar Thomson works more than competitive with any mill anywhere in the country.[22] In 1888, the first bad year after the rail pool was formed, Edgar Thomson rolled 148,000 tons of rails, while eastern, central, and western Pennsylvania mills rolled respectively 106,000, 470,000, and 114,000 tons. In 1898 Edgar Thomson made 561,000 and the other three districts 7,000, 334,000, and 148,000 tons. Illinois output had gone up from 436,000 to 549,000 tons, but its share of national output fell from 31·0 to 27·7 per cent.[23]

READJUSTMENT AMONG THE EASTERN RAILMAKERS

The eastern rail mills had to be reorganized in the face of this competition, and for some of them adjustment proved too difficult. Troy Steel Company had been reconstructed in 1895 and 1896, and had begun to make basic Bessemer steel, thereby lessening its dependence on Lake Superior ore. Half of this was to be transferred to the rail mill. The basic Bessemer process, however, proved more costly than acid practice under American conditions and Troy could not withstand the competition of 1897 and 1898. Its rail mill was soon idle, and in February 1902, when the plant was sold by auction, it was bought by American Steel and Wire to supply pig iron from the blast furnace to its Worcester, Massachusetts works. In 1906 the rest of the works was dismantled, any useful machinery being shipped to the western mills of the Steel Corporation.[24] Bethlehem, much more important, gradually withdrew from the rail business, though for a time it hung on to the production of special low phosphorus rails for which it was able to charge a premium of $2 a ton. Eventually others began to make improved rails with no extra charge so that it was dislodged from this refuge. As late as 1901 it still had a rail mill of rated capacity 205,000 tons, but in such a good year as 1898 the whole of the eastern Pennsylvania district in which it was located produced an output of under 7,000 tons. In 1902 the four-converter Bessemer shop was dismantled, and the company gave up any pretensions as a railmaker. By this time, it had already obtained a wide reputation in forge and ordnance work which contributed to the belief that the district was ill-suited for tonnage lines.[25] Cambria at Johnstown was in better shape, though over the nineties it did little more than hold its own. For coal and ore supplies, and for access to western construction, it had a better location. In addition to rails it had a considerable trade in structurals and a well-established speciality trade.

Howe in 1892 had described the Steelton management as 'able and energetic', but Pennsylvania Steel and its offshoot, Maryland Steel of Sparrows Point, operated well below capacity in the mid-nineties and lost money. After this the plants were enlarged and improved. Steelton, like Bethlehem, developed new high-grade lines. It made special open hearth steels, entered the bridge trade, and by 1907 claimed to be the biggest plant in the country for frogs, switches, and general railway equipment. In spite of these developments, in the first years of the twentieth century the capacity of the Steelton rail mill was very largely increased. Close connection with the Pennsylvania Railroad made the marketing situation rather easier than for an independent concern like Bethlehem. Sparrows Point made only small profits in the late 1890s in spite of large plant extensions. The average profit per ton, after deducting fixed charges but making no allowance for depreciation, was $2·133 on average over the years 1898–1907, but in 1898 and 1899 only $1·136 and $1·199 respectively.[26] In 1900 50 per cent of its rails were exported, and in addition it had the great advantage of shipping to the tidewater terminals of southern railroads.

None of the foregoing had been of anywhere near the stature of Lackawanna. There were two major works in Scranton at the beginning of the nineties, Lackawanna, which had first rolled steel rails in 1875, and the Scranton Steel Company which entered the trade eight years later. Each had a rail capacity of about 200,000 tons a year. In 1891 they merged as the Lackawanna Iron and Steel Company, and the older mill, Lackawanna's, was closed. By 1896, rail capacity was almost 450,000 tons and by 1901 approached 630,000 tons a year. The company suffered severely in the struggle for business in 1897 and two years later took the remarkable decision to move all the Scranton operations to Stoney Point, a largely waste site on Lake Erie just south of the Buffalo city limits. By March 1902 the last rails had been rolled at the old plant.

West of the Alleghenies growth of rail capacity continued, although it too was accompanied by rationalization. In Pittsburgh, Edgar Thomson in 1901 rolled well over twice the tonnage that it had ever made in any one year before 1897. Homestead last rolled rails in 1894 before concentrating on open hearth steel and other mill products, especially plates and structurals. Duquesne, occasionally an important railmaker in the mid-nineties, last turned them out in 1897 before turning to billets and other semi-finished steel. Illinois Steel and its successor, Federal Steel, carried out a similar rationalization programme in Chicago. Rail production ceased at North works, the Union works were idled, and Joliet and, above all, the South works were given their business. By 1901 the capacity of the last two was greater than at all four works five years earlier. Further west Colorado Fuel and Iron extended Pueblo from a rated capacity of 54,000 tons of rails in 1890 to 120,000 in 1896 and 200,000 by 1901. By this time, Tennessee Coal Iron and Railroad Company was building the first rail mill to roll open hearth rails on a large scale. Youngstown, unusually, had a footing in the rail trade at this time though the plant was usually kept on bars and billets. Apparently, its rail production was not particularly competitive.[27]

CAUSES OF THE LOCATIONAL CHANGES IN RAIL PRODUCTION

The westward shift of rail production within Pennsylvania and even the difficulties which Illinois Steel experienced in competing with Pittsburgh may be explained in part by 'natural' trends, the changing nature, sources, and conditions of delivery of materials and to a smaller degree the alterations in marketing. Some of these were outside the control of individual firms, but on the other hand steel company size and initiative determined how completely the advantages of full integration, of ownership of minerals and their means of supply, were taken up. In the cut-throat business of the time, these 'personal' factors could be decisive.

The sharp lowering of Lake Superior ore costs at the mine, and at Lake Erie ports, as mechanization and the physical means of delivery were improved, and as organizational change in the ore trade went on, helped to worsen the relative position of eastern works. These lacked adequate supplies of Bessemer ores of their own and were impeded in use of foreign ores by the tariff and the costs of hauling ore inland. It seems that costs of rail transport of iron ore fell much less in this period than lake transport costs, so that the longer the distance of rail carriage from the lower lakes, the greater the disadvantage in the use of Lake Superior ore. The haul by the Lehigh Valley Railroad from Buffalo to Scranton was over 220 miles and to Bethlehem 60 miles more, while from Cleveland to Pittsburgh was only half as far. In addition, the eastern railroads had a bad reputation for charging high rates per ton mile. In 1897, the most critical year in the rail business, the landed price of upper Lake ore on Lake Erie was at its lowest point to date, so that the long haul eastwards was especially embarrassing. By this time too the Carnegie Company had its own Conneaut–Pittsburgh mineral railway.

There were two exceptions to intrinsic eastern difficulties with ore— Cornwall ore and the situation of Sparrows Point. The Cornwall mines were worked on a large scale and in great open pits on the pattern of Mesabi. Both Pennsylvania Steel, whose Steelton works were only fifteen miles away, and Lackawanna at Scranton, no further from Cornwall than Pittsburgh from Lake Erie, used this ore in rail steel. By the 1890s iron made from Cornwall ore was usually employed as one-third of the Bessemer charge, though sometimes alone. However, as will be seen later, the Cornwall operations were inefficiently organized. Pennsylvania began to build the works at Sparrows Point in 1887, and although the tariff on imported ore had to be paid there was a remission on what was worked up into export articles. Other eastern companies, and especially Bethlehem also used foreign ore. In fuel supply eastern works were penalized by their original location to use anthracite. Lower operating costs with mixed coke and anthracite charges were partly cancelled out by the long coke hauls, coals nearer than those of south-western Pennsylvania being unsuitable for coking by the beehive process. In times of recession the freight bill was cut by more use of anthracite even though this meant lower productivity.

While the conditions of raw material assembly had moved against the eastern producers, it is difficult to maintain this excuse for Chicago. Even by the late 1880s costs of iron production in Chicago were little higher than in Pittsburgh. (Table 32) By the end of the century, the coke rate in the best practice had fallen to 17 or even 16 cwts. a ton of pig iron, while the iron content of the ore had also fallen. Theoretically in Chicago's favour, the obvious implications of these changes were completely upset by the varying readiness with which they were taken up. The character of the Carnegie

Table 32.   *Estimated rail freights on iron ore and coal to*
Chicago and Pittsburgh 1887
(dollars per ton of pig iron)

	Ore	Coke	Total
Pittsburgh	2·50	1·15	3·65
Chicago	–	3·79	3·79

Note: These figures assume a coke rate of 23 cwts., and 60 per cent Fe ore. Lake freights on ore were less to Chicago than to Cleveland.
Based on: *I.A.* 14 Apr. 1887, p.19, 21 Apr. 1887.

Company was undoubtedly an outstanding factor. By connection with the H.C. Frick Coke Company, Carnegie avoided the open market coke price oscillations which other iron companies had to contend with, but in the more efficient organization of the supply of Lake Superior iron ore it was Illinois Steel which had taken the initiative through its 1887 link with the Minnesota Iron Company. In 1893 Minnesota Iron's Duluth and Iron Range Railroad was extended to Mesabi. In the previous year, with the grudging help of Carnegie, the Oliver Iron Mining Company had gained a small interest in Mesabi. Both Minnesota and Oliver were dwarfed by the properties and operations which Rockefeller took over in the depression of 1893/4. The Rockefeller mines were soon linked to the lower lakes by an independent railroad and the Bessemer Steamship Company. In the Carnegie Company Rockefeller's purchases were at first ridiculed, for it was believed that the ore was not only of rather lower quality than on the old ranges, but above all, too powdery to be used with great success. However through the agency of Frick and Oliver and, intially, without Carnegie's approval, Carnegie Steel made a fifty-year agreement with Rockefeller to take at least 600,000 tons of ore a year. Oliver estimated as a result that the cost to Carnegie of ore delivered at Lake Erie ports fell from perhaps $4 to only $2·50 a ton. Within two months the rail pool had broken and Carnegie began the 1897 ore shipping season with this very great ore supply advantage, equal to almost $3 a ton saving per ton of pig iron. The agreement was recognized by all producers as putting Carnegie in an almost unassailable position. *Iron Age* observed that the agreement 'gives the Carnegie Company a position unequalled by any steel producer in the world'.[28] The company sought to go even further.

Carnegie had often threatened to build railroads from Pittsburgh to obtain lower rates than on the independent lines, as with his support of Vanderbilt's plan for a railroad from Reading to Pittsburgh in 1883, or the plan to take over the American Midland in 1889 so as to obtain an independent route to Chicago.[29] Sometimes, but not always, the threat alone secured concessions. (Carnegie recalled for the Stanley Committee that the Pennsylvania agreed to give lower coke rates from Connellsville to Pittsburgh on condition that the Carnegie Company gave up plans for an independent line, but that when the Carrie furnaces were acquired it was found that they had been charged a rate for coke carriage which was below the one Carnegie had been given as a great concession.[30]) After building the Union Railroad in 1892 to connect the various Pittsburgh mills, Frick suggested that Carnegie Steel should run its own mineral trains over the Pennsylvania tracks from Lake Erie to Pittsburgh, but before this project was carried through the Pittsburgh, Shenango and Lake Erie Railroad was acquired. From the ore docks at Conneaut, then being deepened at Federal expense, the line stretched at this time to Butler. An extension was made to link with the Union Railroad at Bessemer, where ore yards were laid out near to the Edgar Thomson works. Renamed the Pittsburgh, Bessemer and Lake Erie, by 1901 this line carried ore from Conneaut to Bessemer for as little as 40 cents a ton—when the rate by independent lines was about $1·25. Two years before this, through the Oliver Iron Mining Company, the Carnegie interests had purchased six ore carriers from the Lake Superior Iron Company to form the Pittsburgh Steamship Company. However, for lake ore carriage Carnegie remained right to the end largely dependent on Rockefeller.[31]

Owning or leasing ore and coal resources adequate for decades—including Gogebic and Vermilion properties turning out the valuable 'Old Range' ore, bought up at the low prices of 1897 and 1898—Carnegie Steel was largely shielded from raw material price fluctuations, and could invest in modernization or expansion with more confidence than its rivals. Illinois Steel, its best-endowed competitor, had 4,500 acres of Connellsville coking coal land by 1890 and from its 1,150 ovens 1,500 company rail cars brought coke to Chicago, but its long haul was over independent railroads.[32] Moreover, in Connellsville Frick owned the finest properties and other coke lands were exorbitantly priced. At the end of the nineties the opening of the Lower Connellsville district improved the situation for other producers than Carnegie Steel.

The unrivalled excellence of its practice was another and vital factor in Carnegie's success and so in the good showing of Pittsburgh in the rail business. Control of materials and excellent methods both had a common root in the organizational structure. While other companies were public concerns accountable to shareholders, and continually pressed for dividend distribution, Carnegie Steel, though dominated by Andrew Carnegie, was a partnership to which were admitted young men who showed outstanding ability in their work. To this small dedicated group, intimately aware of the inside of the steel business, the uncertain conditions of the nineties provided both wonderful incentives and rewards for a policy of scrapping and rebuilding in order to retain the highest efficiency and the lowest costs in the trade. Profits from good times were used to put the plant in better condition for the periods of poor trade so that then, and particularly in 1897 and 1898, it was possible to go for the business at prices which broke other companies. The ruthless pursuit of business efficiency, for years it might seem, merely for its own sake, came later to be identified as 'American practice'. Its roots went far back, beyond W.R. Jones to A.L. Holley, and other companies were important contributors to its reputation, but, even so, it was Carnegie's example at this time and the harsh lash of its power which spread it far wider.[33]

THE REVIVAL OF COMPETITIVE POWER AND THE ASCENDENCY OF ILLINOIS

Illinois Steel was one of the big companies which learned the lesson from Carnegie competition, and in so doing was able again to exploit Chicago's natural advantages for rail manufacture. Rationalization was capped by the formation of Federal Steel, and in 1898 and subsequent years, though it did not out-produce it until 1902, Illinois performed much better in relation to Allegheny County than in 1897. Presumably it was a recognition of this new competitiveness which in 1898 made Carnegie agree to E.H. Gary's suggestion that Federal should have half the rail orders they were jointly able to secure, an agreement which made the production figures less indicative of cost differences than of the cut and thrust of 1897. (Table 33).[34]

Table 33.   *Bessemer rail production, Illinois, Allegheny County, and Edgar Thomson Works 1890—1901*

	Illinois	*Allegheny County*	*of which (Edgar Thomson)*
1890—2 (Average)	447	357	309
1896	311	305	301
1897	436	539	477
1898	549	564	562
1899	588	606	604
1900	605	631	627
1901	704	711	708

Based on *A.I.S.A.* and *I.A.* 14 Apr. 1904, p.18.

The balance of advantage shifted more strongly against Pittsburgh in the next few years. By-product coking, the growing popularity of open hearth

rails after 1907, and, as far as the east was concerned, the abolition of the
iron ore tariff and improvement in its system of delivery, were all important
factors, but perhaps the decisive one was the formation of the United States
Steel Corporation. Within this widespread group orders could be more
rationally allocated (Table 34).

Table 34.   *Production and average mill costs of rails by*
districts 1902—1906
(thousand tons and $ per ton)

	1902		1903	1904	1905	1906	
	Prod.	Cost $	Prod.	Prod.	Prod.	Prod.	Cost $
Chicago and the West	814	19·26	801	619	918	890	21·19
Lake Erie	96	23·01	142	291	511	633	21·13
Valley	314	20·01	319	+	160	249	18·71
Pittsburgh	709	18·46	723	548	717	762	19·93
Eastern	701	24·02	710	456	652	759	24·30

+ No production reported for 1904.
Based on Bureau of Corporations. *The Steel Industry*, 3, 1913, pp.560—1.

When it was decided to build the new Steel Corporation plant on Lake
Michigan, soon to be named Gary, officials of the corporation estimated that
rail production in the Chicago area was 450,000 tons a year too small. Gary
was equipped to roll up to 75,000 tons of open hearth rails a month. There
was also a rapid growth on Lake Erie, largely due to the new Lackawanna
works. Until the 1907/8 depression Pittsburgh continued to do well in rails,
but in 1908, before Gary started work, it was already remarked how poorly
Allegheny County rail mills were performing, a far cry from their aggressive-
ness in 1893 and 1896. By May 1910 Chicago works had order books full for
months ahead, while Pittsburgh and eastern mills were so short of business
that they were having to roll billets and specialities.[35] In that year only two
Steel Corporation plants made heavy Bessemer rails. Edgar Thomson turned
out 314,000 tons, the Chicago works 610,000. Pittsburgh's leadership, which
had owed so much to technical excellence and commercial drive, now passed
back to the area which had from long before had the most advantageous
location, and Chicago re-established its ascendency in rails.

NOTES

[1] *Tariff Proceedings and Documents 1839—1857,* Senate Documents, 1911, Part 3, p.1354.
[2] J.L. Ringwalt, *Development of Transportation Systems in the United States,* 1888, p.133.
[3] J.P. Lesley, *The Iron Manufacturer's Guide,* 1859, *passim.*
[4] A.I.S.A.

[5] A. Nevins, *Abram S. Hewitt*, 1935, pp.234–5.

[6] Ringwalt, op. cit., pp.200–1.

[7] *Eng.*, 15 Aug. 1884, p.163, 29 Aug. 1884, p.208.

[8] M.H. Howe, *The Metallurgy of Steel*, 1892, p.259.

[9] *Eng.*, 12 Sept. 1884, p.258, 16 Jan. 1885, p.67.

[10] D. Brody, *Steelworkers in America; The Non-Union Era*, 1960.

[11] Evidence to Stanley Committee, *I.A.*, 10 Aug. 1911, p.316.

[12] E.C. Potter, 'Rails past and present', *I.A.*, 17 Feb. 1898, p.14.

[13] *Eng.*, 21 Nov. 1884, p.471; *I.A.*, 11 Feb. 1897, p.18; Report of Commissioner of Corporations, *The Steel Industry*, 1911, pp.69–70; A. Berglund, *The United States Steel Corporation*, 1907, pp.34–5, 167.

[14] *I.A.*, 4 Jan. 1894, p.8, 22 Feb. 1894, p.364.

[15] *I.A.*, 11 Feb. 1897, p.18.

[16] The 1897 Crisis recalled by W. Scranton, former President of Lackawanna, in *I.A.*, 30 Jan. 1913, p.315, and by J. Gayley, formerly of Carnegie Steel, in *I.A.*, 22 June 1911, p.1532.

[17] E.H. Gary, Evidence Stanley Committee, *I.A.*, 15 June 1911, pp.1467, 1470; Dissolution Suit, *I.A.*, 5 June 1913, pp.1394–5.

[18] *I.A.*, 30 Jan. 1913, p.315.

[19] Letter of 28 Sept. 1899 (mistakenly referred to by Gary as 28 Sept. 1889), quoted Dissolution Suit, *I.A.*, 5 June 1913, p.1396.

[20] *I.A.*, 1 June 1911, 15 June 1911, p.1470.

[21] *I.A.*, 1 Dec. 1898, p.37.

[22] *Tariff Hearings* 1908–1909, vol.2, evidence of C.M. Schwab, pp.1628, 1639–40; *I.A.*, 18 Jan. 1912, p.197.

[23] *A.I.S.I.* and *I.A.*, 14 Apr. 1904, p.18.

[24] *I.T.R.*, 1 Feb. 1912, p.308; A.I.S.A. *Works Directories*, 1890, 1896, 1901, 1904; V.C. Clark, *History of Manufactures in the United States*, vol.3, p.34.

[25] J.S. Jeans, *American Industrial Conditions and Competition*, 1902, pp.128, 133; A.I.S.A. *Works Directories*.

[26] E.C. Felton (President of Pennsylvania Steel), *Tariff Hearings* 1908–1909, vol.2, p.1608.

[27] *I.A.*, 22 Feb. 1900, p.18.

[28] *I.A.*, 17 Dec. 1896.

[29] *I.A.*, 17 Oct. 1889, p.601.

[30] *I.A.*, 17 Jan. 1912, p.196.

[31] C.M. Schwab, quoted J.S. Jeans, op. cit., pp.92–6, 420–8; J.H. Bridge, *The Inside History of the Carnegie Steel Company*, 1903, p.272.

[32] G.W. Cope, *The Iron and Steel Interests of Chicago*, 1890, pp.24–7.

[33] See H.H. Campbell, *The Manufacture and Properties of Iron and Steel*, 1907. p.470.

[34] I.M. Tarbell, *The Life of Elbert H. Gary*, 1925, p.99.

[35] *I.A.*, 26 May 1910, p.1212.

CHAPTER 6

The Locational Implications of
Changes in Raw Materials, Techniques, and
Economic Organization 1890-1920

The rapid rise of the United States to a predominant position in world output of iron and steel at the end of the nineteenth century was the outcome of complex factors. Growth in national wealth, in population, and the spread of economic development over the whole subcontinent undoubtedly greatly assisted or perhaps called into being a pioneering spirit in matters of expansion and modernization, and, when the growth rate periodically slackened, firms for many years proved ready to engage in cut-throat competition. There was a fresh, new-world readiness to scrap and rebuild, but while adaptation, absorption, and the elimination of the inefficient went on in a grand, social-Darwinian way, the other basis for the growth of the industry was the development of a system of raw material production and a linking transportation system unrivalled in the world. This work too was done with admirable persistence and ability, but the large unified market justified a scale of planning and execution which linkages in Europe could not hope to match. Yet as Campbell pointed out in 1896, 'this wonderful progress has not been the unearned harvest of bounteous nature, for it has been accomplished in defiance of mighty obstacles in the enormous distances through which the raw materials must be carried, and, although the achievement may be a just source of national pride, it involves inevitable expenses and disadvantages which may be lessened by energy, but which can never be swept away.'[1] In the thirty years centred on the turn of the century the mineral supply pattern changed in a number of ways which together upset the balance of locational advantage in different parts of this area and improved the situation of some other districts. Changes in ore supply were accomplished first, but those involving fuel were quite as important.

PROGRESS IN FUEL SUPPLY

As opposed to the large, concentrated yield of mineral fuel, wood had to be gathered laboriously from a wide area, a condition of supply quite at variance with developing American business attitudes. Charcoal was generally too soft to support the heavy burden of a medium to large furnace without crushing, and so ruled out high productivity. Small outputs and uncertainty of supplies made charcoal furnaces unsuitable to feed the big outputs of Bessemer steel plants, and although locally or even regionally still important, charcoal furnaces had by the end of the century been relegated to the production of special irons. (Table 35) Early in the twentieth century Michigan charcoal iron production was expanding to supply the very tough iron for railroad car wheels. With the subsequent advance of electrical and alloy steels this market was lost. Nationally charcoal iron production was still as high as 376,000 tons in 1917, but down to 138,000 in 1929.

Consumption of Ohio block coal declined rapidly in the eighties, displaced by mixed coal and coke charges, and then by coke alone. Anthracite could support a large burden, but was too dense for rapid driving, such a distinguishing feature of American furnace practice. As increased productivity became important, eastern works turned to mixed coke and anthracite charges and in the eighties these came to far surpass the use of anthracite alone. In the nineties the decline in anthracite iron accelerated, and by the early twentieth century it was reckoned that coke would be preferred to anthracite even if it cost up to 25 or 30 per cent more per ton. 1910 was the last year in which anthracite was used alone as a fuel, and in 1923 all use of it in iron smelting ceased. Long before this, conditions of coke or of coking coal supply to eastern works had improved.

Table 35.   *Pig iron production by fuel used 1880–1905*
(thousand tons and percentages)

Type of fuel:	1880 Tons	%	1890 Tons	%	1900 Tons	%	1905 Tons	%
Charcoal	389	(11·5)	593	(6·7)	303	(2·1)	414	(2·5)
Charcoal and coke	—	—	—	—	53	(0·4)	—	—
Anthracite	994	(29·5)	295	(3·4)	46	(0·3)	30	(0·2)
Anthracite and coke	638	(18·9)	1,690	(19·1)	1,796	(12·4)	1,275	(7·7)
Coke and bituminous coal	1,355	(40·1)	6,266	(70·8)	12,254	(84·8)	14,909	(89·6)
TOTAL	3,376		8,845		14,452		16,628	

Source: Bureau of the Census, Census of Manufactures, 1905, Bulletin 78, p.37.

Pennsylvania made 84·2 per cent U.S. coke in 1880, Connellsville alone having two-thirds of the nation's ovens. By 1890 this district made 6·5 million out of the country's 11·5 million tons of coke. There were new districts, but their output was still small, costs often higher, and their product apparently invariably inferior to that of Connellsville. (Fig. 13)

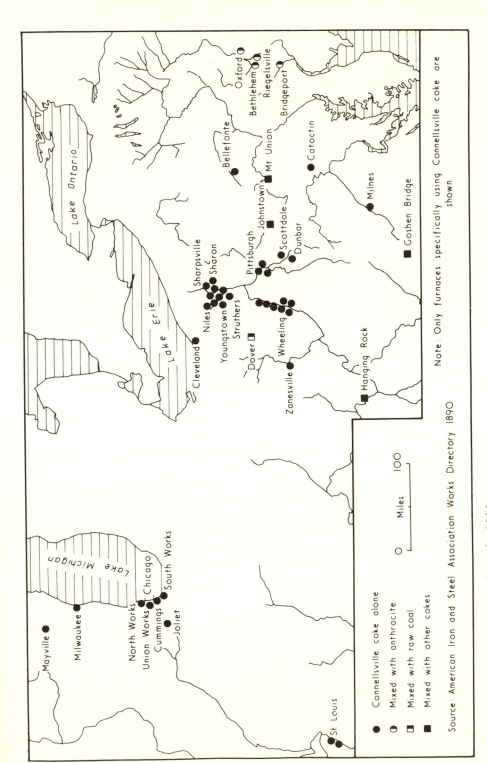

Lake Ontario

Lake Erie

Lake Michigan

Mayville

Milwaukee

North Works
Union Works
Cummings
Joliet
Chicago
South Works

St. Louis

Cleveland

Sharpsville & Sharon
Niles
Youngstown
Struthers

Dover

Wheeling

Zanesville

Pittsburgh

Johnstown
Scottdale
Dunbar

Mt Union

Hanging Rock

Bellefonte

Oxford
Bethlehem
Riegelsville
Bridgeport

Catoctin

Milnes

Goshen Bridge

Note: Only furnaces specifically using Connellsville coke are shown

● Connellsville coke alone
◐ Mixed with anthracite
▣ Mixed with raw coal
■ Mixed with other cokes

○———— 100
Miles

Source: American Iron and Steel Association Works Directory 1890

13. Blast furnaces using Connellsville coke 1890

West Virginia was the second northern coke state, its rather higher-volatile coals coming from the south-western continuation of Connellsville. Further south, the Pocahontas Flat Top region, spanning the border between West Virginia and Virginia, was undeveloped until 1881. In 1890 it shipped 500,000 tons of coke or one-thirteenth the tonnage from Connellsville ovens. Whereas Connellsville coal went direct from the coal face to the ovens, in the Flat Top, Central, and New River districts coal had to be broken and screened and, in some cases, washed before charging. In spite of this beehive ovens generally made an inferior coke from it, though in the Kanawha field, careful screening made possible a coke almost up to Connellsville quality.[3]

To the west, through to Indiana and Illinois, the volatile content of coals went up to 35 or 40 per cent (for Connellsville coal the range was 25–30 per cent). In the west too the sulphur content was often too high for furnace coke. Eastwards volatiles fell to 20–25 per cent, and on the very eastern edge of the coal basin to 16–19 per cent. Coke made from high-volatile coal was too spongy or brittle; the lack of volatiles in eastern coal checked the development of a cellular structure without which coking was incomplete.[4] These coals were not uncokeable—indeed Fulton pointed out that second-grade U.S. coals were superior to the best on the continent of Europe—but they would not give a satisfactory product with standard equipment or with methods as simple as those of Connellsville. Because of this, while Connellsville output was sufficient to meet demand, and while price and freight charges remained low enough to permit satisfactory iron-making profits at a distance from the coke district, the area had no challengers. The 1890s upset this situation.

Demand for coke, essential for the high furnace productivity needed in a time of keen competition, went up so much in the nineties, that by the end of the decade prices were increasing sharply. The opening of the so-called Lower Connellsville district in 1900 eased the position for a while, but the desirability for big companies to obtain new coking coal properties was obvious.[5] In favoured cases they might obtain them nearer to the plant, and out of the control of Frick or Oliver, purchase from whom was often tantamount to a subsidy to Carnegie Steel.

There was little success with attempts to use high-volatile coals. It was suggested that, mixed with dry eastern coals they might perhaps be coked in a broad, horizontal oven, but the plant would cost more than a conventional beehive one, and low-volatile coal would have to be hauled from east of Connellsville. At the end of 1899 166 of the 182 ovens in Illinois and Indiana were idle, though patent ovens said to be able to coke the region's coal were being built at Chicago.[6] In fact little success was achieved for over fifty years.

The problem was more serious for eastern coke users, for they lacked the highly competitive position of Illinois furnaces with respect to ore assembly.

Faced with similar low-volatile coals, Belgian, German and French engineers had pioneered the retort oven. A narrow, slot oven with very much more rapid coking than with beehive ovens made possible production of an acceptable coke from dry coals. Introduction of the slot oven, equipped to recover by-products, improved the prospects of ironworks well away from Connellsville.

The new ovens gave a higher coke yield per ton of coal—the coke yield in Connellsville 1880–96 averaged 66·8 per cent; a 73·6 per cent coke yield was claimed for the Otto Hoffman oven.[7] They also produced valuable by-products, especially gas of high calorific value, tar, and sulphate of ammonia. They were, however, much more expensive, and the coals they used had to be crushed, sorted, and washed. If located on these poorer-grade coals the slot ovens might reduce freight costs as compared with charges on Connellsville coke, but how then were the by-products to be used? If located near the point of coke consumption, physical deterioration of coke in transit could be avoided, but even with a 75 per cent coke yield a market orientation for ovens would increase the tonnage of material moved by rail by one-third, and the freight bill to almost the same extent. In Europe by-products had been a great attraction, but the chief of these, ammonium sulphate, was of little value in the United States where use of artificial fertilizers was as yet of negligible importance.[8] Until about 1905, Fulton's figures suggest that the trend of net ultimate coke cost was against the by-product oven, but even so, the need to improve plant efficiency forced distant concerns to try even difficult solutions.

The decisive factor in the location of by-product coke ovens concerned the heat balance of integrated iron and steel works. Mid-western works frequently used natural gas as an open hearth fuel, or to raise steam for mill drives. Even when this eventually became a costly fuel, they could employ local coal in gas producers or in boiler plants. Where there was no local fuel, or where it was expensive or had other more remunerative outlets, coke oven gas and tar provided excellent substitutes. In 1913 James Farrell revealed that United States Steel had found that, using by-product tar in the melting shops, steel output was in some cases as much as 24 per cent higher than with producer gas, coal, or other fuels.[9] Thus the early by-product ovens were located near to or at steel plants while for a number of years further expansion at Connellsville generally involved more beehive ovens, though the Dunbar Furnace Company in Connellsville did in fact build the industry's second ovens. Not surprisingly, eastern firms bulked large in these developments. Following the Solvay Process Company, Cambria in 1895 was the first American steel company to build by-product ovens. Between then and 1903 batteries were built at seventeen works. Five were in the Connellsville–Pittsburgh–Valleys–West Virginia area, one each at Cleveland, Milwaukee, and Duluth, three to the south and the other six east of Pittsburgh. Some eastern firms lagged behind—the big Reading and Thomas iron companies of the Lehigh Valley were examples, though these were merchant iron firms only—and Bethlehem had no ovens until 1912. Batteries were built at Sparrows Point in 1903 and at Steelton in 1907. Between 1901 and 1904 Cambria increased the by-product plant at Johnstown from 160 to 260 ovens, and cut the number of its Connellsville beehives from 920 to 508.[10]   After building ovens at its Lebanon furnaces in 1903, Lackawanna Steel installed them in the following year at Buffalo, obtaining coal lands in Indiana County much nearer both than Connellsville. (Fig. 14)

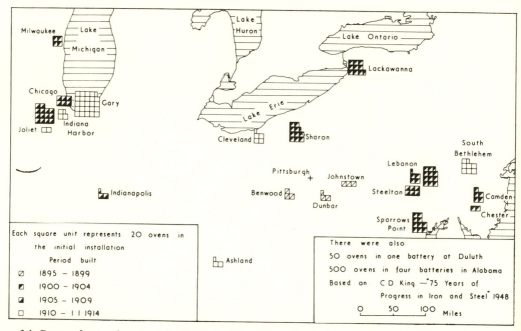

14. By-product coke plants built in the Manufacturing Belt to 1914

Potentially the change in coking method was beneficial to Pittsburgh too, for, in place of a rail haul on coke, firms there could now barge coking coal down the navigable Monongahela. A number of factors delayed this. Firstly Pittsburgh firms, and especially United States Steel, had very large investments in Connellsville ovens. At the end of 1900 when there were 1,085 by-product ovens in the country the Steel Corporation, formed a few months later, controlled none of them. Only Jones and Laughlin as yet made coke in Pittsburgh. It was not easy to find room in Pittsburgh for coal yards and ovens, and U.S. Steel also had the problem of a multiplicity of furnace plants requiring either a number of separate installations or a great central establishment linked by rail with the furnaces and through a gas pipeline network to melting shops and mills. For a number of years coke capacity extensions needed for Pittsburgh continued in the form of new coalfield-located beehive ovens, so that nationally the number of beehives reached its peak as late as 1910, almost 83 per cent of coke output still being made in them. In that year Jones and Laughlin began to build coke ovens at their new Aliquippa Works down the Ohio from Pittsburgh, and these too were of the old type. But in 1918, on a site just north of Clairton Works, U.S. Steel put up the world's largest by-product coke plant. Supplied with 1,250 tons coal a day, barged down the Monongahela, Clairton sent coke on to Pittsburgh and Valley plants, and gas through nine miles of forty-inch main to the Edgar Thomson, Homestead, Duquesne, and Clairton works. Fuel assembly costs in Pittsburgh were reduced, but, as Youngstown area plants had to tranship any coal which they barged down the Ohio and finish the journey by rail, the assembly cost situation of Valley independents relative to Pittsburgh costs was worsened by the change.

In the rest of the United States only two other significant coke-producing areas emerged at this time, and compared with the north-east they were very small. In 1901 seven beehive plants in the coalfields towards the New Mexico border supplied Pueblo furnaces. In eighteen years to 1913 Alabama coke production increased from 839,000 to 2,975,000 tons a year. Beginning with Tennessee Coal, Iron and Railroad Company's Ensley plant in 1898, three other by-product coke plants were built by 1913. Even with by-product ovens Illinois failed to become a major source of coking coal, though by-product ovens of large capacity were soon established in Chicago.[11]

In summary by-product coking lessened the cost of fuel assembly for plants remote from Connellsville, making possible an admixture of poorer coal, sometimes mined nearer the point of coke consumption. It improved their heat economy in integrated operations, or, if there was no steel capacity, such plants could sometimes sell their gas to nearby municipalities. In these ways new coking techniques reinforced the general, long-term trend of fuel economy, helping the dispersal of the industry from its old centre around the upper Ohio. Changes in ore supply assisted the process.

IRON ORE

At the end of the nineteenth century iron ore supply was transformed by a great extension and a thorough-going reorganization of the means of working and delivering Lake Superior ore. Other ore districts declined, while onto the westward drift of the iron industry was superimposed more strongly than ever before a lakewards pull (Table 36). The opening of Mesabi in 1892 was decisive. Here Rockefeller gained control of the largest deposits and exploited them on the rational model laid down on the Vermilion Range by Minnesota Iron Company, low mining costs being matched by reduced rail tariffs to the lake shore and by a superb system of lake transport and terminal or transhipment facilities. Keen competition in the nineties caused ceaseless improvement of this system. Largely because of this, by 1898 it could be claimed that Pittsburgh could produce one ton of pig iron for no more than had been paid for ore alone in 1884. Mesabi Royalties were higher than on the older ranges. There was a longer haul to the lake. In addition, for many years Mesabi fine-textured ore could not successfully be used alone in the furnace but had to be mixed with old range ore. As against this, the ore body was far richer than others, was rationally developed, and was largely worked by steam shovels in open pits. The advantage of the last was outstanding. In Steel Corporation underground mines in 1901 ore output per man day was 4·69 tons, in the open workings 21·53 tons. Open workings additionally had far more flexibility, Gayley reckoning that the nine Corporation pits in Mesabi could have shipped 77 per cent more than they did that year.[13]

Table 36.　　　*Production of iron ore, U.S.A. and states*
1880–1920
(million tons and percentage of U.S. total)

	U.S.	Minnesota		Michigan and Wisconsin		New York, New Jersey, Pennsylvania		Alabama		All other States	
	Tons	Tons	Percentage	Tons	Percentage	Tons	Percentage	Tons	Percentage	Tons	Percentage
1880	7·1	nil	—	1·7	23·6	3·7	52·7	0·2	2·4	1·5	21·3
1890	16·0	0·9	5·6	8·1	50·5	3·1	19·4	1·9	11·8	2·0	12·7
1895	15·9	2·9	24·2	6·5	40·5	1·5	9·3	2·2	13·8	1·9	12·2
1900	27·5	9·8	35·7	10·7	38·7	1·7	6·0	2·7	10·0	2·6	9·6
1905	42·5	21·7	51·1	11·7	27·6	2·5	5·8	3·8	8·9	2·8	6·6
1910	57·0	31·9	56·0	14·4	25·3	2·5	4·5	4·8	8·4	3·2	5·8
1915	55·5	33·4	60·3	13·6	24·5	1·8	3·2	5·3	9·5	1·4	2·5
1920	67·6	39·4	58·4	18·5	27·3	2·1	3·1	5·9	8·7	1·7	2·5

Source: Works Projects Administration, *Technology, Employment and Output per Man in Iron Mining*, 1940.

The rail cars which carried the ore from the pits to the docks were still frequently of only twenty-five tons capacity, but traffic was organized so effectively that by 1901 capacity of some lines was 1,000,000 tons ore a month.[14] Lake ore carriers grew quickly in capacity and changed in style. Only six steel vessels were on the lakes as late as 1886, but by 1891 there were 89, and in 1899 there were 296.[15] In the nineties the 'whaleback' design of McDougall of Duluth made its valuable contribution, though by 1900 it had been superseded. The largest boat of 1890 could carry about 2,500 tons of ore, but by 1901 about 7,000 tons. The *Augustus B. Wolvin*, launched in 1904 at Lorain, had a capacity of 10,000 to 11,000 tons.[16] As vessels grew so efficient company fleets were built up to maintain a shuttle service. Rockefeller's Bessemer Steamship Company was the chief of these, but by the end of the nineties other companies were also extending. By 1899 Rockefeller controlled 70 vessels, Pickands Mather about 35, Carnegie's new Pittsburgh Steamship Company was expected to have about 15 vessels by the following season, and in the autumn American Steel and Wire purchased a fleet. National Steel owned 9 ore boats.[17]

Low-cost ore carriage was matched by rapid improvement of unloading facilities at lower lake ports, as grab buckets superseded hand labour and wheelbarrows. From the lake shore the ore was in the control of independent railroads, except in one or two cases, such as Federal Steel's Elgin, Joliet and Eastern or Carnegie's Pittsburgh, Bessemer and Lake Erie.[18] Through the efficiency of this line Carnegie cut its delivery costs for ore well below those of other iron-makers. By 1898, when the lowest normal movement costs by rail were reckoned about $0·004 per ton mile, the Carnegie line could carry for as little as $0·003 but on Lake ore vessels the ton mile rate was $0·00079.[19] In short, however adequately high plant efficiency seemed to compensate, conditions were clearly improving for lake-shore centres of manufacture. (Fig. 15)

Eastern ore supply was more difficult. Lake Superior ore, like Connellsville coke, came eastwards in very large tonnages, in the process killing off many famous mining enterprises. An eventual improvement of the position of eastern works came in part with by-product coking, but also with the discovery and efficient organization of overseas ore deposits which could be exploited on a large scale and with bulk long-distance transport—an oceanic variant of the model of the Great Lakes. The initiative lay for a time with U.S. Steel. In 1909/10 W.N. Merriam, one of their geologists, surveyed the large ore fields of Brazil. Returning he advised the Corporation to buy, but, as it had no plant near the east coast, his suggestion was ignored.[20] Opportunity then passed to Bethlehem Steel and its Chilean ore supply.

By spectacular reorganization of the Great Lakes ore supply, and, a little later, a breakthrough in eastern deliveries the ground was laid for the expansion and partial relocation of the steel industry in the twentieth century. This reorganization was also affected by important changes in the nature and pattern of consumption. Puddled iron reached its peak importance at the end of the eighties, and after that declined rapidly. With this decline went a fall in the locational attraction of coal of which the forges had been such wasteful users. 'Had iron remained king', remarked Carnegie in 1898, 'the Pittsburgh district would have been his throne and the sole seat of his dominion.' In finished steel, by the late nineties, new lines were expanding much more rapidly than rails, the old specialism. Rail demand was still moving westwards, though renewal demand from the eastern system remained a useful outlet for mills which, with the single

Lake Superior

Lake Huron

Lake Michigan

Lake Erie

Lake Ontario

Ashland

Ishpeming

Newberry

Florence

Black River Falls

Fond du Lac

Mayville

Milwaukee

North Works

Chicago

South Works

Joliet

Brazil

Fruitport

De Pere

Leland

Elk Rapids

Mancelona

Fayette

St Ignace

Ironton

Wyandotte

Detroit

Cleveland

Tonawanda

Rochester

Elmira

Emporium

Troy

Millerton

Wassaic

Canaan

Kent

Oxford

Catasaqua

Riegelsville

Norristown

Bethlehem

Chester

Pottstown

Sparrows Point

Baltimore

Danville

Pottsville

Hollidaysburg

Steelton

Saxton

Kittanning

New Castle

Sharon

Sharpsville

Niles

Youngstown

Leetonia

Pittsburgh

Johnstown

Scottdale

Fairchance

Wheeling

Bellaire

Steubenville

Dover

Columbus,

Zanesville

Hocking Valley

Ironton

1890

exception of Pueblo, were east of the Mississippi and north of the Ohio and Potomac. Structurals were rapidly expanding. Demand for bridge structures was widespread. Steel office construction too was scattered, but there were major concentrations in the big north-eastern cities and above all in congested New York City, a consumption pattern completely different from that for rails. Along with structurals, plate found a big new market in a range of uses including construction, ship-building and rolling stock manufacture. By 1902 one freight car maker alone used 400,000 tons of steel a year, and the biggest centres of this activity were in the western half of the manufacturing belt, where Pressed Steel of Pittsburgh and Chicago, American Car and Foundry of Detroit, and Cambria Steel were important firms.[21]

With the growth of the barbed wire business after the mid-1870s, rapid advance in wire nail production a few years later, and employment of wire for cables and for telegraph purposes, a great surge occurred in rod and wire production—output going up from 250,000 tons in 1875 to 1,200,000 tons 1898. Here too there was a strong westward pull. In 1891 Washburn and Moen of Worcester, Massachusetts, built new works at Waukegan, Illinois. Alabama Steel and Wire built the Ensley works, the first southern rod mill, but rod production remained strongly concentrated in the area immediately around Pittsburgh, with northern Illinois and Indiana as a second major centre.[22] The head of the Ohio Valley and the Valleys district of eastern Ohio had become the chief centre of sheet iron manufacture, and this area also gained a large share of the new American tinplate industry. By specializing on light, flat rolled steel at this time the area's survival and growth was insured when, in the years immediately following the First World War, these were the chief growth lines of the whole industry. Much pipe and tube consumption was scattered throughout the various centres of population. This was especially the case with gas pipe, standard pipe used in plumbing, or tube used by the boiler makers. Demand from the oil and gas industry of Pennsylvania and the Middle West brought the main producing units west of the Appalachians, but when the new oilfields of California and the Gulf Coast were opened up early in the twentieth century pipe and tube works remained centred in the north-east, Pittsburgh, Wheeling, the Valleys, and soon Lorain being especially prominent centres. The seamless tube industry also obtained a footing in the United States at this time, its early growth and its location being associated with the popularity of the bicycle in the eighties and nineties. (The index of a trade journal such as *Iron Age* was full of references to bicycle patents and developments until about 1897. By 1900 this section was shrinking and entries under the heading 'automobiles' were expanding.)

By the end of the nineteenth century, for manufacturing as a whole, although the north Atlantic area kept its lead and had the largest increases, the most rapid growth was occurring in the north-central division. Growth in steel consumption was already much slower in the Pittsburgh area than in the hinterland of Chicago, where by 1899 value added in all manufacturing industry was almost three times that of Pittsburgh and its chief satellite towns, Allegheny City, Braddock, Homestead, and McKeesport. Primary metal already bulked dangerously large in the total manufacturing output of Pittsburgh. (Table 37) Even before the great expansion of the nineties this

Table 37.   *Value of Pittsburgh area manufactures 1905*
(million $)

	All industry	Iron and steel
Pittsburgh	164·4	89·4
McKeesport	23·0	17·0
Allegheny City	45·8	4·6

Source: Bureau of Census, *Census of Manufactures,* 1905 Bulletin 60, Pennsylvania.

had been recognized as a serious problem. In autumn 1892 T. Oliver warned the Pittsburgh Chamber of Commerce that, helped by low rates on semi-finished steel (or on steel rather than on the products fabricated from it), the area was making slow progress in fabricating.[23] (Table 38)

Table 38.   *Rail freight rates on semi-finished and finished*
steel from Pittsburgh 1892
(Dollars per ton)

Pittsburgh to:	Semi-finished steel		Finished steel	
Cleveland	Billets	1·00	Wire and nails	2·13
Chicago	Wire rods	2·40	Wire nails	4·15
St. Louis	Wire rods	2:75	Wire nails	5·22

Based on *I.A.* 3 Nov. 1892, pp.342—3.

At the end of the century it seemed that the United States might become a great steel exporter. Maryland Steel had major interests in foreign markets, in which trade it secured a rebate on the iron ore import duty, Alabama iron began to supply markets in Europe in the mid-1890s, but the biggest fears of European firms were connected with the plans of Carnegie Steel and the early U.S. Steel. By 1902 the freight rate from Pittsburgh to the seaboard was $2·10 to $2·40 a ton, though Kennedy later claimed that, by using the Bessemer and Lake Erie, and interchanging with the major east-west routes, export steel could be delivered in Jersey City for as little as $1·50.[24] Yet although Carnegie obtained orders for shipbuilding material at well below European prices, it quickly lost this market when the Europeans cut their prices to compete. Carnegie billets were delivered in the English Midlands for $18·50 a ton, but again this proved a trade difficult to hold.[25] In fact although exports grew they failed to become a major element in shaping the pattern of American steel-making. With the exception of the eastern plants and the relatively small developments in Alabama the industry remained pivoted on the Great Lake/Ohio river flows of ore and coal, though the changing balance of ore and coal was laying the foundation for important locational shifts within this area. Looking at it at this critical time, Carnegie failed to see the changes implicit in this situation and painted too static a picture. Above all, he failed to make due allowance for the dynamism of Chicago or the revival of the East.[26]

> The present centre of steel [he wrote] is in the square made by a line drawn from Pittsburgh to Wheeling, northward to Lorain, eastward to Cleveland, and south again to Pittsburgh . . . As far as the writer sees there is little chance of this region being soon displaced. Colorado will, no doubt, expand as the western coast is developed. Chicago's position as a steel manufacturer is assured. There is no sign of the great South West making steel to any extent. The transfer of the great Lackawanna Iron and Steel Works of Scranton, Pennsylvania to Buffalo, and the splendid triumphs of the Bethlehem Steel Company, in Pennsylvania in armour, guns and forgings as specialities, which give it a unique and commanding position, are proofs that, for the making of ordinary steel the East is not a favourable location. The history of the steelworks at Troy is another case in point. There is one exception to this march westward, at Harrisburg in Eastern Pennsylvania, which remains a prosperous and important centre of manufacture. The Maryland Steel Company has advantages for export, but probably more important for the future of that company is its development in shipbuilding, for which its plant is well located. So far as the writer sees there is nothing to change the centre of steel manufacture in this country in the new century; it is in the Central West already described, and there it is likely to remain. (Fig. 16)

ORGANIZATIONAL CHANGE WITHIN THE STEEL INDUSTRY

Into the 1890s the iron and steel industry consisted of a large number of relatively small firms, buying raw materials—pig iron or blooms or, more commonly, iron ore or coke,—in the open market. Mines and transport facilities were independently owned. Competition was keen, for in all but the exceptional year over-capacity was already a leading characteristic. At the end of the century a group of factors, some technical, some organizational, began to change the situation. The replacement of wrought iron by steel increased the optimum size and complexity of the production unit. The puddling forge was of small capacity, easy to build and duplicate and, operating on cold metal, not disadvantaged as far as technical considerations were concerned if there were no blast furnaces on the same site. Steel converters or furnaces were of larger capacity, much more costly, and, it was gradually realized, could operate most economically on hot metal. Big integrated units, combining smelting, steel-making, and rolling required large

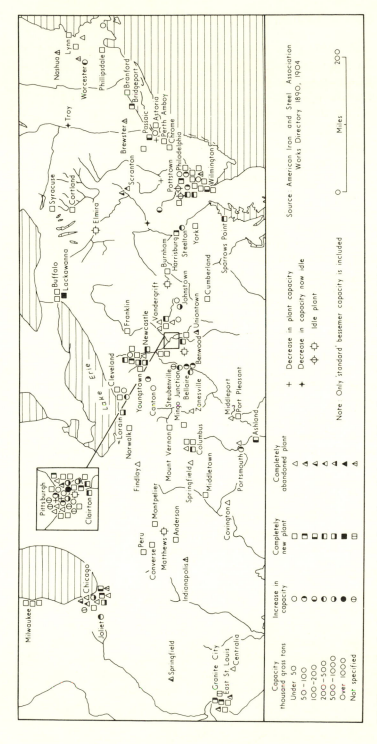

16. Changes in the ingot capacity of steelworks in the Manufacturing Belt 1890—1904

supplies of coal and iron ore and thus made full control over these more attractive. At the same time high-level operations were desirable to cover the huge commitment of capital which they represented. The very uncertain trade conditions of the last twenty years of the nineteenth century threatened their position, and provided much of the motive behind the backward integration of the period and the proliferation of pools and great consolidations. There was an overlap in time between these various organizational forms, but broadly, up to 1898 many producers competed for business, their cut-throat relationships being moderated by pools or gentlemen's agreements which usually broke down when the strain was especially great. As John Gates put it, looking back, 'Well in those days we used to have a few agreements. The boys would make them and Andy would kick them over.'[27] 1898 and 1899 were the years of the great consolidations, largely horizontal, and of a scope indicated by the word 'American' which usually prefixed their title. The two most powerful groups were, however, vertical combinations, renamed or extended at this time, Federal Steel, successor to the Illinois Steel Company, and the Carnegie Steel Company. There followed the logical backward integration of the finishing groups, and the forward integration of the others, till then concerned with pig iron, steel, and semi-finished mill products. This seemed likely to lead to over-capacity in all lines of trade, and the incorporation of the United States Steel Corporation in February 1901 was an attempt to avoid this by bringing a large part of the industry beneath one gigantic overarching wing. Too big to behave in the same irresponsible way as its predecessors without receiving public condemnation, the Steel Corporation perhaps inevitably became a conservative force. As a result of this, outside firms could expand more rapidly or wholly new ones find a profitable line of business. Each new form of organization was a recognition that earlier ones had been inadequate to deal with the problem of ever-expanding capacity and highly variable demand, and yet produce the desired results of 'satisfactory' operating rates for the majority of companies and a 'fair' profit margin. Locational values altered along with the structure of the industry.

POOLS AND AGREEMENTS

Agreements—'gentlemen's agreements' as they were euphemistically called—were informal, pools were much more precise, providing for quotas, fixing penalties for over-production and occasionally even paying firms to withdraw from a particular trade. The pools helped to preserve plants in poor locations, or, more accurately, preserved the less efficient producers, which was not always the same thing. The conditions within the rail pool and the results of free competition after its collapse in February 1897 show this clearly. Until this time the worsening position of the eastern rail mills had been disguised. The annual payment to Maryland Steel of Sparrows Point showed that pools could also subdue a well-located competitor. The structural steel pool of 1897–1904 illustrates the same principle. By the late nineties the New York area had become the chief market for structural steel, but the pool paid $5,000 a month to the excellently located New Jersey Steel and Iron Company of Trenton not to produce beams and channels.[28]

The horizontal groupings, like the pools, helped restrict competition, so that prices rose. This provided a temptation for the establishment of new firms. In other respects results were different and on the whole led to greater operating efficiency than with pools. A better division of the market, more plant specialization or, alternatively, reduction of cross-hauling were important advantages. The locational effects were considerable. Gates reckoned that, by cutting out cross-hauling from formerly independent and competing plants, American Steel and Wire could save perhaps $500,000 a year, while Guthrie of American Hoop, which produced a wide range of steel—bars, hoops, bands, cotton ties, skelp—and had a capacity of 700,000 tons a year, reckoned that mill specialization rather than the attempt to roll all sizes might save $1 million a year. Plants were closed on a big scale—after the American Tinplate Company was formed in October 1898 dismantling of small or badly located works went on, so that, whereas the A.I.S.A. 1896 *Works Directory* listed 69 completed tinplate works, 1 rebuilding and 4 being built, the 1901 *Directory* showed only 55 completed works and 7 building. One of the conditions imposed by American Tinplate when purchasing plants was that the old owners should not take up the trade again in less than 15 years within a radius of 1,500 miles.[29]

THE FORMATION OF THE UNITED STATES STEEL CORPORATION

By virtue of unequalled competitiveness Carnegie Steel occupied a key position in reorganization at the turn of the century. Largely through Frick's initiative, it acquired a dominating position in Mesabi ore supply, its royalties to Rockefeller being only a litle over 38 per cent of the going rate. The cost of landing ore at lake shore and furnace was cut to a new low level. Through the H.C. Frick Coke Company fuel was also obtained cheaper than for most, perhaps for all, rivals—the contract of January 1899 guaranteeing Carnegie all the coke needed for five years at $1·35 a ton delivered on rail cars at the works. A year later the market price was at least $3·50 a ton.[30]

Expanding ore and coal lands, installing new furnaces, entering new trades, the various consolidations seemed likely to founder in a grand display of industrial Darwinism. The initial contacts for a further rationalization movement were not particularly successful. Then in December 1900 at a dinner given in his honour, C.M. Schwab sketched out his version of a new, ideal steel world. Though it was jogged along further by Carnegie's plans for more forward integration, announced in the following month, the prospect of a rationalized, efficient steel industry dominated by a giant amalgamation was itself sufficient to dazzle the susceptible and to intrigue those of more calculating disposition. Morgan, involved with National Tube and other steel concerns, and more deeply still in railroads, was vitally interested in stability and in the prospects which Schwab outlined. Schwab reckoned that the steel industry had about reached the limit of cost reduction from a metallurgical or mechanical point of view—a mistaken impression which probably explained part of the progress which smaller, independent companies subsequently made. Economy was now to be won from a reorganization of the industry. Transport costs were a major factor in the trade and most existing plants were badly located, being survivals of the early days of iron-making.[31] Plant specialization was a key factor in future success. Instead of a mill making ten, twenty, or more products, the greatest economy would come from each making one product only. It would be desirable to build separate mills to roll angles or beams exclusively, even, perhaps, six different mills to roll beams of different sizes. Good though they were in part, some of these arguments were undermined by the changes of the time; others were mutually incompatible. If plants specialized to a very high degree, process costs might fall, but those for marketing would rise sharply. At this time freight rates seem to have reached their lowest levels, and the case for extreme specialization was invalidated by their subsequent increase. More-over, a 'balanced line' of production lessened the danger of a heavy local incidence of depression which the uncertainties of some trades would cause under Schwab's proposals. On the other hand, a great amalgamation would weed out inefficient or badly located plants. Mergers also provided important financial benefits and especially those of stock inflation. In February 1901 the organization of a giant amalgamation was decided, in the following month the work of putting the new concern together was under-taken, and on 1 April 1901 the United States Steel Corporation began business. (Fig. 17)

UNITED STATES STEEL AND ITS COMPETITORS

Carnegie, Federal and National Steel, and the chief horizontal consolidations, American Steel and Wire, National Tube, American Tinplate, American Steel Hoop, and American Sheet Steel were included in the Steel Corporation from the start. Very soon after, American Bridge was brought in, Rockefeller's Lake Superior Consolidated Iron Mines were acquired for $80 million, and his Bessemer Steamship Company—with 56 vessels and a single trip capacity of 228,600 tons of ore—for $8·5 million and stock in the Corporation. The capacity of the various constituent companies was estimated by the Bureau of Corporations as 7·4 million tons of pig iron, 9·1 million tons of ingots, 7·7 million tons of finished products per year. J.S. Jeans's figures, probably greatly inflated, especially in respect of finished products, are given below: (Table 39)

The Steel Corporation quickly made moves to acquire a still firmer grip on raw materials. By December 1901 50,000 acres of good Pocahontas coking coal land had been leased, but, from the start, emphasis was placed on

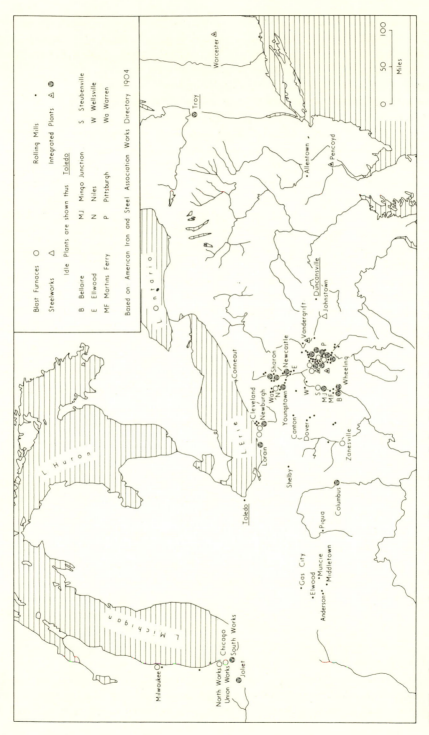

17. Blast furnaces, steelworks and rolling mills of the United States Steel Corporation 1901

Legend (within map):

Blast Furnaces ○
Steelworks △
Rolling Mills ·
Integrated Plants △ ◎

Idle Plants are shown thus Toledo

B Bellaire
E Ellwood
MF Martins Ferry
MJ Mingo Junction
N Niles
P Pittsburgh
S Steubenville
W Wellsville
Wa Warren

Based on American Iron and Steel Association Works Directory 1904

Map labels:

L Huron
L Michigan
L Erie
Ontario

Conneaut
Cleveland
Newburgh
Lorain
Toledo
Shelby
Canton
Dover
Zanesville
Columbus
Piqua
Gas City
Elwood
Anderson
Muncie
Middletown

Milwaukee
North Works
Union Works
Chicago
South Works
Joliet

Sharon
Newcastle
Youngstown
Vandergrift
Duncansville
Johnstown
Wheeling
Allentown
Pencoyd
Troy
Worcester

Miles
0 50 100

extending control of iron ore. A month after the Corporation's formation Lake ore reserves 'in sight' were put at 500 million tons. Fourteen months later the estimate was 700 million tons.[32] At this time U.S. Steel reckoned

Table 39.   *Iron, steel and finished products capacity of*
United States Steel Corporation plants
April—May 1901
(thousand tons)

	Pig iron	Bessemer ingots	Open hearth ingots	Finished products
Carnegie Steel	2,740	2,000	1,900	3,866
Federal Steel	1,855	1,760	240	2,360
(Lorain Steel)	(400)	(550)	—	(506)
National Steel	2,325	2,100	110	2,000
National Tube	605	480	—	996
American Steel and Wire	1,030	935	365	2,645
American Steel Hoop	500	—	10	730
American Sheet Steel	—	—	247	920
American Bridge	—	—	230	—
American Tinplate	—	—	—	534[1]
	9,455	7,825	3,102	14,557

[1] Black plate. Source: J.S. Jeans, *American Industrial Conditions and Competition,* 1902.

its ore fields were worth twice as much as its manufacturing plant and represented almost half the value of all its properties. Purchase of the Union Steel Company in November 1902 added large new properties on Mesabi, in 1903 the ore holdings of the Chemung Iron Company of Duluth were added, and in 1904, with Clairton Steel, there were bought in not only 2,644 acres of Connellsville coking coal land but 20,000 more acres of Mesabi ore. The biggest remaining Mesabi holdings, those of the Hill interests, were acquired on a royalty basis in October 1906, the high price paid confirming the fears of many that the Steel Corporation was seeking to maintain its predominating position by preventing rivals from obtaining adequate ore supplies. However, in fact the steel Corporation's share of Lake ore shipments fell from 61·6 per cent in 1901 to 51 per cent in 1910.

By 1906 U.S. Steel owned or had a controlling interest in 65 mines in the Lake ore districts, 113,000 acres of coking coal land, 93 blast furnaces, and over 700 iron and steel works located on 1,100 miles of company railroad. [33]

As with earlier, smaller amalgamations, a good deal of rationalization of these properties was needed. The process proved difficult and was not carried to anywhere near its logical completion for over a quarter of a century. Even so between 1901 and 1927 over 60 plants were dismantled.[34] Early plant abandonment gave little indication of important changes in locational values. (Table 40) Most of the plants closed were in the Pittsburgh

Table 40.   *Active blast furnaces and rolling mills of the United States Steel Corporation. 1901, 1904*

| Companies | 1901 | | | 1904 | |
	Blast furnaces	Rolling mills		Blast furnaces	Rolling mills
Carnegie Steel	19	6	merged		
			March	43	24
National Steel	18	6	1903		
Federal Steel	18	4		19	3
Lorain Steel	2	2		2	1
National Tube (Ohio)	—	—			
National Tube	4	9		5	7
American Steel and Wire	11	15		12	16
American Tinplate	—	28	merged		
American Sheet Steel	—	21	merged	—	41
American Steel Hoop	3	14	After March 1903 included in Carnegie Steel		
American Bridge	—	1		—	1
Shelby Steel Tube	—	5		—	4
Union Steel	—	—		4	6
Clairton Steel	—	—		3	1
Total	75	111		88	104

Based on A.I.S.A. *Works Directories*, 1901, 1904.

or Valleys districts, but the bulk of U.S. Steel properties had been in the same areas in April 1901 and remained there (Table 41 and Fig. 18). Nor did early plant expenditure give much indication of what the Steel Corporation judged would be the major metallurgical growth areas, for it was widely known that Carnegie properties needed little improvement, while those of Federal Steel required a good deal of investment to bring them to the same standard. (Table 42) The large expenditure at National Tube's McKeesport plant suggested that Carnegie's claim that a Conneaut mill could have produced tubes for $10 a ton less owed as much to better plant as to better

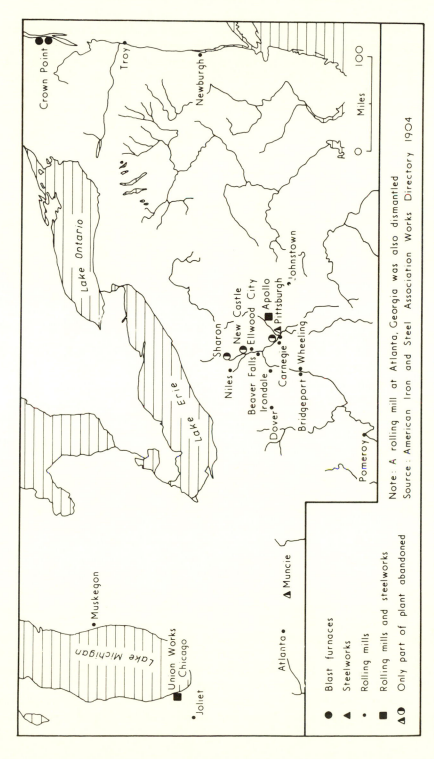

18. Iron and steel plants of the United States Steel Corporation abandoned or dismantled 1901–1904

location. Clearly the Steel Corporation's central planning group shared Carnegie's low evaluation of eastern prospects. Apart from Pencoyd and Worcester it had no important plants there, and in 1902 American Bridge,

Table 41.   *Location of blast furnaces and open hearth furnaces owned by the United States Steel Corporation 1906*

Area	Blast Furnaces	Open Hearth Furnaces
Within 50 miles Pittsburgh	48	111[1]
The Valleys	9	18
Within 50 miles Chicago	21	24
Lake Erie Shore	10	6
East of the Alleghenies	2	19
Other	7	2
Total	97	180

[1] 60 open hearth furnaces were at Homestead. Based on *I.A.* 16 Aug. 1906, pp.406–8.

Table 42.   *United States Steel Corporation plant expenditure announced March 1903*

Company	Works	Investment (thousand dollars)
Carnegie Steel	Homestead	1,135
	Edgar Thomson	275
	Duquesne	330
Illinois Steel	South Chicago	5,075
	Joliet	1,470
National Tube	McKeesport	9,255
	Lorain	8,646
All other manufacturing plants		8,207

Source: *I.A.* 5 Mar. 1903.

which operated the former, began a large new works on the Ohio river below Pittsburgh south of the village of Economy, the foundation of the new community of Ambridge. A lakeward shift of interest was shown in developments at Lorain, where big new tube-making capacity was built onto the existing rail and billet plant, while the Conneaut site remained undeveloped. Both Rockefeller and Carnegie seem to have contemplated building works in the Chicago area in the late 1890s.[35] In 1905 the announcement of the Gary project provided conclusive evidence of the attractions of a Chicago location, attractions hidden until this time largely by reason of the greater technical efficiency of Carnegie Steel.

The effect of the formation of the United States Steel Corporation on other firms was naturally very great, and indeed in some respects worldwide. Rival companies began to overhaul their own structure, and to strive to acquire bigger mineral properties. Jones and Laughlin bought some ore in the open market until 1901, but after that expected to be able to supply all their own needs.[36] Pennsylvania Steel and Bethlehem reorganized Cuban ore properties, and Republic Iron and Steel Company, formed only two years before the Steel Corporation, extended its ore and coal holdings. Although its achievement in this respect was minute as compared with that of the giant new concern, a good deal was none the less achieved. When it was organized, Republic had 1·5 million tons of coal reserves, but by July 1903 14 million tons of Connellsville coal. At its formation, northern ore reserves were 2·5 million tons, but by February 1903 they were 19 million.[37]

A number of men of talent and ambition were displaced or soon came to feel their initiative too restricted. Leaving, they built new plants or took over and revived old concerns. The outstanding example of this was the resignation of C.M. Schwab from the presidency of the Steel Corporation in August 1903, and the transfer of his business genius to the apparently moribund plant at Bethlehem which he made the catalyst in the reorganization and revival of the eastern steel industry. His successor in the presidency, W. Corey, also resigned in 1911 and built up a concern that for a time appeared likely to be a revival for Bethlehem in the leadership of the east, and which had Midvale Steel as its core. J.W. Gates, who had played a significant role in the formation of U.S. Steel, became an important factor in Republic Steel, for a short time was associated with Colorado Fuel and Iron, and then with the Tennessee Coal, Iron and Railroad Company.

Survival and extension of old firms and the growth of new ones was helped by three aspects of Steel Corporation policy—its co-operative, almost paternalistic stance, the associated price stability, and the development of Pittsburgh Plus pricing. The first two provided a suitable climate for growth, the latter gave extra incentives for new capacity at a distance from Pittsburgh.

Co-operation was only adopted by the Steel Corporation after the conversion of the old Carnegie Steel men to E.H. Gary's principles. In the early days Schwab remained for a time thoroughly committed to the old way. Gayley, also from Carnegie Steel, and a Vice-President of the Steel Corporation, recalled, 'I was brought up in the school of keen competition . . . I had been trained under the old competitive conditions. I was finally won over to the new plan.'[38]

Price stability now replaced the wild oscillations of the recent past, the most outstanding example being rails. In the month the Corporation was formed, the price was raised $2 to $28 a ton; it was maintained at that level for fifteen years. New entrants to such a capital-intensive trade as rail manufacture were rare, but in other lines, where plant construction was cheaper, price stability encouraged the growth of new concerns, and helped to account for the Steel Corporation's dwindling pre-eminence. Between 1901 and 1910 its share of the rail and structural steel businesses fell respectively from 65·4 and 57·9 per cent of the national total to 60·2 and 51·3 per cent, but in plates and sheets—the two unfortunately not being separated at this time—its share went down from 64·6 to 48 per cent, and from 1906 to 1910 alone its part in the tin and terne plate business declined from 72 to 61 per cent.[39] In sheet and tinplate a number of new concerns, of great importance in the future, got an important footing at this time, including Youngstown Sheet and Tube, the American Rolling Mill Company, Weirton Steel, and Inland Steel.

NOTES

[1] H.H. Campbell., *The Manufacture and Properties of Structural Steel*, 1896, p.v.

[2] J. Fulton, *Coke*, 1905, pp.326–8. C.D. King, '75 Years of Progress in Iron and Steel', *T.A.I.M.E.*, 1948, pp.9–10.

[3] Fulton, op. cit., p.44.

[4] Fulton, op. cit., Chapter 10.

[5] *I.A.*, 4 Jan. 1906, p.9, 26 July 1906, p.231.

[6] Fulton, op. cit., p.9; *I.A.*, 13 Jan. 1898, p.15; J.S. Jeans, *American Industrial Conditions and Competition*, 1902, p.17; U.S. Geological Survey *Annual Report, 1899–1900*, part v.1; G. Thiessen and others, *Coke from Illinois Coals*, Illinois State Geological Survey, Bulletin 64, 1937.

[7] Fulton, op. cit., p.251.

[8] Letter to A.J. Moxham quoted Fulton, op. cit., pp.277–88.

[9] *I.A.*, 22 May 1913, p.1242.

[10] A.I.S.A. *Works Directories* 1901, 1904.

[11] Editorial, 'Illinois Coke', *I.A.*, 13 Jan. 1898, p.15; *I.A.*, 22 May 1913, p.1242.

[12] *J.I.S.I.*, 1898, 1, p.576.

[13] Jeans, op. cit., pp.33–4.

[14] Jeans, op. cit., p.106.

[15] *Report of the Industrial Commission*, 1900, IV, U.S. Industrial Commission Reports, *Transportation*, testimony of the Secretary of the Lake Carriers Association, p.718.

[16] Jeans, op. cit., pp.106–7; F. Popplewell, *Some Modern Conditions and Recent Developments in Iron and Steel Production in America*, 1906, p.58.

[17] 'The Iron and Steel Trade of the United States', in *Summary of Commerce and Finance*, August 1900, p.229; *Report of the Industrial Commission*, 1900, I, *Trades and Industrial Combinations*, p.944.

[18] *Summary of Commerce and Finance*, August 1900, pp.228–31; Popplewell, op. cit., pp.59–62. On this railroad and Conneaut ore docks, see J.T. Odell, Vice-President of the railroad, in *Summary of Commerce and Finance*, pp.231–2.

[19] Ibid., pp.226, 235.

[20] *I.A.*, 9 Oct. 1913, p.818; also *I.A.*, 25 Apr. 1912, p.1037.

[21] Jeans, op. cit., p.315.

[22] Jeans, op. cit., p.166, lists wire rod mills at the end of 1901.

[23] *I.A.*, 3 Nov. 1892, pp.342–3.

[24] *I.A.*, 29 May 1913, p.1329; Jeans, op. cit., p.112.

25 Jeans, op. cit., p.278; *I.A.*, 4 Apr. 1912, p.890.

26 A. Carnegie, 'Steel Manufacture in the United States in the Nineteenth Century', *New York Evening Post*, 12 Jan. 1901, Review of the Century Number, reprinted in *The Empire of Business*, 1912, pp.236–7.

27 *I.A.*, 8 June 1911, p.1407.

28 W.C. Temple, Stanley Committee Hearings, *I.A.*, 17 Aug. 1911, p.367.

29 *I.A.*, 13 June 1912, p.1492.

30 *I.A.*, 1 Mar. 1900, pp.22–3.

31 B.J. Hendrick, *The Life of Andrew Carnegie*, 1933, pp.487–489 and T.N.E.C., Monograph 13, *Relative Efficiency of large, medium and small businesses*, p.413.

32 C.M. Schwab, in *Report of Industrial Commission*, 1901, XIII, p.464. Bureau of Corporations, *Report on the Steel Industry*, I, 1911, p.203.

33 A. Berglund, *The United States Steel Corporation*, 1907, p.88.

34 E.D. McCallum, *The Iron and Steel Industry in the U.S.A. A Study in Industrial Organisation*, 1931, p.181.

35 Nevins, *A Study in Power*, 1933, vol.2, p.266; *I.A.*, 17 Nov. 1898, p.10.

36 W.L. King, evidence in *Report of The Industrial Commission*, VIII, May 1901, p.499.

37 *I.A.*, 5 Feb. 1903, p.21, 13 Aug. 1903, p.22.

38 E.H. Gary, *I.A.*, 5 June 1913, p.1396, C.M. Schwab, *I.A.*, 10 Aug. 1911, p.316; J. Gayley, *I.A.*, 22 June 1911, p.1532.

39 Report of the Commissioner of Corporations, *The Steel Industry*, Part I, 1911, pp.360–3.

CHAPTER 7

Varying Patterns of Development 1890-1920.
I. Pittsburgh and Chicago

The iron and steel industry in both Pittsburgh and Chicago began near the present city centres. The first Pittsburgh foundries, naileries, and mills were in the streets and along the river edges near the Point. When the North Chicago Rolling Mill was completed in the summer of 1858 its site on the north branch of the Chicago River, 3 miles inland from the lake, was open prairie just outside the town. In both places early plants remained growth points; in Chicago until the end of the century, in Pittsburgh until much later.

THE PITTSBURGH AREA

From the Point works were built up river along both the Monongahela and Allegheny, though the greater availability of big sites meant that the biggest works were concentrated on the former. In the early days some of meander core sites seemed ample enough, indeed the original Edgar Thomson plant was 2,000 feet from the rivers edge, so that difficulty was experienced in obtaining process water. Expansion and the need in many cases to build open hearth steel plants, with their greater space requirements than Bessemer shops, began to fill up some of these areas. Major new plants within the city boundary, as at Homestead in 1879—81 and at Duquesne in 1889 were also placed on large river meander slip-off slopes, and their success and growth extended them impressively along the river front until stopped by residential building, other industrial plant, the railroad marshalling yards or the approach of the river and bounding cliff. In the nineties the process continued, from Duquesne up river to McKeesport and then, in 1901, to the integrated works at Clairton built to produce semi-finished steel for the members of the new Crucible Steel Company. (Fig. 19) From there development went on beyond the county line, the first furnace of American Steel and Wire's Donora works blowing in 1905. Three years before, the Pittsburgh Steel Company had begun development at Monessen, a site which was to become a most extreme illustration of the problems of congested works.

There was growth, though it was both slower and smaller, out along the Allegheny and its tributaries. Beyond the city the valley of the Kiskiminetas River was already important in sheet-making and was added to on a large scale in the last years of the century. At Tarentum on the Allegheny the Carnegie Company contemplated building the giant works which would mark its entry into the tube trade, but lack of room made it decide for

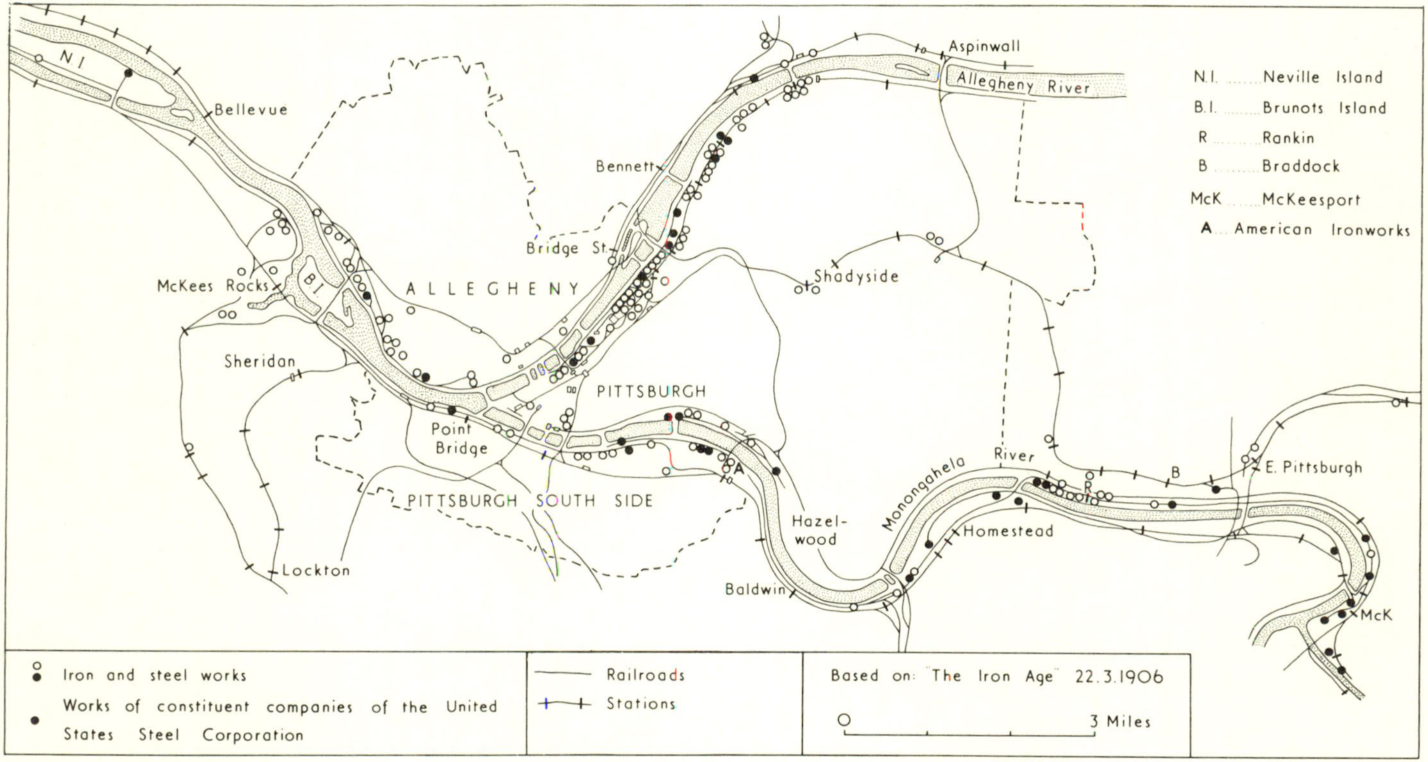
N.I.
Bellevue
McKees Rocks
B.I.
Sheridan
Lockton
Point Bridge
ALLEGHENY
Bridge St.
Bennett
PITTSBURGH
PITTSBURGH SOUTH SIDE
Shadyside
Aspinwall
Allegheny River
Monongahela River
Hazel-wood
Baldwin
Homestead
A
R
B
E. Pittsburgh
McK
N.I. .... Neville Island
B.I. .... Brunots Island
R .... Rankin
B .... Braddock
McK .... McKeesport
A ... American Ironworks
Iron and steel works
Works of constituent companies of the United States Steel Corporation
Railroads
Stations
Based on: "The Iron Age" 22.3.1906
0
3 Miles

19. The Pittsburgh district, 1906

Conneaut instead. A mile up river the Allegheny Steel and Iron Company made its small beginnings at Brackenridge in 1901. By the early twentieth century construction was moving down the Ohio with steelworks at McKees Rocks and Coraopolis and blast furnaces on Neville Island. Just beyond the county line, American Bridge built structural fabricating works and a small melting shop at the company site of Ambridge in 1904/5, and Jones and Laughlin bought Woodlawn Park with its two-mile river frontage. There in 1907 it began the Aliquippa works. Beaver Falls was already a metal centre, and after selling Clairton to the Steel Corporation in 1904, two years later Crucible Steel built a new works at Midland. This, and the new sheet mills at Follansbee, built by an old Pittsburgh tinplate dipping firm, caused growth by Pittsburgh area firms to merge with the sprawl of riverside works centred on Wheeling and those creeping up and down the Mahoning and Shenango Valleys. By 1906 the area whose bounding river points were Charleroi on the Monongahela, Kittanning on the Allegheny, and Beaver Falls on the Ohio contained 253 separate iron and steel plants, the United States Steel Corporation alone having eleven separate blast furnaces plants in this area.[1] Impressive though this figure and the physical setting was, there were many site deficiencies which pushed up operating costs. These factors aggravated the difficulties caused by changing raw material supply and market situations.

Sites were everywhere rather small. Homestead, before Lackawanna and Gary the biggest works in the country, covered only 156 acres in 1904, yet into this Carnegie Steel squeezed 0·7 million tons of iron, 2 million tons of steel and 1·4 million tons of rolling mill capacity, as well as some ancillary plant. New departments caused additional problems, as with the need, when natural gas supplies became unreliable, for gas producer plants, or, later still, by-product coke ovens. As integration of operations became more desirable the disadvantages of having main railroads pass through the centre of the works, as at Edgar Thomson, or of separation of the various units by the river became obvious. The piecemeal acquisition of works which it later became desirable to integrate increased the problems, though this was often tackled resolutely. Jones and Laughlin built a bridge across the Monongahela from the Eliza furnaces to the American Ironworks as early as 1877 though it was not used for molten iron until 1899 or 1900. In 1900 Carnegie Steel linked the Carrie furnaces and Homestead steelworks in a similar manner. Even before this 700 to 800 tons of hot metal a day were being railed five miles from the Duquesne blast furnaces to Homestead.[2]

In 1895 Carnegie Steel bought one mile of additional river frontage between Homestead and Duquesne and built there a major steel extension of sixteen open hearth furnaces. The last gaps suitable for development began to disappear. The conflict between Carnegie and Frick at the end of the nineties was deepened by suspicion that Frick was planning to develop one of the last big open sites. Jones and Laughlin's Pittsburgh works, the oldest major plant in the area, provides an excellent illustration of the area's problems. Use of fine Mesabi ore in furnaces designed for coarser, old range ore created clouds of red dust which, settling in neighbouring residential areas, caused a good deal of litigation and was believed to be one factor in the development of Aliquippa.[3] By 1917 a new melting shop at the Pittsburgh works had to be built on the north side across the river from older open hearth furnaces.

The same physical difficulties hindered the development of railway facilities on a scale necessary to deal with movement at peak periods. Traffic grew at a startling rate in the years around the turn of the century, and the boast that Pittsburgh was far and away the world's leading originating and terminating point in tonnage terms began to pale when this led to hopeless congestion. (Table 43) Carnegie's acquisition and development of the Pittsburgh, Bessemer and Lake Erie line relieved the burden of ore delivery,

Table 43.   *Pittsburgh area originating and terminating*
freight 1897–1912
(million tons)

	Railroads	Rivers	Total
1897	37	7	44
1904	78	10	88
1905	92	11	103
1912	—	—	177

Based on *I.A.* 14 Dec. 1905, 26 June 1913.

and as early as 1890 Jones and Laughlin made coal shipments by barge down the Monongahela, but most coke for Pittsburgh ironworks was still made in the coalfield beehive ovens and shipped in by rail. At this time, through control over the Baltimore and Ohio, the Pennsylvania Railroad acquired a temporary monopoly over the traffic. In the boom conditions of 1902 it proved quite unable to move the necessary tonnage of coke. By the end of the year coke was piled high in the yards of Connellsville, and at least a score of blast furnaces in the Valleys had been put out of blast or banked. Early in 1903 leading railroad officials came to Pittsburgh personally to unravel the tangle, but by March U.S. Steel had as many as 11 furnaces out of work; 250,000 tons of coke were piled up at the ovens. To a greater or less extent the problem remained serious through to the end of 1904.[4] Reacting to this general congestion the City Council swung from its opposition to the entry of the Wabash Railroad, and in July 1904 regular running into the city began on that new line. The Wabash was, however, built across country at extra-ordinarily high cost, there was bad organization, and U.S. Steel delayed in providing it with the traffic Carnegie Steel had promised. In the depression of 1907 the Wabash failed hopelessly.[5] Another development, anticipating later, bigger ones, was a move to improve navigation. Soon after the Wabash episode the Ohio was improved to Beaver Falls, and in 1911 congress authorized improvement of the river throughout its whole length to a minimum depth of nine feet. This was not achieved until 1929. A private company was chartered at the same time to build a canal of fourteen foot depth from the mouth of the Beaver River to Indian Creek on Lake Erie at a cost of $53 million. This vision of an effective water link from the Ohio Valley and the Valleys to Lake Erie was never to prove more than a mirage.

Inadequate transport facilities were already blamed for the slow growth of Pittsburgh's finishing capacity as well as for lack of further extension in crude steel. (Fig. 20) In the next fifteen years or so the switch to by-product coking shifted much of the burden of carrying fuel from the railroads to the river Monongahela, but at the same time re-emphasized the lack of room for new plant in Pittsburgh and the disadvantages of scattered operations.

Before long outsiders were often willing to write off Pittsburgh as a finished place for steel expansion. In this they were to be proved woefully wrong, but growth was now won only at extra cost. As early as 1907 Campbell of Steelton observed that almost every piece of level ground along the river front in the immediate vicinity of Pittsburgh had been taken up. Sixteen years later Crolius contrasted riverside conditions well away from Pittsburgh with the central plants and concluded '... it is almost an axiom that the further we remove from Pittsburgh ... the greater will be found the freedom for balanced development and the cleaner will be found the general atmosphere'. Freight problems still persisted. 'Several days frequently elapse before car loads find their way into or through the intricacies of the local Pittsburgh labyrinth.'[6]

CHICAGO

Whereas Pittsburgh area works, developing out from the centre, became more widely scattered but shared similar site conditions, in the case of Chicago the industry became concentrated in a new area and with much better sites than in the early days. The North Chicago works was built to reroll rails, but in 1864 another rail mill equipped with puddling furnaces was installed, blast furnaces followed five years later, and a Bessemer plant in 1872. By 1878 layout was already cramped, the converters were inconveni-ently placed in relation to other sections of the works, and, though the plant was river-based, most materials had to be brought in by rail. Two of the three other early locations also turned out to be poor. Built as an iron rail mill in 1863, Union works on the South Chicago river in the south-west of the town was making steel by 1871 and a little later pig iron as well. Joliet

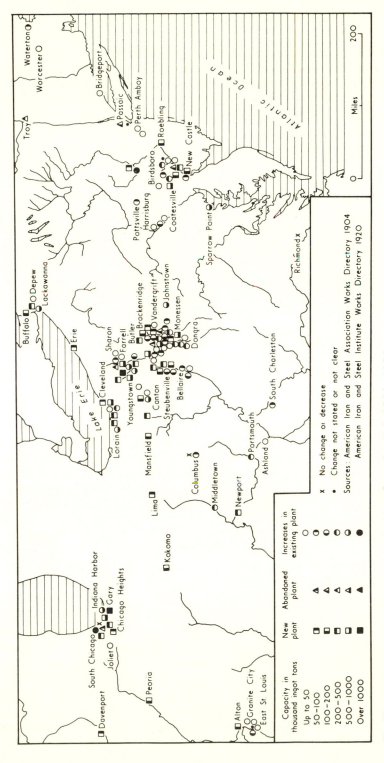

20. Changes in steel ingot capacity in the Manufacturing Belt 1904–1920

on the Illinois–Michigan Canal, where the same product evolution occurred, had to rail in its ore from Chicago. The Chicago works performed well enough, especially in rail manufacture, but it can scarcely be claimed that they were favoured at this time for raw material assembly, or in their layout. Pittsburgh plants certainly had easier assembly conditions and, on balance, probably better sites as well.

In the last twenty years of the nineteenth century new growth points were established in South Chicago and Calumet; after 1900 plant construction centred on the sand barrens straddling the Illinois–Indiana stateline. (Fig. 21) The Calumet plants survived, but the older city sites were given up. By 1901 the rail mill at North Chicago had been dismantled and at Union works it was idle. At both the steel plant and mills were abandoned 1901–04.

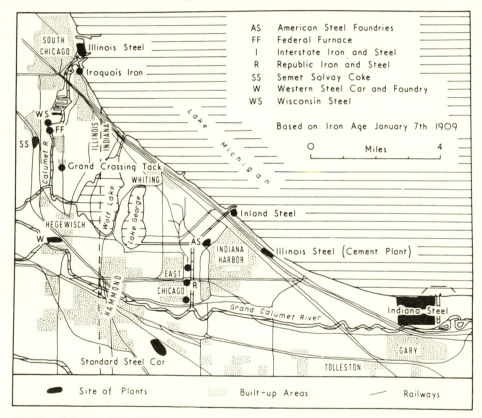

21. The Calumet Area 1909

Having enlarged resources after the acquisition of the Milwaukee Iron Company in 1878, and excellent rail-making prospects, the North Chicago Rolling Mill Company decided to build a new, lakeside rail mill. This was not done at Milwaukee works, and between there and Chicago there was no favourable site, rather high ground cutting off the lakeshore, while urban communities were already growing along it. To the south were the Calumet lowlands, sparsely populated yet near enough to the city to tap its labour supply. There were no natural harbours in this area, even the mouth of the Calumet River being too shallow for small boats. In 1880 the company bought a sandy lake beach three-quarters to one mile long and extending half a mile up the north side of the Calumet River. Already four main railroads were within a mile of the site and another two less than two miles away. Four blast furnaces were built in 1881, and next year the Bessemer plant and rail mill began work. All the coke came by rail from Pennsylvania, the dolomite flux from the quarry at Hawthorne 8 miles west of Chicago on the Chicago, Burlington and Quincy railroad. South Chicago proved its competitive strength very quickly, and in less than ten years was the chief works in the area. After the formation of the Illinois Steel Corporation in May 1889 its ore docks also supplied Joliet and the Union works. This justified the construction of new docks to handle 5,000 tons a day. Through the nineties there was ample room for growth, and in 1890 four new blast furnaces, an open hearth melting shop and plate mill were added. In 1880 the site had been only 70 acres, but by 1897 258 acres had been acquired, and by 1904 the area was 330 acres.[9]

Apart from Illinois Steel's fifteen blast furnaces, there were only two iron-making plants in the Chicago area in 1890. The Calumet Iron and Steel Company of Cummings, South Chicago, had water transport along the Calumet River. The Iroquois Furnace Company operated some way above the mouth of the same river. Iroquois improved on its location in the late nineties by building a wholly new plant with a lake frontage. By 1903 three other plants had been built on the Calumet. The Federal Furnace Company was a small blast furnace concern, but opposite it a much larger growth began when the Deering Harvester Company bought a small, hand-filled furnace on a large site two miles from Lake Michigan. There in 1901 it put up a rolling mill to supply part of its steel requirements, and when, the same year, International Harvester was formed, work began on the fully integrated Wisconsin Steel Works. The third plant was the Grand Crossing Tack Company, named from the major railroad crossing where South Chicago Avenue crossed 79th Street. Integrating backwards Grand Crossing put up melting furnaces and mills in 1902/3 on the Calumet one mile above Wisconsin Steel and five miles south-east of its old plant. All these works were better located than those of central Chicago, but they had disadvantages enough.

Except for South Chicago and the new Iroquois furnaces all were river plants with draught limitations as ore boats increased in size, and with rail and road bridges as obstacles to movement and to improvement. (Fig.21) Proximity to the southern fringes of Chicago provided them with a good local labour supply when the journey to work could only be short, but on the other hand urban growth pushed up land costs and threatened to restrict their expansion. Room for slag dumping or for slag processing plants was more difficult to find than on the lake shore. Major new mills or branches of operation, by-product ovens for example, were more difficult to fit in. Yet Wisconsin Steel was able to bring ore and coal boats to the works, found room for winter ore stockpiles, and in 1919 found room for its first coke battery. Beyond the Calumet lay sparsely populated townships which were, in large part, still a wilderness at the end of the nineteenth century.

The southern rim of Lake Michigan is fringed with dunes, belts of ill-drained soils, and glacial moraines. By the end of the nineteenth century it was already crossed by the main railroads running into Chicago, and at the crossing points in this system very small settlements had grown up. There were in addition a few small towns, Whiting making fence posts and paint, working lumber, and establishing the basis of its future prominence in oil refining, Hammond with a wider range of industry, and Crown Point the capital of Lake County.[10] In 1901 steel-making made its first move into this wasteland. The Michigan Land Company offered a fifty-acre lakeshore site to any firm which would undertake to build a steelworks and spend at least $1 million on plant. The offer was taken up by Inland Steel Company, established eight years before to reroll scrap rails in Chicago Heights. Its new site on the lakeshore north of East Chicago had at the beginning of 1901 only 'a railroad tower house surrounded by a vast expanse of seemingly useless sand'.[11] Sheet and bar production was begun in 1901. Within a few years the first blast furnace was built, and structural, plate, and rail-making began. As a result some of the site's potentialities and limitations began to appear. Iron-making involved a much greater movement of material, and in 1906 the first harbour—the beginning of Indiana Harbor—was created. Until this time a lake-shore location had had no transport value. Expansion also showed that railroads constricted the plant, and although expansion on made land in the lake was begun in 1908, it was slower and not quite as satisfactory as a large land site. Plant layout was, therefore, not ideal, a problem which was to become worse in the thirties when two hot strip mills with all their ancillaries had to be accommodated. There was plenty of land around the plant in 1901, but the growth of East Chicago, and, after 1916, of the new works of the American Steel and Tube Company on the western side of the harbour made it both scarce and costly.

The 1905 Annual Report of the United States Steel Corporation contained the announcement of major New Chicago works.

> Although the capacity of the producing furnaces and mills located at Chicago and vicinity has been materially increased from time to time, it has not kept pace with the increased and rapidly increasing consumption tributary to this location, and therefore a large percentage of this tonnage is now supplied from Eastern Mills. In consequence of these conditions it has been decided to construct and put into operation a new plant to be located on the south shore of Lake Michigan, in Calumet Township, Lake County, and a large acreage of land has been purchased for that purpose.

Several other locations had been considered, including the existing works at South Chicago. Further lake infill would have been necessary there, and it may well be doubted whether this could reasonably have been on a big enough scale for the new capacity. At any rate, permission for lake infilling was refused until a delegation of local interests petitioned the legislature at Springfield. By the time the ban had been removed it was too late, for the Gary project was under way.[12]

U.S. Steel bought 1,500 acres of sand barrens in 1905 and in the following year added a much larger block acquired from meat packing interests which had held it for development if the Chicago city authorities had decided that the Union Stockyards could not be continued near the heart of a great metropolis. The Gary site was uneven, sandy hillocks alternating with swampy sloughs and much of it covered with thickets, but it was found that, packed and confined laterally, sand made an excellent foundation for heavy plant.[13] Because of its size and influence the Corporation was able to secure re-location of the Baltimore and Ohio and Lake Shore railroads, an interesting contrast with Inland, as yet an almost insignificant newcomer, whose expansion left the plant divided by the Pennsylvania, New York Central, Baltimore and Ohio, and Elgin, Joliet and Eastern. On this large, almost completely empty site Corporation engineers put down a plant with an unusually rational layout, making provision for future growth but without in the early years increasing the length of the journey between consecutive stages of production. By this means heat loss and internal haulage were both reduced—cost savings which might match those derived from a favourable raw material assembly position. The excellence of Gary's plan and internal material flow patterns has scarcely if ever been exceeded. There too land has been reclaimed from the lake for some later expansion. The Gary site also had sufficient room for other U.S. Steel subsidiaries to build plants to work up its steel. American Sheet and Tinplate began sheet mills to the west in 1909, and tinplate mills in 1915, American Bridge opened fabricating yards in 1911, and National Tube built a major plant. West of these, along the foreshore, there was space for another subsidiary, the Buffington Portland Cement Works, using Gary slag. In the agglomeration of all these operations Gary was the very antithesis of the scatter of Steel Corporation operations in the Pittsburgh area. Eventually other U.S. Steel Chicago plants, except the South works, ceased to operate in the primary lines of iron and steel-making. Bay View works, Milwaukee, closed in 1928; Joliet, although it retained its mills and coke ovens, was abandoned as an iron and steel-making plant in 1932.

After the American Steel and Tube Indiana Harbor works was built in 1916, no new integrated plant was built in the sand barrens. At the time of the Pittsburgh Plus hearings Jones and Laughlin bought a large site, strangely located, over two miles from the lake on the western arm of the Indiana Harbor Canal, but decided not to build on it. In 1925 Bethlehem Steel issued a formal denial of rumours that it had bought 2,000 acres near Michigan City as a steel plant site.[14] Four years later the newly formed National Steel Corporation acquired a large area at Portage well to the east of Gary, but depression stopped plans for its early development.

NOTES

[1]*I.A.*, 22 Mar. 1906, p.1022.

[2]*J.I.S.I.*, 1899, 1, pp.409–10.

[3]*I.C.T.R.*, 15 Jan. 1907, p.547.

[4]*I.A.*, 1 Jan. 1903, 29 Jan. 1903, p.37, 19 Mar. 1903, p.26, 22 Dec. 1904, p.41.

[5]F.H. Spearman, *The Strategy of the Great Railroads*, 1905, pp.93–109; W.Z. Ripley, *Railroad Finance and Organization*, 1923, pp.521–4; Carnegie, at Stanley Committee Hearings, *I.A.*, 18 Jan. 1912, p.196.

[6]H.H. Campbell, *The Manufacture and Properties of Iron and Steel*, 1907, p.469. *B.F.S.P.*, Jan. 1923, p.70.

[7]A. Holley and L. Smith., *Eng.*, 18 Jan. 1878, pp.39–41, 8 Feb. 1878, pp.97–9, 22 Mar. 1878, p.227.

[8]G.W. Cope, *The Iron and Steel Interests of Chicago*, 1890, p.12; H.E. Kaufman, '75 Years of Blast Furnacing at South Works', *B.F.S.P.*, Dec. 1952, pp.1457–62; *I.A.*, 30 Oct. 1890, p.756; *Iron*, 19 Mar. 1880, p.209; *I.A.*, 15 July 1897, p.6; *J.I.S.I.*, 1904. 2, pp.398–9.

[9]J.B. Appleton, *The Iron and Steel Industry of The Calumet District*, Illinois Studies in the Social Sciences, 13, 1925.

[10]See Tolleston, Indiana quadrangle of the 1:62500 topographical map, surveyed 1897.

[11]J.L. Block, 'The Growth of the Inland Steel Company', *B.F.S.P.*, Sept. 1928, p.1200; *Fortune*, July 1958.

[12]*I.A.*, 17 Dec. 1908, p.1814; Brody, *Steelworkers in America: The Non-Union Era*, 1960, p.112.

[13]Appleton, op. cit., p.30; *B.F.S.P.*, Aug. 1937, p.880; Illinois Steel Company, *The Gary Works of the Illinois Steel Co.*, 1922.

[14]*I.A.*, 20 Aug. 1925, p.477.

CHAPTER 8

Varying Patterns of Development 1890-1920.
II. Decline and Revival in the East

At the end of the nineteenth century the metal centres east of the Appalachians suffered severely in competition with the west. Rail mills were driven to the wall, foundry iron furnaces closed, and, as Mesabi ore began to move down the lakes, large numbers of eastern ore pits were abandoned to become covered with scrub or fill with water. Charcoal and anthracite were both discredited fuels, and surviving works had long hauls on coke. Many eastern plants were small and antiquated, unfitted for the new era, and even the big concerns like Bethlehem, Pennsylvania, and Lackawanna were soon eclipsed by the new groupings in the west. As there seemed little hope of anything better, enterprise languished, was too small, or proved misguided. By 1898 Abram Hewitt, whose firm had once been a national pacemaker, was writing an epitaph on the east. 'For the manufacture of crude iron in the East I can see no hope of better times . . . The great Lehigh region is ruined and our great plant in New Jersey is useless. We have a plant which we could have sold at one time for $3 million but now not a furnace is in blast.'[1] Hewitt looked back to a day when Trenton Works was worth $3 million. Within little more than three years the United Steel Corporation was popularly known as the 'Billion Dollar Trust'. About 1900 Hewitt gave a dinner for Carnegie at which he reckoned that, because of a failure to relocate their Trenton operations, net profits for the whole period 1858 to 1900 had not equalled the amount of Carnegie's charitable gift to the Cooper Union which the dinner celebrated.[2] Others too thought little of eastern prospects. Carnegie observed in 1898, 'in the East Pittsburgh has nothing to fear. Steel cannot be manufactured on the Atlantic seaboard under any circumstances in competition with her.'[3] Within less than a generation the obsequies for the eastern industry proved to have been premature.

As late as 1890 the Lehigh, Schuylkill, and Lower Susquehanna Valley districts made 2 million tons of iron or 500,000 tons more than Allegheny County; Illinois and Indiana produced only 800,000 tons. In 1902 output was 1·7 million in the three eastern districts, 4·7 million in Allegheny County, and 1·9 million tons in Illinois and Indiana. As early as 1888 three large furnaces on the edge of the anthracite coalfield in Berks and Lebanon counties were running wholly on coke. Opening of the mid-Appalachian field by the efficient 'coal roads' running down to Hampton Roads provided a lower-cost supply of coal which could be used for coking. The development of by-product coking, already going ahead rapidly in Europe, promised a breakthrough for the east, by removing any disadvantage in coke quality and enabling an admixture of the 'drier' coals of the Appalachians along with Appalachian plateau coal (Table 44). The slowness of by-product coking in catching the imagination of American pig iron makers in general is to be

Table 44.   *New York and New Jersey pig iron production*
by   fuel used—1880, 1890
(thousand net tons)

	1880	1890
Charcoal	21	16
Anthracite	173	36
Anthracite and coke	119	176
Coke	—	117

Source: *I.A.* 8 June 1893, p.1290. These figures differ considerably from *A.I.S.A.*

attributed to the excellent coking coal available in Connellsville. In the east, where the advantage would have been greatest, the lack of fully integrated units or of enterprise on the part of those who managed the big works hindered a competitive iron industry.

Eastern ores were deficient in both quantity and quality, and productivity was low. Output of merchantable ore per shift in 1880 was 0·83 tons in New York, New Jersey, and Pennsylvania, and in the three Upper Lake states 1·14 tons. By 1900 the respective figures were 1·61 and 3·75 tons.[4] There were a few big ore bodies like the Cornwall Banks, but even there ownership was divided. Concentration of eastern lean ores was either a costly failure or pushed the price up too high. Smelting eastern magnetites involved technical problems not solved till the 1930s and high coke rates in the meantime.[5] The opening of Mesabi and the improvement of ore handling sharply cut the cost of ore in areas fringing the Great Lakes, and though large tonnages of this ore were railed to eastern furnaces, their competitive position was worsened relatively by these developments. The obvious resort of eastern companies was to smelt more foreign ore. Bethlehem took interests in orefields fifteen miles from Santiago, Cuba, in 1882, and two years later Pennsylvania Steel also began to work Cuban ore. Between 1884 and 1897 3 million tons of ore passed through Santiago to the U.S.A. Yet while their own costs were falling, western interests managed to ensure that a tariff on imported ore burdened their eastern competitors. The tariff was 75 cents a ton when in 1892 imported ore was $2·46 at the point of shipment, and Bessemer ore at the lower lake ports averaged $5·50. With freight from Santiago, the short land haul to the eastern valleys, and the high cost of transporting coke, iron costs in the east would probably have scarcely been competitive even if the furnaces had been of equal efficiency. The 1897 act retained this level of duty and not until the Payne-Aldrich tariff of 1909 was it cut to 15 cents generally, and to 12 cents for Cuban ore. There were some advantages in the use of foreign ore—in years of large grain shipments to Europe ballast rate cargoes of Mediterranean ore continued to come west— but on the other hand the supply was not always reliable in quality or in quantity, and the Spanish–American War cut Cuban shipment by more than half. With rationalized foreign supplies handled in bulk, and smelted on or near the coast in big new coke furnaces, the prospects of the east were by no means as dismal as Hewitt and others believed. In evidence to the McKinley Committee, the President of the Maryland Steel Company had already bravely claimed that if the duty was removed his company would undertake to make pig iron against the world—and could prove from their working results that they could do so. The switch from Bessemer to open hearth steel favoured the east, but the benefits showed themselves only slowly because of the delay in the changeover in the United States. The east was short of Bessemer ores and the high molten iron requirement of that process—now that cupola practice was rapidly being abandoned—placed the works of the region at an increasing disadvantage by virtue of their higher iron costs. On the other hand its great wealth of scrap from processing industries and large centres of population made it more favourable for the open hearth; in addition this process could profitably be conducted on a smaller scale than with Bessemer converters and so located nearer to market outlets. The acid open hearth furnace did not require such stringent control over phosphorus content in the ores as the acid Bessemer process, and in the 1890s there occurred the large-scale adoption of the basic open hearth furnaces for which the range of suitable iron ores was wider still. (Table 45) Long-continued emphasis on rail production slowed the advance of the melting shops so that as late as 1890, when 44 per cent of British steel was open hearth, in the United States melting shop output was 574,000 tons but Bessemer 4,131,000 tons. A host of small melting shops, very largely market orient- ated, were established in the east in the nineties but the big firms struggled on in the old trades until the change was literally forced on them by competition from the west. That a certain lack of foresight was present may be surmised, though a full consideration of it may be deferred.

Another eastern weakness, widely commented on at the end of the nineteenth century, was its high rail transport charges. As early as 1889 Carnegie reckoned eastern roads could still make a profit if their rates were only two-thirds as high, and *Iron Age* then and later painted a sad picture of the decline which high rates induced. 'Let any intelligent man visit the Hudson River Valley, travel along the Lehigh and the Schuylkill. He will witness signs of decay which, unless it is soon arrested, will leave among the wrecks only a few flourishing concerns, which the railroads have been unable to kill, which will live in spite of them.'[6] A number of reasons to explain these high charges may be adduced. Hauls, as compared with those in the west, were relatively short and there was such an abundance of other freight that the railroads could afford to charge low-grade material heavily. Another

Table 45.   *Steel plant and output, east and other districts, 1901*

District	Bessemer plant (standard size) Number	Bessemer plant (standard size) Average capacity (tons)	Open hearth furnaces[1] Acid Number	Open hearth furnaces[1] Acid Average capacity (tons)	Open hearth furnaces[1] Basic Number	Open hearth furnaces[1] Basic Average capacity (tons)	Steel output 1901 All kinds (thous. tons)
Steelton	3	10	3	45	9	40	427
Johnstown	4	12	2	35	8	40	656
Lehigh Valley	4	7	6	30	2	40	69
S.E. Pennsylvania	–	–	12	30	33	40	629
New York and New Jersey	–	–	2	25	8	20	150
New England	–	–	5	15	6	40	173
Sparrows Point	2	20	–	–	–	–	352
Pittsburgh	30	10	35	30	84	40	7,317
Illinois	5	11	3	25	9	40	1,750
U.S.	58	–	84	–	204	–	13,474

[1] Excludes steel castings plant.
Based on H.H. Campbell, *The Manufacture and Properties of Iron and Steel,* 1907, pp.442–3.

factor related to the structure of the eastern steel industry. No company was big enough or aggressive enough to build its own line to bring the railroads to heel. Crippling freight rates were in this respect another illustration of the organizational weaknesses of the eastern steel industry. The most vigorous reaction was Lackawanna Steel's move from Scranton to Buffalo.

The regional marketing situation of eastern works was improving by the turn of the century. Growth and rebuilding in the coastal cities provided big new outlets for merchant steel, for pipes, and for structurals and plates for bridges and now increasingly for use in steel-framed buildings. Export possibilities, especially in Latin America, began to be realized. The ship-building market was tempting but proved rather disappointing. Iron shipbuilding began in the United States with Harlan and Hollingsworth at Wilmington in 1843/4. The nearby firms of John Roach of Chester and William Cramp of Philadelphia also built early in iron. By the seventies the lower Delaware was already being called 'the American Clyde'.[7] In the early

1880s Roach built the first big American steel ships. Progress was slow for American costs were high, and not until 1891 did annual merchant construction in iron or steel exceed 100,000 tons; not until 1900 was half the shipping built and registered of steel or iron. In the 1890s construction of a new navy was begun, and the wider horizons opened by the Spanish–American War encouraged further growth. Yet in 1900 the United States built only 365,000 tons of merchant vessels, one-eight of the British output in that year. Less than half this was on the Atlantic and Gulf coast, and in that section it was focused on the Delaware.[8] Shipbuilding was only to a very small degree linked with steel-making. In the 1870s the Delaware River Iron Shipbuilding and Machine works had co-operated with the Philadelphia and Reading Railroad in establishing the Chester Rolling Mill Company to roll plate and angles, but this was a slight and unusual venture. Almost thirty years later Charles Cramp noted he had tried without success to improve operations by commercial links with Bethlehem Steel, Midvale or even Carnegie. Such connections as there were, were tenuous or often not with local firms at all, the New York Shipbuilding Company, formed in 1899 with a magnificent new yard on the Jersey shore opposite Philadelphia, had links through H.C. Frick, one of its leading shareholders, with Pittsburgh steel interests and not with local ones.[9] The lack of steel-maker's initiative with respect to shipbuilding highlights that organizational deficiency which seems to have been at the root of the problem of the eastern industry in the 1890s. It may be illustrated from two of the leading companies, Pennsylvania and Bethlehem.

In July 1883 the tariff on iron ore was reduced from 20 per cent *ad valorem* to 75 cents a gross ton, and the following year the Pennsylvania Steel Company began to ship Cuban ore, railed on to Steelton from Baltimore. A logical step seemed to be a coastal smelting works, and in 1887 it decided to built a plant at Sparrows Point on Chesapeake Bay, fifteen miles south of Baltimore. Although the blast furnaces were blown in first, a steel plant was also planned, and the early reports spoke of rail and plate mills. Sparrows Point works was laid out on a 1,000 acre site, soon had a 28 foot channel to the ore docks, and developed a profitable export trade in rails, for which there was a rebate on the ore duty. In 1900 45 per cent of its steel and as much as 50 per cent of its rail output was for export. An important shipyard was also errected there, which at the beginning of 1901 had 54,000 tons of shipping being built.[10] Yet in spite of its low costs and on-the-spot market for plate, Sparrows Point had no plate or angle mills, its shipbuilding materials being obtained from the works of the parent company at Steelton, over seventy miles away.

Bethlehem Steel, as John Fritz exposed in his autobiography, frittered away the advantages of location by wrong choices or the unwillingness to make decisions. It lost the opportunity to control large supplies of Bessemer ore, and its Cuban ore operations were inefficiently conducted. In the early nineties the directors refused to adopt Fritz's suggestion that the firm should go into the structural steel trade in a large way, and so lost to Pittsburgh the opportunities which proximity to New York and Philadelphia gave it. Even more tragic was the way in which it bungled its entry to the plate trade. For a number of years Bethlehem had made armour plate and forgings for naval vessels, and by the beginning of 1897 it was completing a plate mill which it was understood was the first step in a comprehensive plan for the production of merchant steel. By all accounts the mill was of high quality. Market conditions were good, national output of plate and sheets—the figures were not then separated—going up from 1·2 million tons in 1897 to 1·4 million in 1898 and 1·9 million in 1899, but by the beginning of 1899 the mills had been sold to Carnegie Steel, transferred to Homestead, and were operated there with great success. The explanation of this strange development seems in part to have been due to the high steel-making costs at Bethlehem, for it was said that other works could not produce basic open hearth steel at prices competitive with Homestead.[11]

Bethlehem reverted to a belief that special products would give it the best return, so that by 1902 Jeans observed, 'the steel-producing plant at Bethlehem is of more or less special character, and adapted to more or less exceptional work, such as the manufacture of armour projectiles and large forgings.'[12] In 1902 Bethlehem became part of another abortive project. the United States Shipbuilding Company. Within a year the group was in diffi-culties, and even of Bethlehem, the soundest part of the whole, it was said that its valuation of $30 million was excessive, one employee claiming he could duplicate the plant for $7 million. Languishing while others expanded, in 1901, a record steel year nationally Bethlehem made only 69,000 tons of steel, 0·5 per cent of the national total.[13]

The key significance of the entrepreneurial factor in the east may be seen also in the early twentieth century, when raw material and marketing conditions were improving. The ore duty was 40 cents a ton but the fuel position had improved with by-product coke oven construction getting under way. By 1901 it had been decided to build 200 ovens at Sparrows Point and by 1904 there were ovens building for Pennsylvania Steel. As new markets for steel developed so open hearth steel made greater headway. As late as 1898 melting shops turned out just under one-third the tonnage of the converters, but by 1903 two-thirds, and by 1907 the tonnages were 11·66 million gross tons Bessemer steel, 11·54 million tons open hearth steel. In 1903 the market price for No. 1 heavy melting scrap in east Pennsylvania was $17·66 per gross ton or 78 cents a ton less than in Pittsburgh. For the next three years eastern prices ranged higher than in Pittsburgh, but by 1907 they were lower. Only in one subsequent year, 1912, were eastern scrap prices as high as or higher than those of Pittsburgh.

There followed a host of proposals for new tidewater works in the east. Some merchant blast furnaces were built. By 1908 there were three on the lower Delaware and three in the New York area, at Newark and at Secaucus, on the Hackensack Meadows behind Jersey City. Reconstruction of the Erie Canal, it was suggested, would cut delivery charges on Champlain ore to the New York area by two-thirds. In 1913 a plant site on the Lower Bay was said to have been chosen, and by the following year Witherbee and Sherman of Port Henry had done considerable planning for a blast furnace and coke plant between Perth Amboy and Sewaren, New Jersey. By 1920 the three New Jersey coastal furnaces and one of the three on the lower Delaware had been abandoned. Exposure to foreign, Great Lakes and southern competition accounted for their failure.[14] Proposals for new coastal steel-works were equally unfruitful. The New York Shipbuilding Company was said to be planning Delaware river steelworks in 1902/3, and in 1905 there were persistant rumours of an integrated works on New York harbour, later crystallizing into a steel plant and four furnaces at Carteret west of Staten Island. By 1911, when it acquired smallish Cuban ore properties, there was speculation as to whether the Steel Corporation was planning an east coast works, and three years later the Industrial Bureau of the Merchant's Association of New York published a thirty-three page booklet promoting New York harbour as an advantageous location for iron and steel manu-facture.[15] Eastern revival and expansion was not to be at major new sites.

EASTERN REVIVAL

At the beginning of the new century there could be no doubt that Pennsylvania Steel and Maryland Steel were the most successful of the eastern firms. Yet over the next dozen years production at Bethlehem boomed while that at Sparrows Point and Steelton grew very slightly. (Table 46) Looking back it seems that the Pennsylvanian interests made two closely

Table 46.   *Output of finished products—leading eastern works*
1901, 1913
(thousand tons)

	1901	1913	
Bethlehem	18	704	
Cambria	467	1,193	
Lackawanna	333	544	
Pennsylvania and Maryland	515	539	(1911)

Source: *I.A.* 22 Oct. 1914, p.951.

related mistakes. One was to concentrate expansion at Steelton, the other was to fail to realize the potentialities of Sparrows Point. (Table 47)

Steelton, it is true, had a good location for eastern markets, and the Pennsylvania Railroad presumably continued to treat it kindly in the matter of freight charges on raw materials. Low-cost production possibilities at

Table 47.   *Ingot capacity at Steelton and Sparrows Point*
1901, 1904, 1916
(thousand tons)

	1901	1904	1916
Steelton	500	650	1,515
Sparrows Point	400	500	720

Based on A.I.S.A. and A.I.S.I. *Works Directories.*

Sparrows Point were not fully exploited, concentration on rails and billets burdening it with low-value products, for which there were usually cut-throat prices in international markets, though it retained its advantageous situation in much of the south. (By 1913 the rail freight charge, Birmingham to Galveston, was $3·40 a ton but at that time the Baltimore—Galveston water rate was never over $2·50. [16]) Meanwhile, its shipyards continued to obtain material from Steelton. In 1912 two-thirds of the plate used came from there, the rest from Lukens and Worth in south-eastern Pennsylvania or from United States Steel.[17]

In 1905 C.M. Schwab took over full direction of Bethlehem Steel, rationalized the company's organisation, travelled indefatigably to get business, and poured in capital resources, taking little or nothing back for a number of years. Most important, he overhauled raw material supply and Bethlehem's product line. By-product coke ovens were built at Bethlehem in 1912. Ore supply presented greater difficulties. Reorganized, the Juragua Iron Company in Cuba was shipping up to 350,000 tons of ore a year by 1913, but this was less than a quarter of company requirements. There was some connection of Bethlehem's name with Texas ore, but this never came to anything. By 1911/12 an agreement had been drawn up for the delivery of 3 million tons of Swedish ore over the next 9 years, and by late 1913 Kiruna ore was arriving at a rate equal to 400,000 tons a year. In 1913 Bethlehem made an offer, which was refused, to buy Witherbee Sherman's Adirondack ore for $1 a ton delivered at Port Henry.[18] At the beginning of the same year the company carried off the decisive ore deal, one which helped to make it the king-pin in the reorganization of the whole eastern iron and steel industry. This was the purchase of the El Tofo orefield in Chile. Schneider of Le Creusot had owned this ore, but imminent completion of the Panama Canal alone gave it practical significance, and then especially for a coastal works in the United States. By the end of 1913

Bethlehem was planning to strip 1·5 million tons of ore a year, rail it four and a half miles to Cruz Grande Bay and then 4,400 miles to Philadelphia. By March tenders had been invited for ten ore vessels of 15,000 tons carrying capacity.[19] Schwab reckoned that 68 per cent Fe ore could be delivered in Bethlehem for $4·08 a ton—made up of $1 a ton on board at the Chilean port, $2·25 to $2·50 freight cost and canal dues and the rest rail charges. In that year Mesabi non-Bessemer ore averaged only about 51 per cent Fe and the average price at lower lake ports was $3·40 a ton.[20]

Product reorganization was achieved even more rapidly. The decision was taken to roll basic open hearth rails and structural steel on a large scale. In rails the basic process removed the penalty of lack of Bessemer ore. The choice of structurals was far more important. A major local market existed in the New York–Philadelphia belt. Henry Grey's universal beam mill produced a lighter section than the older structural mills. Leading U.S. companies had turned Grey's idea down, but he had build successful mills at Differdange, Luxembourg. Schwab as a mill man realized the practicability of Grey's scheme and had reputation sufficient to overcome all obstacles to its realization, including the scepticism of structural engineers and doubt on the part of Bethlehem colleagues. In the crisis of 1907 finance for the project, already spiralling from its original $4 million estimate to $12 million, ran out. At this stage personality became a locational factor of first rank. Schwab obtained credit for the project, induced contractors to work for no immediate payment, and even persuaded the Lehigh and the Philadelphia and Reading railroads to accept freight on credit. By 1909 Bethlehem could make 1,000 tons of structurals and 500 tons of rails a day. With its new plant, and favoured by short hauls to market, it prospered in the new trades, while Sparrows Point, better located for raw material assembly, did less well, and other, smaller firms remained very much in the doldrums as compared with those west of the Appalachians. As late as 1911 Bope, the Vice-President of Carnegie Steel, claimed that 'the hand writing is on the wall for the plants located east of the Allegheny Mountains.'[21] In fact conditions were swinging in favour of the east but it required business acumen of a high order to realize, and dynamism and a touch of organizational genius to seize the new opportunities.

MERGER AND RATIONALIZATION IN EASTERN STEEL

In spite of the great success of the section mills of Bethlehem Steel's Saucon division, it was war material which gave that company the opportunity to become the focus for the wholesale reorganization of the eastern steel industry. In 1913 it purchased the Quincy Massachusetts, yards of the Fore River Shipbuilding Company. Along with the older Bethlehem ordnance trades this led to early recognition of the company as a major source of supply for the allies in the First World War. As Schwab put it to his Annual Meeting in April 1915, with a fine commercial turn of phrase, '. . .

while the year so far has been very bad for the general steel business, the Bethlehem Company has been fortunate in being engaged in the manufacture of lines which are in strong demand.' Net profits in 1915 were $18 million, and early that year $20 to $30 million plant extensions were under way. By the end of 1917 90 per cent of Bethlehem orders were on government account.[22] From 1915 to the end of 1921 the company earned about $150 million, but preferred and common stockholders received only $44·6 million in dividends, the rest was reinvested. By August 1916 *Iron Age* was recognizing how war production and profits had changed the position of the east '. . . the European War has done more in a year to bring in the long looked-for day of the eastern steel trade than twenty-five years of promotions and theoretical discussions of the Atlantic Seaboard's advantages as a situs of profitable steel-making.'[23] The first moves in regional consolidation and reorganization came from the Midvale Steel and Ordnance Company.

The Midvale Steel Company had been established in Nicetown, Philadelphia, in 1866. By 1890 its annual capacity was 25,000 tons of tires, axles, castings, and other special steels, and, like Bethlehem, it had ordnance and forging sections.[24] In 1911 Schwab's successor as President of United States Steel, William Corey, resigned as a result of a clash of personality with E.H. Gary, and, like Schwab, went east, in this case to control Midvale. In October 1915 Midvale acquired the Coatesville steelworks and plate mill from Worth Brothers and subsequently largely extended them. Four months later Cambria Steel was purchased. Cambria's Johnstown plant had 1·78 million tons capacity, and there were large Lake Superior ore holdings, a Great Lakes ore fleet, and coking coal properties. In November 1917 Midvale acquired the 144,000 ton new plant of the Wilmington Steel Company. Expanding all its works, by 1922 the group had a steel capacity of 2·7 million tons. Bethlehem Steel early in 1922 had steel capacity only 500,000 tons greater, but held the ace for eastern development, big foreign ore properties.

In February 1916, nine days after Midvale took in Cambria, Bethlehem acquired Pennsylvania Steel and Maryland Steel, the Pennsylvania Railroad fearing that under new antitrust legislation it would not be allowed to operate manufacturing plant. By mid-1916 Bethlehem was committed to spend $30 million at Sparrows Point, and $10–12 million at Steelton, as well as $30 million at Bethlehem.[25] In May 1922 Bethlehem acquired Lackawanna Steel. This gave it access to the Great Lakes market for rails, plate, and structurals, new coal properties, and fifteen years' supply of Lake ores, Lake Champlain ore deposits, and the Lackawanna share of the Cornwall ore banks, whose rational redevelopment Bethlehem could now embark on. The three Bethlehem works then had ingot capacity of 3·2 million tons, Lackawanna of 1·8 million. Midvale and Bethlehem were direct competitors in many fields, but with Bethlehem much better located than

Johnstown for structurals, and with nothing to rival Sparrows Point, it was clear where the initiative lay. Early in 1921, hoping to find business for its war-extended plants, Midvale began cutting prices. Over the next eighteen months it featured in various wide-ranging merger proposals, but in November 1922 it was agreed that all the Midvale properties except the ordnance plant should be brought into Bethlehem Steel.[26] With a capacity of 7·6 million ingot tons, Bethlehem now was almost exactly one-third as big as the Steel Corporation. Rationalization was essential to make its regional empire fully viable. Over the next four years $167 million was spent on plant development.[27]

Integrated, coastal, or lakeside operations, in locations well placed for marketing, were most desirable, and like some of the older Bethlehem properties, many of the Midvale works did not measure up well to those standards. (Table 48) Wilmington was well located, but was very small, and had no blast furnaces. It was closed down in the 1920s. Coatesville was a little way inland though it had iron capacity, Worth Brothers having built the first furnaces in 1909, and Midvale adding two others in 1917. At the time of the merger Coatesville plate capacity was 310,000 tons, as compared with only 200,000 at Sparrows Point. In 1931 the works was abandoned, the

Table 48.   *Steel ingot capacity of Bethlehem Steel Company plants 1916, 1922, 1930*
(thousand gross tons)

	1916	1922	1930
Bethelehem	1,129	1,412	1,699
Steelton	1,515	788	678
Sparrows Point	702	850	1,753
Lackawanna	1,600	1,840	1,984
Coatesville	425	550	336[1]
Wilmington	–	144	–[2]
Johnstown	1,780	2,016	1,825
Total	7,151	7,600	8,275

[1] Coatesville abandoned 1931.
[2] Wilmington abandoned in the 1920s.
Based on A.I.S.I. *Works Directories.*

plate mill being moved to Sparrows Point, where costs were lower and there were shipyards next to the mill. Johnstown was the biggest plant in the group at the outset, but, although it provided a foothold in the west, it was inferior in this respect to Lackawanna, which as a low-cost iron-maker and with cheap scrap, and now benefiting further from growth of the Detroit market and lake shipping of finished steel, was to prove, along with Sparrows Point, Bethlehem's most successful plant. By 1926 about $35 million had been spent on plant improvement at Johnstown, but it remained a rather high-cost producer. Gradually it concentrated on lines for which it had a local outlet, such as mine and railroad supplies, or in which the costs of iron and crude steel production were relatively unimportant in the value of the finished product, as with specializations built on its long-established Gautier division.

Rail production at Bethlehem was given up in favour of the rail mills at Sparrows Point, Lackawanna, and Steelton. The latter had a combination rail and structural mill built in 1915 but its location raised costs, so that at this time it turned more and more to billets, Lackawanna taking over much of the other trade. Eventually, as the rail market declined still further, and fewer producers remained in the field, Steelton, located centrally in the eastern market, again became the chief Bethlehem rail mill, leaving the bigger, better-located plants free for expanding lines of business. Lackawanna was equipped with Grey mills in the twenties, enabling it to exploit its position in lakeside markets for structurals, and to divide with Bethlehem outlets east and west of the Alleghenies. Ten years later, with strip mill development, the same pattern was to be followed, this time by Lackawanna and Sparrows Point.

EASTERN REVIVAL IN PERSPECTIVE

Below the colourful battles, negotiations, and changing affiliations of the major concerns, the east continued to show its former uncertainties and high failure rates. The old merchant iron trade withered, and many small steelworks, built at the end of the nineteenth century, quickly disappeared. On the other hand, one or two plants survived at indifferent or bad locations, locations which, if the company had been taken into the bigger groupings, might well have been abandoned. Coatesville shows the pattern very well. Lukens Steel had descended from a Brandywine Creek rolling mill of 1810 which had used charcoal pig iron and water power. In the eighties it began to roll steel plate, and in 1891 built its first open hearth furnace. Worth Brothers built an iron plate mill in the same town in 1881/2, but in 1885 began to roll steel, and in the nineties, like Lukens, built a steelworks. Both built new plate mills in 1902/3, each mill being wider than any other in the United States. Worth provided their own pig iron capacity, while Lukens continued to purchase it, or to bring iron from a distance, having control of two Virginia iron companies by 1920. Lukens built a new melting shop in the First World War, but at this time Worth Brothers had been swept into the Midvale group and from there was taken on into Bethlehem Steel and the works was closed in 1931. Lukens, very similarly endowed, survived and continued its slow growth. The small Phoenixville plant, established in 1783, first made steel in 1889, and now survived as a cold metal works. On the other hand, Alan Wood Iron and Steel Company, formed in 1901 as a reorganization of another old concern, had 250,000 tons steel capacity at Conshohocken in the outskirts of Philadelphia by 1907, and in 1911 put a bridge across the Schuylkill to the Swedeland blast furnace, so making hot metal practice possible. By 1920 Conshohocken had 500,000 tons ingot capacity, though it stagnated for some time after that. If it had fallen to the big groups it too would probably have not survived.[28]

There was a high mortality among the small, market orientated steel casting plants of the lower Delaware, and in the twenties and thirties two of the three bigger works there, Wilmington and the Newcastle works of the former Pennsylvania Seaboard Steel Company failed. Only at Worth's Claymont works was there expansion. (Table 49) New England's steel-making significance declined still further. The Worcester works, the biggest, stagnated, there was small growth at Bridgeport, but the Pennsylvania Seaboard Steel Company's New Haven works was abandoned.

Table 49. *New England steel ingot capacity 1920–30*
(thousand gross tons)

	1920	1930
Connecticut	160	140
Massachusetts	313	286
Rhode Island	39	39

Based on A.I.S.I. *Works Directories.*

Pittsburgh Plus to 1924, and, for many grades of steel, the multiple basing point system through to 1938, gave eastern firms 'phantom freight' as well as their natural protection from competition on local deliveries at a time when the eastern market was growing rapidly. With by-product coke, foreign ore, and favourable differentials on scrap, raw material costs sometimes compared very favourably with those of Great Lakes area producers. By 1940 58 per cent Fe El Tofo ore was put down at Sparrows Point for $4·70 a ton when the open market price for 51·1 per cent Mesabi non-Bessemer ore at lower lake ports was $5·12.[29] By 1940 United States Steel was said to be actively studying the relative advantages of Wilmington, Norfolk, and Richmond as locations for a new integrated steel plant. For the big concern, able to organize long-distance bulk movement of material, the east had become a highly desirable location. But the organizational factor was decisive, and for the small firm regional conditions continued to present many difficulties.

Even so the east had come a long way since the slighting evaluations of Carnegie or Bope. In 1926 Schwab spelled out his own faith in the east. He saw the growth of the eastern market, the increasing cost of Lake ore as compared with the delivered price of Latin American ore, bulk-mined, and efficiently transported, as decisive factors. From the point of view of material costs

> ... study of the iron ore situation would be a deciding factor. Looking toward Lake Superior, I saw costs steadily advancing and the quality of the ore steadily declining. In the old days I could get ore properties there at royalties ranging from 10 to 25 cents: today those royalties have gone to $1·25 and $1·50. Taxes on lake ore have increased several fold, and there have been successive advances in labour costs. On the other hand are the low mining costs in Chile, which have been stable over a long period, and substantially that is the case in Cuba. Our Chilean ore runs 67 per cent in iron, whereas Lake ores that once averaged about 60 per cent in iron are now down to between 50 and 55 per cent. Chilean ore hauled 4,400 miles can be laid down at Sparrows Point at $6 for the ore required to make a ton of pig iron ... While Pittsburgh has its advantage in fuel, the ore advantage of the eastern seaboard plant is such that pig iron can be made more cheaply at Sparrows Point than at any point in the Middle West. The marked advance in rail transportation cost is another factor in the greater future of the eastern steel industry. When I started making steel at Pittsburgh we had a $2 freight on finished steel to the Atlantic seaboard. Today the rate is $7 a ton ... the higher cost of railroad operation is here to stay. If you speak of markets east of the Alleghenies, think of the scale of railroad operations ... and of the large track and car and locomotive requirements. The number of metalworking plants in the east and the range and number of their products are well known. Almost the entire output of our wire mills at Sparrows Point is taken up by spring works and other industrial users of wire in New England and the Middle Atlantic seaboard. Speaking of structural steel, where has there been such a use of it as we have seen in New York City and other points on the Atlantic seaboard in this present wave of new construction? Here the steel construction of the next ten years will far exceed what are called high records today.[30]

Table 50.   *Steel company capacity in certain heavy and light*
lines 1933
(capacity as percentages of all finished lines)

| | Heavy rolled steel | | | Light rolled steel | |
	Plates	Shapes	Rails	Sheet and strip	Tinplate
Bethlehem	14·5	27·6	13·4	below 5·0	below 5·0
United States steel	15·3	11·2	10·8	c. 10·0	c. 5·0
Republic	c. 3·0	none	none	24·7	c. 3·0
Youngstown sheet and Tube	13·5	none	none	9·5	c. 4·0
Inland	12·5	14·2	12·3	24·3	none
National	below 10·0	below 10·0	?	61·8	14·0

Source: *Fortune* Sept. 1933.

He expressed the hope that eventually Sparrows Point would be the greatest steel plant in the world, an ambition realized within thirty years.

Although Schwab spoke of a successful reorganization of eastern steel, he was thinking largely in terms of the old, heavy trades. (Table 50) The depression hit these hardest of all, and in the depths of 1932 no area was operating at a lower rate than eastern Pennsylvania—as little as 10 per cent of capacity at the end of the year—and few major companies were in worse straits than Bethlehem. Still further product diversification was essential.

NOTES

[1] *I.A.,* 6 Jan. 1898, p.19.

[2] *A.I.S.I. Yearbook,* 1930, p.527.

[3] *I.A.,* 17 Nov. 1898, p.10.

[4] Works Project Administration, *Iron Mining,* 1940, Table A.4.

[5] Axel Sahlin in J.S. Jeans, *American Industrial Conditions and Competition,* 1902, pp.406–8; H.H. Campbell, *The Manufacture and Properties of Iron and Steel,* 1907, p.495; J.T. Whiting, 'Microscopic and Petragraphic Structure of Blast Furnace Materials', *A.I.S.I. Yearbook,* 1938; F.L. Nason, 'The Importance of the Iron Ores of the Adirondack Region', *A.I.S.I. Yearbook,* 1922, p.183.

[6] *I.A.,* 4 Apr. 1889, p.515; see also *I.A.,* 28 Mar. 1889, 17 July 1890, p.100, 24 July 1890, p.261.

[7] H.G. Smith, 'Shipbuilding and its relation to the Steel Industry', *A.I.S.I. Yearbook,* 1931, p.84; D.B. Taylor, *The American Clyde,* 1958.

[8] Jeans, op. cit., p.250.

[9] C.H. Cramp, *Report of Industrial Commission,* XIV, 1901, p.419. Jeans, op. cit., pp.250–1.

[10] Campbell, op. cit., p.489; *I.T.R.,* 28 Mar. 1901, p.23.

[11] *I.A.,* 21 Jan. 1897, pp.3, 7; Jeans, op. cit., p.555; *I.A.,* 12 Jan. 1899, p.18.

[12] Jeans, op. cit., p.180.

[13] Campbell, op. cit., p.443.

[14] *I.A.,* 23 Oct. 1913, p.1916; 30 Oct. 1913, p.984, 12 Feb. 1914; U.S. Tariff Commission, *Iron in Pigs,* 1927; R.H. Sweetser, *Blast Furnace Practice,* 1938, p.326.

[15] *I.A.,* 2 Feb. 1905, p.399, 2 Mar. 1905, pp.746–7, 18 May 1905, p.1595, 27 June 1912, p.1593, 11 June 1914, p.1474; Report of the Commissioner of Corporations, *The Steel Industry,* part I, 1911, pp.373–4, 381.

[16] J. Farrell in Steel Dissolution Suit, *I.A.,* 22 May 1913, p.1240.

[17] F.W. Wood (President of Maryland Steel), *I.A.,* 16 Oct. 1913, p.859.

[18] *I.A.,* 23 Jan. 1913, p.249, 27 May 1913, p.1327.

[19] *I.A.,* 20 Mar. 1913, p.746, 16 Oct. 1913, p.858.

[20] *I.A.,* 27 May 1913, p.1327.

[21] *I.A.,* 2 Feb. 1911, pp.304–5.

[22] *I.A.,* 8 Apr. 1915, 21 Jan. 1915, p.201, 1 Nov. 1917, p.1036.

[23] *I.A.,* 3 Aug. 1916, p.256.

[24] Campbell, op.cit., p.493.

[25] *I.A.,* 20 July 1916, p.146.

[26] S.L. Goodale and J.R. Spear, *Chronology of Iron and Steel,* 1931, pp.289–90.

[27] *A.I.S.I. Yearbook,* 1927, p.225.

[28] *I.A.,* 5 Nov. 1903; H.R. Wood, *A. Wood, A Century and a Half of Steelmaking,* published by the Newcomen Society in North America, 1962.

[29] Report on an English Steelmakers Visit, Spring 1941, (privately communicated).

[30] *I.A.,* 7 Jan. 1926, p.11.

CHAPTER 9

Varying Patterns of Development 1890-1920.
III. Steel-making along the
Lake Erie Shore and in the Valleys District

Between 1890 and the First World War two major new steel districts emerged—the Lake Erie shore from Lorain to Buffalo and the Mahoning and Shenango Valleys, commonly known as the 'Valleys', and later as the Youngstown district. Even though they are near together the reasons for their growth and many of their characteristics are different.

THE LAKE ERIE SHORE

As plant grew in size and complexity and integration of iron-making, steelworks and mills at the same place was seen to be more desirable, so big, flat sites with access to process water in very large volumes increased in value. With the exception of Cleveland, where the narrow lake front was already crowded and the Cuyahoga flats becoming congested, for a number of years the chief Lake ports could offer this advantage. Site conditions were, however, of minor significance as compared with fuel economy. By 1895 a large furnace plant in Pittsburgh was using 12 per cent less coke per ton of iron than in 1887. By-product coking, and the possiblity of using a proportion of nearer, even if inferior, coal to that of Connellsville, and with an increased coal/coke yield, was to prove important later, but was of no account through most of the 1890s when Lake furnace plants bought their coke. One furnace was abandoned at Cleveland in the 1890s, but seven new coke-using stacks were built along the Lakes, two at Lorain, two at Tonawanda, and three at Buffalo. (Fig. 22) From 230,000 tons in 1890, lake-shore coke iron capacity increased to 1,085,000 tons by 1900. From nothing in 1888, Buffalo ten years later could produce 250,000 tons of iron and steel.[1] Most of the new capacity was merchant iron plant, supplying local foundries or shipping iron eastwards or to Pittsburgh, still short of pig iron in boom years. At an early stage the advantages of the Erie shore for steel-making also became clear. To large markets and excellent communications there was now added the advantage of locally available iron. (Fig.22) (Table 51). The first big new steel-making development set the pattern of so many others, the movement lakewards of interests already located in the interior.

In the early 1880s T.L. Johnson of Cleveland developed a rail for urban tramways, and about 1885 Cambria began to roll this section at Johnstown. As demand grew, Johnson determined to roll the rails himself, and in 1888

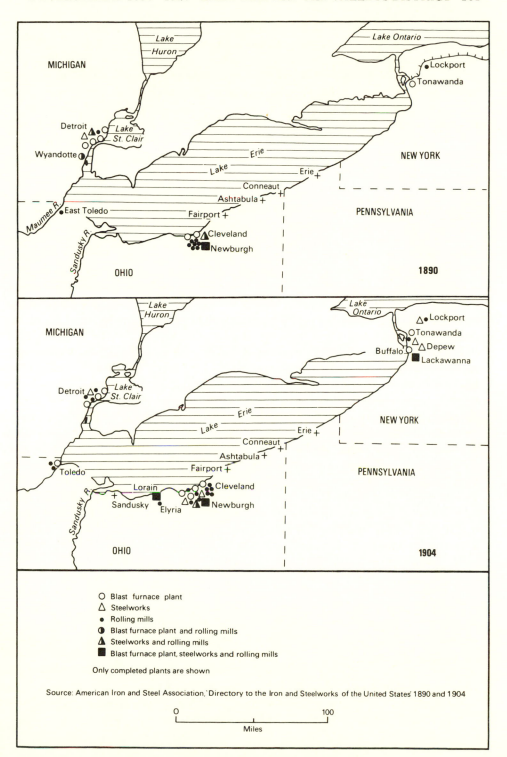

22. The iron and steel plants of the Lake Erie shore 1890–1904

his mill at Moxham near Johnstown began work on purchased Cambria blooms. In 1894 he obtained a large tract at Lorain, and, with a 1,200 tons per day Bessemer plant, started to roll street rails there in the following year. Production of higher-quality lines—frogs, switches, and so on— was left at Moxham. In 1898 the company was reorganized as the Lorain Steel Company, and work was begun on two blast furances and coke ovens. By-product coking was an important factor from the start in this case, John Fulton, the coke expert, being hired as early as 1895 to experiment with and compare by-product coke made in the ovens of the Solvay Process Company at Syracuse with Connellsville beehive coke.[2]

Table 51.   *Value added in manufacturing in Lake Erie shore districts and Pittsburgh 1899–1914 (million $)*

	1899	1904	1909	1914
Pittsburgh Area[1]	108·0	102·8	125·8	129·2
Cleveland	62·0	74·0	117·0	153·0
Lorain	2·9	4·7	14·7	11·1
Toledo	12·5	19·0	27·1	44·5
Detroit	41·0	61·0	122·0	178·0
Buffalo	39·0	59·0	82·0	89·0

[1] Pittsburgh, Allegheny City, Beaver Falls, Braddock, Butler, Carnegie, Charleroi, Connellsville, Homestead, Latrobe, McKeesport, McKeesrocks, Washington. For some of the smaller communities figures are not available for the earliest dates.

Source: Bureau of the Census, *Census of Manufactures,* 1914.

The next lake-shore project never passed on to construction but the case is instructive. As the ore traffic handled by Lake Erie ports increased, the two biggest ports in this traffic, Ashtabula and Cleveland, kept their lead but lost ground relative to a number of others. (Table 52) Toledo, Huron, Lorain, and Erie all had high growth rates, but Conneaut was outstanding. In July 1896 Carnegie Steel let the first contract for the extension of the Shenango and Lake Erie Railroad from Butler to the Monongahela Valley. At the beginning it was expected that 70 per cent of the line's total freight would move southwards from Conneaut, so that over half the returning trucks would be empty. By 1898 the iron ore traffic amounted to more than twice all the other freight on this line, and, in the hope of achieving a better balance between up and down cargoes. Carnegie Steel decided in December 1898 to examine prospects for a blast furnace at Conneaut. Early in 1899 a committee reported that there was room for a steelworks as well. At this time Carnegie engineers began work on a scheme for a Conneaut tube works, and it was later suggested that, but for negotiations with the Moore group which controlled National Tube, this plant would have been built in 1899.[3]

Table 52. *Iron ore receipts at Lake Erie ports 1892, 1893, 1895, 1897 and 1903. (thousand tons)*

		1892	1893	1895	1897	1903
Conneaut	below	10	203	244	495	3,904
Toledo		139	145	260	416	652
Sandusky		49	4	12	79	130
Huron		65	137	146	198	486
Lorain		190	165	214	355	990
Fairport		866	792	914	1,008	1,434
Erie		645	469	811	1,311	1,258
Buffalo and Tonawanda		197	308	719	797	2,150
Cleveland		1,950	1,260	2,312	2,456	4,434
Ashtabula		2,555	1,845	2,474	3,001	4,242
Total		6,660	5,333	8,112	10,120	19,681

Based on *A.I.S.A.*

At the end of 1900, after an attempt to find an adequate alternative site on either the Monongahela or the Allegheny, a $12 million plant was announced for Conneaut.[4] Schwab observed that coke would now occupy hundreds of trucks which so far returned empty to Conneaut every day. With the formation of United States Steel the plant was never built, but in February 1903 it was announced that a new tube mill would be built at Lorain which was at this time turning out 1,200 to 1,400 tons more steel a day than its rail department used, and had been selling this as billets.[5]

The decision to build at Lorain rather than Conneaut provides an interesting illustration of the range of cost factors. The Corporation owned no railroad from Lorain, so could not gain the advantages of a balanced inward and outward movement of material, and the traffic from Conneaut remained unbalanced. On the other hand, capital expenditure was saved by extending a large existing works. No major metallurgical works was ever to be built on the Corporation's 5,000 acre site at Conneaut; Carnegie some years later argued it should have been used in preference to Gary.

There had been rumours in 1895 of plans by Lehigh interests to build furnaces and a 1,200 tons a day Bessemer plant at Buffalo, but nothing came

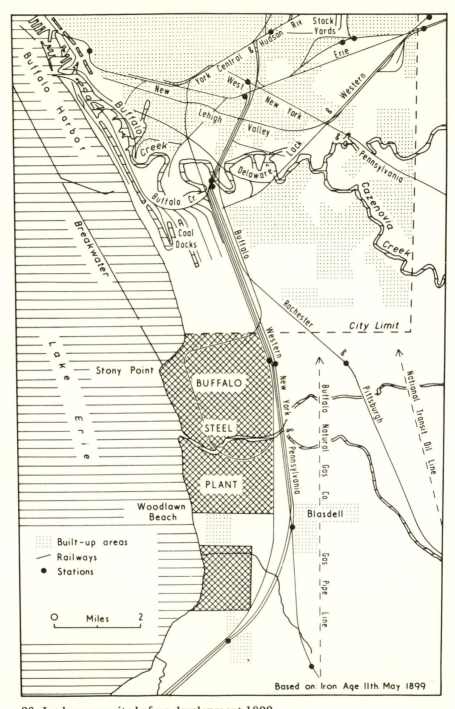

23. Lackawanna site before development 1899

of this.[6] Worsted in the 1897 and 1898 rail struggles, Lackawanna searched for a better location. For some time the possibility of a steel plant and billet mill near the company's Lebanon blast furnace and its Cornwall ore properties was considered, but this would scarcely have solved Lackawanna's problems.[7] Early in 1899 a new concern, the Buffalo Steel Company, bought 1,600 acres of land at Stoney Point on the south side of Buffalo. Billet and rail mills and a steel capacity of 800,000 to 1,000,000 tons a year were planned.[8] Soon it was known that it was planned to remove the Scranton works, but although the rail mill and two blast furnaces were moved from Scranton this snowballed into a very much bigger scheme. In site, layout, and location the new Lackawanna works had impressive advantages (Fig. 23).

Apart from Lake Superior ore brought to its own deepwater harbour, Lackawanna had interests in the Port Henry area and for a time thought of using returning grain boats along the Erie Canal to deliver this. By-produce ovens were built in 1904, supplied from company-owned mines in Indiana and Cambria Counties, Pennsylvania, much nearer than Connellsville yet producing coal which, washed and crushed before shipping, and compressed before charging, made an acceptable furnace coke. Eastern markets remained accessible by rail—or as contemporaries believed by the projected improved Erie Canal—while Ohio valley, lake-shore, and western markets could now be served much more easily than before. The Company's South Buffalo Railroad Company made connection with the main lines.

Both Conneaut and Lackawanna, and even to a large extent the Johnson development at Lorain, were developments by outsiders, not home-grown enterprises. The latter seem to have been largely concerned with merchant iron plants, though in one or two cases these grew into something bigger.

THE MERCHANT IRON TRADE ON LAKE ERIE TO 1914

Much Lake iron-making capacity supplied foundries, forges, or independent steelworks. By-product coking did not affect the unintegrated concerns at all, the only ovens on the lake by 1914 being at Lackawanna, Cleveland, and Lorain. But, helped by fuel economy, Lake ironworks could now more than equal the delivered costs of eastern Pennsylvanian plants in their local markets—though they were in turn sometimes hard-pressed there by southern competition (Table 53)—and there were other major outlets through to the Ohio. Furnace after furnace was built for this trade, mostly in Cleveland or Buffalo, but also at Toledo and Erie. In 1902 the first coke furnaces in the Detroit area were built at the mouth of the Rouge River. Differences in ore assembly and in marketing costs were less important in their long-term profitability than control over minerals.[9] Controlling ore lands and boats, the Hanna interests bought the old Union Iron Company properties at Buffalo in 1892, and next year built a new works on the same site. The most interesting and most important case of all is that of Corrigan McKinney, where expansion proceeded all the way to steel-making.

James Corrigan built up a fleet of Lake ore vessels in the early eighties, and in 1889, as the trade proved highly profitable, bought Lake Superior mines as well. In the 1893 panic the concern went under, to be reorganized

Table 53. *Estimated cost per ton of pig iron supplied to Philadelphia 1908/1909*

Producing district	Production cost	Freight	Total
East Pennsylvania	18·79	0·60	19·39
Buffalo	16·45	2·45	18·90
Virginia	14:17	2·80	16·97
Alabama	11·00	4·00	15·00
Germany	8·71	2·50	15·21[1]

[1] Including $4 duty.
Based on J.G. Butler, *Tariff Hearings*, 1908–1909, pp.1428–9.

by the receiver, Price McKinney. Ore shipments increased but at no time was it possible to sell the full capacity of the mines, so that it was decided to make pig iron for sale. Furnaces were acquired or leased at Josephine and Scottdale in west-central and south-western Pennsylvania, at Charlotte near Rochester, at one other point in New York State, and in 1897 from the Upson Nut Company in Cleveland itself. In 1907 the lease on this last furnace ran out, and the following year Corrigan McKinney obtained a large site on the Cuyahoga and built a new furnace there and shortly a second. The merchant iron business was a speculative one, with a large number of new producers, so that there were years of bad trade like 1907/8 and 1911. By 1912 Corrigan McKinney had decided to put up a steel plant. The rail trade was considered, but it was eventually determined to make billets. By 1916 the plant was at work. It eventually became clear that it was necessary to go on from semi-finished to fully finished products. Long before this Corrigan McKinney had gained a reputation for aggressive, independent action which later made it refuse to adopt Pittsburgh Plus pricing. In 1904, on behalf of the Steel Corporation, James Gayley recommended the ore firms not to cut prices in spite of recession. Corrigan McKinney went so far as to announce that they would take business at whatever price was necessary to get it. 1904 ore shipments were 2·5 million tons down on 1903, but their shipments rose by 700,000 tons.[10]

FURTHER EXPANSION AND THE EMERGENCE OF SITE PROBLEMS

By the early twentieth century good sites were becoming rare in the neighbourhood of some of the chief cities of the Lake Erie shore, and there were none left to equal those on the southern edge of Lake Michigan. As early as 1904, the President of Buffalo Chamber of Commerce observed that there was no extensive uncontrolled tracts with water access there, and a few years later the only suitable areas were said to be along the Niagara River between Buffalo and Tonawanda. An example of movement to this area was the outcome of the 1906 search of the Union Drawn Steel Company of Beaver Falls for a source of bars for its cold drawing mills. Shunning its local area, long considered ideal for low-cost assembly, it started up a blast furnace, two open hearth furnaces and a blooming mill on a Niagara River site in 1907. The following year Wickwire Brothers, with a small wire works at Cortland, New York, brought blast furnaces and a steelworks into production at Tonawanda.[11]

The increase in size of ore boats cut the biggest of them off from the upstream furnaces on the Cuyahoga flats. Until about 1900 all lake carriers could navigate the heavy industry section of the Cuyahoga; by 1913 only about 55 per cent of them and by 1924 32 per cent. Piecemeal development gave undesirable patterns of interplant linkage. Already by 1904 molten metal was carried five miles from U.S. Steel's central furnaces to Newburgh Steelworks through the congested Cuyahoga belt.[12] Upson, making nuts and bolts in Cleveland since 1872, subsequently acquired an old blast furnace, and now built a small steelworks on the east side of the river, hemmed in between the railroad tracks. As *Iron Age* put it, 'the engineers had to make the plans fit the site'. In the late seventies, Holley and Smith had observed that Otis Steel had a fine potential blast furnace site next to the mills and melting shop, but in 1912, with plans for three blast furnaces, a new melting shop and additional primary and finishing mills, the company applied to the government engineer in the city to fill in fifty acres of lake. Permission was refused, and instead Otis had to split its operations, building a new works in the upper Cuyahoga flats.[13]

Table 54.   *Steel ingot capacity of Lake Erie shore counties*
1890–1920

1890–	242
1896–	802
1901–	1,260
1904–	2,544
1920–	6,509

Based on A.I.S. and A.I.S.I. *Works Directories.*

Between 1900 and 1908, 19 blast furnaces were built on the Erie shore. At the end 2 more were building, and another 4 projected. Of the total of 37 existing blast furnaces, 18 were integrated with steel operations, the others were merchant plants.[14]   Their competitive strength was great as may be seen by the way in which Buffalo came to dominate the iron trade of New York State. At the beginning of 1905 the Buffalo area had 11 of the state's 22 furnaces, by the middle of 1914, 19 out of 26, and by early 1926 21 out of 25. There was a big increase of steel capacity as well, though from 1904 to 1920 the Lake Erie share of national capacity went up only from 10·1 to 11·1 per cent. The advance of by-product coking during and after the war, further fuel economy, and booming local markets soon made the belt more important still. (Table 54) Between 1920 and 1938 increase in ingot capacity was 26·5 per cent of the national total. In this period the centre of most rapid expansion moved from Buffalo and Cleveland to Detroit.

THE VALLEYS

Valley competitive conditions for pig iron manufacture worsened in the last twenty-five years of the nineteenth century. As early as 1879/80 84 per cent of the ore used in the six main works of the Mahoning Valley was from Lake Superior or Canadian mines. Raw coal was not available in tonnages sufficient for the growing output; more important it was too soft for big furnace burdens. By 1869 one Youngstown furnace was using Connellsville coke, and mixed coke and coal charges were common in the next twenty years, with coke increasing its relative importance, and eventually becoming dominant. (Table 55) By the nineties a once self-sufficient district had

Table 55.   *Raw materials used in Valley furnaces 1890*
                     (by number of furnaces)

| | | Fuel | | Iron Ore |
	Coke	Coke and Raw Coal	Iron Ore Lake Superior	Lake Superior and Local Blackband
Mahoning Valley (15 furnaces)	15	–	7	8
Shenango Valley [1] (23 furnaces)	16	5	23	–

[1] The fuel supply of two furnaces in the Shenango Valley is not indicated. Based on: A.I.S.A. *Works Directory*, 1890.

become dependent on long—distance transport; in this situation the Valleys' location between the coking district and the Lake ports placed it at a freight cost disadvantage. (Table 56)

In process costs too the situation of the Valleys was deteriorating. The legacy of small, outdated furnaces, built to use coal, was one adverse factor.

Table 56.   *Mineral freight rates and estimated raw*
material assembly costs of Valley, Cleveland, and
Pittsburgh furnaces, 1887
(dollars)

	Assembly costs (per ton of raw material)		Total assembly costs per ton of pig iron
	Connellsville coke by Pennsylvania Railroad	*Ore from Cleveland Docks*	
Cleveland	2·20	0·30	3·07
Pittsburgh	0·60 (?)	1·50	3·06
Valleys	2·65	0·85	4·47

Consumption of 31·4 cwts. ore and 23·7 cwts. coke per ton of pig iron is assumed.
Based on *I.A.* 14 and 21 Apr. 1887.

(Table 57) In the middle of 1884 the ten furnaces at work in Pittsburgh had a weekly average capacity of 773 tons, the ten Shenango Valley furnaces of 539, and eight in the Mahoning Valley of only 382 tons. By 1890 Pittsburgh advantages were great. Production figures from the depressions of the mid-1870s, of 1884/5, and of the 1890s, show that the Shenango Valley was not, as might be supposed from the table, more resilient in bad times than the Mahoning Valley but less so, and that both were terribly vulnerable to

Table 57.   *Date of construction of Valley and Pittsburgh*
blast furnaces existing in 1890

	Before 1850	*1850–9*	*1860–9*	*1870–9*	*1880–9*	*Total*	*Building*
Allegheny County	–	1	5	4	11	21	4
Shenango Valley	1	–	7	11	4	23	–
Mahoning Valley	2	3	6	2	2	15	–

Based on A.I.S.A. *Works Directory* 1890.

the trade cycle. Pittsburgh steel mills could sometimes get Bessemer iron more cheaply from the Mahoning Valley than from local works, in spite of the freight of 40 to 50 cents a ton, because of the keen competition between Valley firms to find some outlet for their production.[15]

The position could be improved in two ways: either by lowering costs of production through better equipment, perhaps lower wages, and lower freight costs, or by carrying manufacture through to a more highly finished state, under which condition high furnace costs and even high freight charges might be covered by the price for the product sold. Essentially it amounted to either freight rate reduction or entry into steel-making. Singly, none of the Valley firms was big enough to influence railroads as the giants of Pittsburgh were able to do. At the beginning of 1891, as pig prices fell, Valley operators threatened to suspend iron production in protest against high freight rates, and a month later the Mahoning and Shenango Valleys Iron Manufacturers' Association, whose main declared aim was to secure 'equitable freight rates', met the railroads' agents. They asked for rate reductions equal to 10 cents per 100 lbs. on finished iron, or $1 a ton on pig iron from the Valleys to Chicago. Delaying till November, the railroads then refused this petition. Meanwhile inability to compete closed most of the furnaces from January to June.[16] Nationally, 1891 pig iron production was 90 per cent that of the previous record year, but in the Mahoning Valley only 71·3 per cent. Apart from enabling a higher-value product to be sold, steel-making was desirable in order to replace the puddled iron trade, which, after uncertainty in the late eighties, declined sharply in the 1890s. (Table 58) In

Table 58.   *Pig iron, muck bar and finished iron production*
           by members of the Mahoning and Shenango Valleys Iron
           Manufacturers' Association 1895, 1896, 1897
           (thousand tons)

	1895	1896	1897
Pig Iron	896	463	932
Muck Bar[1]	170	172	25
Finished Iron	378	251	325

[1] Muck Bar was the rough, unfinished bar which needed piling, reheating and rerolling into finished form.
Source: *I.A.* 21 Apr. 1898, p.20.

1890 the Valleys had three small, special steel plants including an idle open hearth works at Youngstown. They were owned independently of the ironworks, and their annual capacity was 10,000 tons. That year Allegheny County made 1·54 million tons of steel. (Fig. 24).

VALLEY STEEL-MAKING

On the day the railroad companies turned down the request for lower freight rates, the new Shenango Valley Steel Company of Newcastle contracted for a Bessemer plant, the first in the Valleys. Eleven months later it began rolling billets and slabs. It was planned that about half this

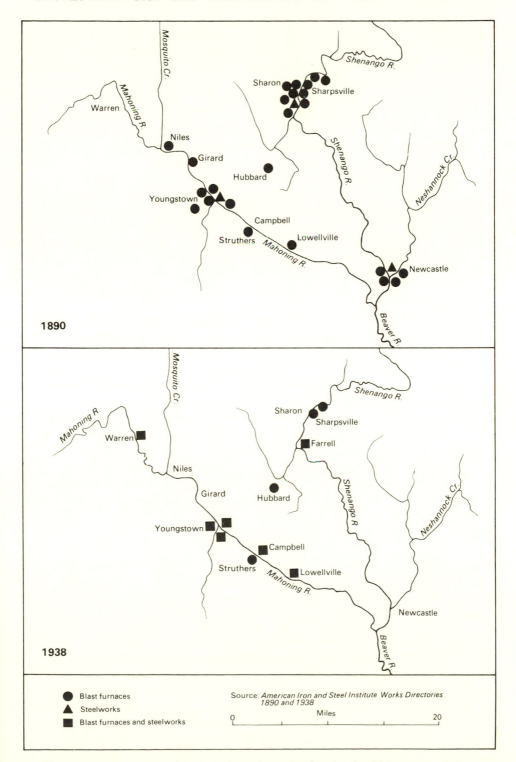

24. The changing pattern of iron and steel production in the Mahoning and Shenango valleys 1890–1938

semi-finished material should be sold to local mills, which until then had mostly rolled wrought iron.[17] Seeing the inevitability of the swing to steel, a number of Mahoning Valley iron rolling mills formed the Ohio Steel Company in 1894, and in February 1895 made that valley's first Bessemer steel in their Youngstown works.[18] Again output was mostly of semi-finished products—billets, slabs, sheet, and tin bars. By 1896 the Shenango and Ohio works had annual ingot capacities of 180,000 and 400,000 gross tons respectively. Steel capacity was still less than one-third that in pig iron in the district, but the steelworks soon became the nucleus around which further reorganization and rationalization of Valley's trade crystallized. Shenango Valley Steel became closely connected with a number of Newcastle area tinplate works, and Ohio Steel, also supplying them, figured prominently in a number of proposals for wider Mahoning Valley associations with finished iron firms, themselves often the product of linkage between smaller mill concerns suffering declining business.[19]

In the hard conditions of the mid-1890s, when control over raw materials proved so significant in the success of bigger companies, unintegrated Valley steel-makers were still at a disadvantage. Their semi-finished products were often in keen competition with those from these bigger concerns. Carnegie's Duquesne works, for instance, left the rail business in 1896/7 to concentrate on semi-finished steel. Backward and forward integration was clearly desirable in the Valleys.

Backward integration involved acquisition of existing furnaces or construction of new ones, and in either case removed an attractive local outlet for merchant pig iron firms. As early as 1894 Shenango Valley Steel controlled two separate furnace plants with an iron capacity equal to some two-thirds its capacity for finished steel. By 1897, it was rolling billets down into more valuable sheet and tin bars. Ohio Steel was already producing these. The latter's steel capacity provided a market for 1,200 tons of molten iron a day at full work by 1897, and it was contemplating, though it did not build, an open hearth shop. By 1901 its ingot capacity was 650,000 tons a year and it had four furnaces capable of 600,000 tons of iron.[20] From 1899 Ohio Steel was the main steel plant of the National Steel Company, third in importance to Carnegie and Illinois Steel, and with its strength firmly based in the Valleys. (Table 59)

The blast furnace companies continued to lose outlets for iron both locally and in other districts; for instance in 1898 Lorain Steel, which had till then bought from them, built its own furnaces. In spite of this, in good times, they could still find markets, and it was the beleaguered iron mill firms which took the initiative in entering steel production. Three firms which survived as major concerns into the 1970s were established with a Valley base at this time.

Republic Iron and Steel was formed as a rolling mill trust in 1899, its properties mainly old puddled ironworks and rolling mills. Except for its small interest in northern Alabama, all these lay west of Pittsburgh. It had twenty-eight mills, five blast furnaces—none west of the Valleys—and melting shops at Minneapolis and Birmingham. For a time it was believed that a new

Table 59.   *Pig iron and steel ingot capacity of Valley Works*
1890, 1896, 1901, 1904
(thousand tons)

		1890	1896	1901	1904
Pig Iron					
	Mahoning Valley	576	919	1,725	1,840
	Shenango Valley	713	1,001	1,682	2,080
Steel — Bessemer					
	Mahoning Valley	—	400	1,000	1,200
	Shenango Valley	—	180	500	600
	Open Hearth				
	Mahoning Valley	—	—	—	4
	Shenango Valley	9	13	464	620

Based on A.I.S.A. *Works Directories.*

Cleveland steelworks might be built, but, in a very extensive rationalization, Republic made the Valley district the heart of its operations, so using to best advantage its iron capacity there. Seven rolling mills in the manufacturing belt had been abandoned by 1904, but only one was in the Valleys, which retained six of twenty mills still active. Steelworks at Columbia Heights, near Minneapolis, at Moline, Illinois, and at Covington, Kentucky, were dismantled. Late in 1899 Republic acquired the rolling mills and steelworks of the Union Steel of Alexandria, Indiana, and the Springfield Iron Company in Illinois. It had a finished iron capacity of 100,000 tons at its Brown Bonnell works, Youngstown, and this plant was now converted into an important steel plant, converters from both Alexandria and Springfield, and the Alexandria blooming mill being transferred to it.[21]

Sharon Steel Company was established in the late nineties working up its own semi-finished material on tinplate, rod, and wire mills until these were sold to the Moore group. One of its new, independent clients was the Sharon Steel Hoop Company, founded in 1900. When Sharon Steel became part of the United States Steel Corporation, the Hoop Company built its own open hearth plant. Until 1918, when it purchased its first blast furnace, Sharon Steel Hoop worked on purchased pig and scrap. The old Sharon steelworks, from which it had purchased until 1902, became the United States Steel Corporation's Farrell works.

In November 1900 the Youngstown Iron Sheet and Tube Company was formed, and, against the trend of the last ten years, decided to make puddled iron on a new, cheaply acquired site outside Youngstown. In 1903, 90,000 tons of pipe and 32,000 tons of black and galvanized sheet were made, and in the following year output more than doubled. All the pig iron was bought at first, but in 1902 it leased the Sharpesville furnace of 72,000 tons, and cold iron was brought across from there into the Mahoning Valley. Bessemer production began in 1905, but not until 1908 was the first blast furnace built. By 1910 Youngstown Sheet and Tube had three blast furnaces, and in 1912 took over the Ohio Iron and Steel Company's Lowellville stack.

By 1901 Valley steel capacity was almost 2 million tons, though the district still exported large tonnages of pig iron, especially to Pittsburgh. Production costs in the Valleys remained high. Two medium-distance rail hauls for minerals were involved, rather than one long one, in the case of Cleveland, or one longish rail haul and a short bulk water movement, as with mills in the Monongahela Valley or on the Upper Ohio. Being land-locked, the area was subject to high rail rates with no redress or check on excessive charges until truck transport became important. In 1922, for instance, in the pipe business for which the Mahoning Valley alone had seventeen mills at work, the freight from Youngstown to the oil and gas fields of the south-west was $7 when it was only $4 from Chicago.[22] As works became bigger, so even recently acquired sites like Youngstown's Campbell works became congested.

Partly to cover high assembly and marketing costs, partly by reason of the time of its development as an important district, and in part too because of its old iron specialisms, Valley works went into lines of business in which the amount of work put into rolling operations was at its maximum. In plates, structurals, and rails production was highly mechanized and the palm went to big, cheap producers. Fortunately, sheet, tinplate, and pipe were products in rapidly growing demand. Tinplate consumption increased as standards of living improved, sheet with the needs of the oil industry, the shift to steel rail passenger cars, and, by the First World War the demand from the automobile industry. Pipe demand boomed with the opening of the new oil bonanzas of Kansas, the Gulf coast, and California.

Whereas demand grew more slowly in the older, heavier trades, the booming market in Valley lines carried it, despite its higher iron and scrap costs, to higher levels of expansion and often of operation than its rivals. By 1913 plates and sheets—overwhelmingly the latter—and tube mill skelp totalled 53·2 per cent of Valley production; in the United States as a whole their proportion was only 33·2 per cent. New companies continued to appear, as with the consolidation which created Brier Hill Steel Company in the Mahoning Valley in 1912. Stemming from the initiative of the displaced president of one of the constituent companies there followed the foundation, a little further up the Mahoning at Warren, of the sheet and tinplate works of the Trumbull Steel Company.[23]

Swept up into the post-war boom which was so marked in sheet, the Valleys continued to do well. In 1920 2·36 million tons of their 4·14 million tons of finished products were plates, sheets, and skelp, 18·8 per cent of the national total. Rolled iron and steel production fell in the next depression year to only a shade more than 50 per cent of this figure, but Allegheny County managed only a 44·8 per cent operating rate. The latter had undoubtedly retained its production cost advantage, but had a much less suitable product specialization than the Valleys under the market demand conditions of the twenties. True, as early as 1922 Youngstown Sheet and Tube was said to be contemplating a Chicago plant and had the land under option, a sign of that lakeward and westwards pull later to become such a theme in the district. In the following year it acquired the Indiana Harbor works of the Steel and Tube Company of America. At the same time, however, the Brier Hill Steel Company was brought in, and Youngstown became the headquarters of an iron and steel company of major, national stature. In 1928 the Valley district, as defined by the American Iron and Steel Institute, rolled more steel than Allegheny County.

NOTES

[1] *I.A.,* 15 Oct. 1908, p.1082, 26 Feb. 1903, p.17.

[2] R.W. Hunt, 'The Development of the American Rolling Mill Industry', quoted in *Iron,* 4 Dec. 1891, p.538; *I.A.,* 31 Mar. 1898, p.22; J. Fulton, *Coke,* 1905, p.277.

[3] J.H. Bridge, *The Inside History of the Carnegie Steel Company,* pp.269–71, 281, 288, 358; *I.A.,* 9 Mar. 1911; J. Kennedy, *I.A.,* 4 Apr. 1912, p.892; C.M. Schwab, *I.A.,* 29 May 1913, p.1326.

[4] Carnegie and Schwab Accounts, *I.T.R.,* 17 Jan. 1901, p.19, 10 Jan. 1901, p.9.

[5] *I.A.,* 5 Mar. 1903.

[6] *I.A.,* 31 Jan. 1895, p.228.

[7] *I.A.,* 9 Mar. 1899, p.18; *Bethlehem Review,* autumn 1956, p.23.

[8] *I.A.,* 27 Apr. 1899, p.33.

[9] 'The Field of Lake Front Furnaces', *I.A.,* 1 Jan. 1914, p.103.

[10] P. McKinney in Dissolution Suit, *I.A.,* 9 Oct. 1913, p.819; *I.A.,* 13 Mar. 1904, p.666; Mussey p.144.

[11] H.G. Moulton, *Waterways versus Railways,* 1912, p.431. Republic Steel Corporation, *Steel Making: Buffalo District,* (undated pamphlet).

[12] L. White in *Dennison University Bulletin,* 1929, pp.81–95; *J.I.S.I.,* 1904. 1, p.371.

[13] *I.A.,* 2 Nov. 1911, p.960; *Eng.,* 27 July 1877, pp.61–2; *I.A.,* 29 Feb. 1912, p.541, 2 Jan. 1913, p.101.

[14] *I.A.,* 5 Jan. 1908, p.28.

[15] *I.A.,* 10 July 1884, p.17, 13 Oct. 1892, p.676.

[16] *I.A.,* 8 Jan. 1891, p.61, 12 Feb. 1891, p.296, 3 Dec. 1891, p.980.

[17] *I.A.,* 13 Oct. 1892, p.676.

[18] *I.A.,* 16 Aug. 1894, p.258.

[19] *I.A.,* 28 Jan. 1897, p.11, 18 Nov. 1897, p.20, 21 Oct. 1897, p.3, 14 Apr. 1898, p.31.

[20] *I.A.,* 15 Feb. 1894, p.308, 28 Jan. 1897, p.11; A.I.S.A. *Works Directory,* 1901.

[21] A.I.S.A. *Works Directory,* 1904; A.I.S.A., 1899, p.50; *I.T.R.,* 30 Aug. 1900, p.11, 4 Nov. 1900, p.11.

[22] *I.A.,* 1 June 1922, p.1569, 24 May 1923, p.1503.

[23] *I.A.,* 18 Jan. 1912, p.213, 25 Jan. 1912, p.259, 2 May 1912, p.1089.

CHAPTER 10

The Structural Steel Trade

Early railroad bridge building employed wood, but as traffic grew, loads became heavier and speeds greater so that stronger structures became essential. Cast iron beams were sometimes used, though a large number of firms which began to build with them had extremely short lives. By the 1850s some wrought iron was employed: in 1859 the New York Central built the first bridges wholly of wrought iron. Five years earlier, the first iron beams for fireproof buildings had been rolled. From this period dates the establishment of the structural iron business both east and west of the Alleghenies.

Cooper and Hewitt began rolling iron beams at Trenton in 1853. Phoenixville turned out its first beams in 1855. Five years later the Keystone Bridge Company was established in Pittsburgh, and in 1864 Carnegie and his partners built the Cyclops mills. Cyclops rolled the first iron beams made in Pittsburgh in 1867. By the late 1870s Union mills—the wider group of which Cyclops was a part—were already rolling some products—bars, angles, tees, as well as beams—in steel as well as in iron, the steel billets coming from the Edgar Thomson works.[1] They were already the leading U.S. mills for structural shapes.

The amount of structural material rolled was too small to be separately recorded in the 1870 census; even in 1880 when production was 87,000 tons, most of it was still of iron. National structural output increased threefold in the eighties, and by 1890 over half was of steel. It was at this time that the centre of production shifted west of the Appalachians and Carnegie Steel became the chief producer. By 1893 Abram Hewitt was writing of beams and girders '. . . whose manufacture I propose to abandon because we are driven out of business by western competition and not by foreign products'. Carnegie Phipps and Co.'s *Pocket Companion* first listed steel beams in 1884. These saved material and weight as compared with wrought iron, rapidly gained popularity, and by 1893 all data on wrought iron were eliminated from the booklet.[2] In 1881 Homestead mill began work with an annual capacity estimated at 50,000 tons of rails and 30,000 tons of structurals. In 1883 it was acquired by Carnegie. The following year the Union Mills first rolled beams of 12 and 15 inches width, but rapid advance followed, so that by 1889 Homestead was already offering 20 and 24 inch beams.

Markets were now rapidly expanding. One estimate puts the steel bridge work done in the United States in the nineties at about 1 million tons.[3] Within a few years structurals were also being used on a large scale in rail car construction—each car built by the Schoen Pressed Steel Company for the Pittsburgh, Bessemer and Lake Erie Railroad in 1898 contained twelve tons of plates and shapes. In the 1880s high buildings with a framework of structural steel, the prototype skyscrapers, first began to poke above the skylines of the biggest cities. Land values in city centres were high, insurance rates burdensome, and in the 1890s the high-speed electric lift made a multi-storey building acceptable. In factories, the adoption of electrical travelling cranes and the use of heavier machinery increased the need for steel frames. Some of this demand was scattered through the country, some rather concentrated. Bridge building was widespread, but the old eastern beam mills had a large regional market—there were 500 bridges built before 1897 and mostly of steel on the Pennsylvania system east of Pittsburgh.[4] Some eastern mills made their mark in foreign bridgework too. Rail car demand was more localized in long-established railway equipment centres such as Philadelphia, Altoona, Chicago, or the Pittsburgh area. Consumption in office building was much more localized. In 1885 the ten-storey Home Insurance Building in Chicago was the pioneer skyscraper and Maitland's *American Slang Dictionary* defined a skyscraper in 1891 as 'a very tall building such as now are being built in Chicago'.[5] In the nineties the most active development shifted to New York, where by 1898 the 29-storey Ivins Syndicate Building was the nation's tallest.

Phoenix, Pencoyd, and Passaic, the main eastern structural firms, had good marketing conditions but they were late in going into steel-making, and built only small melting shops—in 1890 their respective annual capacities were, for all finished products, 50,000, 35,000, and 200,000 tons.[6] They had no blast furnaces to back up their steel plant. Process costs were high as compared with Pittsburgh. Pennsylvania Steel Company had a small structural plant at Steelton, but not until the nineties did Cambria built up the structural business, this time with Bessemer steel. Carnegie, with hot metal, big mills, a good location for marketing east or west, and unrivalled business acumen, was left master of the field. To the west the only potentially important rival was Illinois Steel, but it concentrated so heavily on rails that the structurals for the Home Insurance building were rolled in Pittsburgh. By 1890 North works could make 50,000 tons structurals a year but only at the expense of rail tonnage, and in the year 1889/90 Illinois Steel rolled only 5,161 tons of beams and channels.[7]

The structural trade grew rapidly in the nineties; from one record rolled steel year, 1892, to the next, 1899, structurals increased their share of the total from 7·3 to 8·2 per cent. Pittsburgh remained in the lead. In the structural pool agreement of 1897 Carnegie obtained 49·37 and Jones and Laughlin 12·87 per cent of the total allocation.[8] Allegheny County's output was 61 and 64 per cent of the country's total in 1896 and 1899 respectively. The American Bridge Company was formed in April 1900 as an amalgamation of twenty-six bridge building and fabricating firms. It was dependent for up to two-thirds of its raw material on purchase, and, although it had financial links with Federal Steel, the location of its plants made Carnegie its chief supplier. In 1901 American Bridge was brought into the United States Steel Corporation. As with rails, this marked the end of Pittsburgh's overwhelming predominance. Its share of national structural output fell from 60 per cent that year to 45·8 per cent by 1907 and as little as 37·5 per cent by 1911. Changing organization, marketing, and production conditions were involved in this decline.

NEW STRUCTURAL STEEL CENTRES

Between 1899 and 1912 U.S. structural shape output went up threefold (Table 60) The increase in Allegheny County fell well short of double. In part this was due to the reorganization of the trade carried through by the

Table 60.   *Production of heavy and light structural shapes*
1899, 1909, 1912
(thousand gross tons)

	1899	*1909*	*1912*
New York and New Jersey	N.A.	177	123
Pennsylvania	847	1,642	2,051
: of which Allegheny County	586	907	1,062
Alabama and Ohio	?	60	73
Indiana, Illinois, Wisconsin Colorado and California	?	396	600
United States	906	2,275	2,846

Based on *A.I.S.A.*

Steel Corporation. To save shipping from Pittsburgh to its Chicago area operations, or to independent buyers there, a structural mill was installed at South Works soon after the Corporation was formed. In 1909 American Bridge began new fabricating yards at Gary. U.S. Steel's Pittsburgh area works rolled 60 per cent of its heavy structurals in 1910, Chicago works 17 per cent. Ten years later the two were becoming more nearly matched, Carnegie Steel shipping 517,000 tons of heavy structurals, Illinois Steel 372,000 tons.[9] It was possible to split the northern markets east and west between the two, but Illinois Steel had much the bigger consumption near at hand, while, eastwards, Pittsburgh now met stronger opposition than in the nineties. By 1920 Allegheny County share of U.S. structural output was down to 31 per cent. By this time Pittsburgh structural mills had become high-cost producers—the mill cost of structurals at the Corporation's Pittsburgh works was $52·2 per ton in 1920, when in Chicago the cost was only $42·8.[10] In the 1921 depression Illinois Steel structural output held up better, being 46·6 per cent of the 1920 level as compared with 38·3 per cent for Carnegie Steel.[11] (Fig. 25)

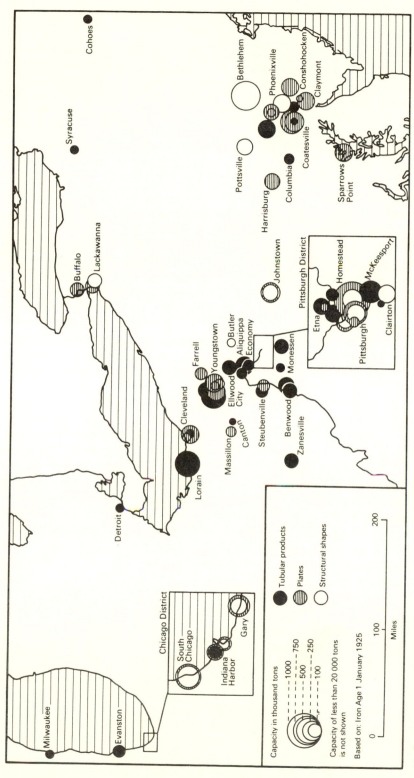

25. Product specialisation—Plate, structural shape and tubular product capacity 1922

In 1901 Lackawanna built structural mills at Buffalo, well located for both eastern and western markets. At Bethlehem much larger capacity, a high quality, new-style product, and a key location proved much more troublesome to Pittsburgh. By the turn of the century it was said that in one year alone 125,000 tons of structural steel were used in New York City for building purposes, excluding bridge work.[12] Yet, though urged on by John Fritz in the early nineties, Bethlehem Steel's directors declined to build a heavy structural mill which might have given them control of this market. Presumably Schwab, formerly superintendent of Homestead, where Carnegie Steel's structural business was concentrated, might have built conventional structural mills there, and with market advantages carved out a profitable trade, but instead he installed Grey universal beam mills. The universal beam mill principle for structurals had been experimented with since the mid-1870s, but not until the mid-nineties did Grey, former general manager for the Cleveland Rolling Mill Company, achieve technical success with a mill built for the Ironton Structural Steel Company at Duluth. Partly because of crude mill design, partly because of shortage of capital, but also by virtue of its location this was not a commercial success. Grey himself later recalled the Ironton operation, 'I hardly think that in the three years 1893–6, there was, all told, as much steel passed through this mill as is passed through the Bethlehem Steel Company mill in 5 or 6 hours.'[13] After overcoming initial opposition from structural engineers, the Bethlehem beams—lighter in section but equal in strength to old-type structurals—became fully acceptable. Backed by new steel capacity and a highly enterprising new management team, Bethlehem exploited its market advantages to the full. By 1920 it had the nation's largest structural capacity.

With Pittsburgh Plus, Carnegie Steel prices long remained competitive in the east, but as rail charges went up, Bethlehem phantom freight on local sales increased. In autumn 1901 the Pittsburgh to New York freight on structurals was $2·60 a ton; by 1908 the general rate on finished steel to New York was $3·20. By the twenties freight rates were rising still more rapidly to the greater disadvantage of the long-distance shipper. Overall from 1903 to 1927, the rail freight rate from Bethlehem to New York increased 161 per cent, that from Pittsburgh 234 per cent.[14] By 1928 the New York metropolitan area was estimated to consume annually about 570,000 tons of structural shapes, 28·4 per cent of the national total.[15] Bethlehem's advantage there was overwhelming, so that by 1929 71 per cent of the product was delivered to consumers less than 100 miles from the plant.[16] In the seven years to 1926 Carnegie Steel sales of plates, bars, and structurals to eastern markets—a market area whose boundaries were not unfortunately clearly defined—were said to have fallen almost 160,000 tons. This is not decisive, for 1920 production of these three classes was 1·8 million tons more than in 1926, but in 1926 it was claimed that Carnegie provided only 74,000 of the 969,000 tons sold there.[17] Westwards too marketing for

Pittsburgh mills worsened as rail charges increased, and in 1921 Chicago became a structurals basing point. Bethlehem built two new Grey mills at Lackawanna in 1927 and the following year introduced weekly boat shipments for structurals to Chicago and Milwaukee. The whole 45,000 tons of structurals for the Marshall Field Merchandise Mart, built in 1928, were shipped direct from Lackawanna to the site on the Chicago River. United States Steel met Bethlehem penetration of the Great Lakes area by building its own wide flange beam mills—seven old type structural mills at Homestead were replaced by the new mills. In January 1927 it began to offer so-called 'Carnegie Beams'.[18] Bethlehem increased its competitiveness by acquiring the widespread fabricating facilities of McClintic Marshall in 1931. Established in 1900 in the shadow of American Bridge, it had bought material for such early jobs as the Marshall Field Store in Chicago or the fabrication of the Panama Canal locks from U.S. Steel. By 1929 over one-third of McClintic Marshall fabricating capacity was in the Pittsburgh area though it had yards at Buffalo, Chicago, Pottstown, and on the west coast. Even at the end of the twenties, when it split the contract for 58,000 tons of structural work for the Empire State Building with American Bridge, it still bought its steel froms the Corporation. Now, however, it became an added outlet for Bethlehem.[19]

NOTES

[1] *Eng.*, 10 Jan. 1879, p.21, see also S.L. Goodale and J.R. Spear, *Chronology of Iron and Steel*, 1931, *passim.*

[2] A.S. Hewitt to W.L. Wilson, 15 Nov. 1893, quoted A. Nevins (ed.), *Selected Writings of A.S. Hewitt*, 1937, p.359; Goodale and Spear, op. cit., pp.216, 213; L.H. Miller, 'Steel Construction', *A.I.S.I. Yearbook*, 1925, p.85.

[3] Ibid., 1926, p.296.

[4] Ibid.

[5] Quoted A.M. Schlesinger, *The Rise of the City 1878—1898*, 1933, p.282.

[6] A.I.S.A., *Works Directory*, 1890.

[7] Ibid.; W. Cope, *The Iron and Steel Interests of Chicago*, 1890, pp.24—7.

[8] W.C. Temple to Stanley Committee, *I.A.*, 17 Aug. 1911, p.367.

[9] Bureau of Corporations, *Report on the Steel Industry*, part III, p.476; *I.A.*, 19 Feb. 1925, p.543.

[10] T.N.E.C., *Hearings*, Part 27, 1939—40, pp.14544—5.

[11] *I.A.*, 19 Feb. 1925.

[12] J.S. Jeans, *American Industrial Conditions and Competition*, 1902, p.315.

[13] H. Grey, letter to *I.A.*, 21 Jan. 1909, p.240; see also on the development of Universal Beams, *I.A.*, 14 Jan. 1909, p.160; *I.T.R.*, 13 Dec. 1928, pp.1542—3; *B.F.S.P.*, Jan. 1941, p.37.

[14] Jeans, op. cit., p.368; *Tariff Hearings*, 18 Dec. 1908, p.1697; *I.A.*, 1 June 1911, p.1298, 24 Mar. 1927, pp.854, 855.

[15] *I.T.R.*, 19 July 1928, p.138, 21 Feb. 1929, p.518.

[16] *I.T.R.*, 10 Jan. 1929, p.169.

[17] *I.A.*, 28 Apr. 1927, p.1232.

[18] *I.A.*, 28 Apr. 1927, p.1236; *I.T.R.*, 3 Jan. 1929, p.83; Goodale and Spear, op. cit., p.310; *I.T.R.*, 13 Dec. 1928, pp.1542—3; *B.F.S.P.*, July 1927, pp.331—2.

[19] *Fortune*, Sept. 1930, p.124, Dec. 1930, p.136.

CHAPTER 11

Steel-making in the South

Hopes for southern steel-making were institutionalized in the names of its new industrial towns—South Pittsburg, Sheffield, Bessemer—but for long eluded realization. At least twenty-one different attempts to make steel in Tennessee and Alabama ended with laboratory success but commercial failure.[1] Technical problems were serious enough; their solution pushed up production costs and so partly, sometimes wholly, cancelled out the advantage of cheap pig iron. High assembly costs, inadequate local markets and costly access to distant ones, shortage of capital, operations on too small a scale, and often ill-equipped—all these problems continued to plague the promoters of southern steelworks.

Alabama coal had a high sulphur content, its ore too much silica and phosphorus. The last ruled out acid steel-making, but the phosphorus content was hardly sufficient for basic Bessemer practice. The basic open hearth process seemed a better proposition. In theory, as H.H. Campbell showed in 1896, Birmingham could make open hearth steel at highly competitive prices, but in fact, real costs were much less favourable (Table 61). As Carnegie put it two years later, 'manufacturers of steel in Alabama make very cheap steel on paper but they have only made it there yet.'[2]

Table 61.   *Estimated cost of stock per ton of steel*
Pittsburgh and Birmingham 1895
(dollars per ton)

Pittsburgh	Acid Bessemer	$12·22
	Acid Open hearth	$12·32
	Basic Open hearth	$12·32
Birmingham	Basic Open hearth	$ 8·19

Based on H.H. Campbell, *The Manufacture and Properties of Structural Steel*, 1896, pp.164, 166.

Lacking scrap, the area had to use much larger tonnages of inferior iron, ore, and lime in its melting shop charges. Productivity was therefore lower, furnace wear greater, iron yields smaller, and slag volumes much larger than in the north. Duplexing—using Bessemer's process to remove silicon and to undertake the early conversion work in bulk and at high speed, followed by finishing in a basic furnace—was attractive technically but involved more capital, which southern firms could not easily command. Campbell, who had worked on their design, advocated employment of large tilting open hearth furnaces in which steel-making was continuous, not all the charge being drawn off at one time so that slag could be removed more easily and frequently than with a fixed furnace.[3] This permitted refining of siliceous irons like those of the south without the usual high proportion of scrap. Benjamin Talbot, who developed the Talbot tilting furnace, was brought over from England in 1889 to work with the Southern Iron and Steel Company.

Chattanooga made the first important strides to southern steel-making but, as with iron, quickly lost the initiative. Roane Iron Company built an open hearth plant in 1878 which used Georgia ore, English iron, and rail crop ends. For four years it made steel which was finished as rails. It then went under. In 1886 the same company built Bessemer works and in 1887 rolled the first southern Bessemer rails. Its ore came from Cranberry, North Carolina, where, as seams were thin, extraction costs were high. Freight charges to Chattanooga were considerable. Some of the iron was too phosphoric and the rails were brittle. The plant was soon broken up and sold. Two other small Bessemer rail plants operated at the same time in Chattanooga and Richmond.[4]

The Henderson Steel Manufacturing Company, established in North Birmingham in 1888, was the first Alabama steelworks. Using the basic process it made good steel, which was employed for tool manufacture and even in Massachusetts for razors. Henderson contracted with a local rolling mill to supply rail steel. To the visiting Percy Gilchrist in 1890 it appeared to have such good prospects that he was amazed it had only one furnace when he reckoned it should have had forty.[5] In fact, though associated with the De Bardeleben blast furnaces at Bessemer, the Henderson works lacked adequate supplies of basic iron and had high operating costs. Birmingham Chamber of Commerce which owned the plant could not raise money for expansion from local iron firms. In the panic of 1893 Henderson went under, like so much of the achievement and ambitions of the south.[6]

TECHNICAL PROGRESS AND COMMERCIAL PROBLEMS IN THE USE OF SOUTHERN IRON

The Tennessee Coal Iron and Railroad Company made its first experimental basic iron in 1891. Four years later it first produced it commercially. By this time the South could for the first time produce low-silicon, low-sulphur pig at a satisfactory price.[7] In 1897 Tennessee Coal and Iron and the Louisville and Nashville Railroad each agreed to invest $100,000 in a steel project by the Birmingham Rolling Mill Company; the citizens of Birmingham also raised a large amount. Excellent steel was made in two small furnaces which ran mainly on scrap, though T.C.I.'s loan appears to have been made in the belief that the plant would be a local outlet for its basic iron. T.C.I.'s Ensley mills rolled some of the steel.

Technically successful, the project was yet another economic failure.[8] In 1899 the Birmingham Rolling Mill Company was brought into the Republic Iron and Steel Company.

Failure clearly resulted from commercial rather than technical difficulties. T.C.I. basic iron was being used successfully in steel-making in the north, and it was there treated in exactly the same way as the product of local furnaces. T.C.I., far and away the biggest southern iron company, fully integrated from extensive mineral properties through furnace plants to iron rolling mills, seemed the best hope of southern steel-making success. Alice furnace was put on basic iron in 1895, and in that year large contracts were received from Carnegie. In 1896 a report on use of T.C.I. iron in the north was accompanied by the suggestion for a Birmingham open hearth plant, an especially attractive proposition when T.C.I. pig iron commanded $6 a ton at the furnace while rails were selling at $28 a ton at Pennsylvania mills.[9] T.C.I. that year made 75,000 tons of basic iron; it seriously contemplated a steel plant, but among other difficulties, could not raise the funds. In addition market prospects were not encouraging, and it was clear that the under-use of equipment could quickly cancel out the advantage of cheap pig iron and the supposed advantage of cheap labour.

The South was still overwhelmingly rural and unindustrialized. New Orleans had in 1890 a population of 242,000, Atlanta only 65,000, while Birmingham itself, so anxious to be a metallurgical colossus, was still a small town of 26,000. Consumption of rolled iron and steel was small, scattered, and largely satisfied from the north. In 1896 rolling mills in Virginia, Tennessee, Georgia, and Alabama turned out 66,000 tons, and much even of this was rolled from northern semi-finished steel.

Two big southern outlets were cotton ties and rails, but neither was enough to keep a works fully employed, and neither would be wholly within the control of a local mill. Cotton ties took 40,000 to 50,000 tons of steel a year, but northern producers had a firm grip on the trade, and by rail or water Pittsburgh's access to much of the cotton belt was as good as Birmingham's. Rail prospects were more exciting, but no less uncertain. Some southern roads promised financial aid to southern promoters on condition that a rail mill was included in their steel schemes. New lines were still being built—track in Mississippi, Louisiana, and Tennessee alone was increased by just under 2,000 miles in the nineties. For the south-west Birmingham seemed well placed to compete with the northern works, but established mills had good rail communications too, or, as with Pittsburgh and Sparrows Point, water routes to a very wide area of the southern network. They also had the advantage of producing Bessemer rails. Basic open hearth rails were for many years too costly to be competitive, and in addition melting costs in the South were inevitably rather higher than in northern plants. It was the opinion of outsiders that the South could not count on sufficient demand to keep a large rail mill at work.[10] If, following its success with sales of pig iron, the South sold billets and other semi-finished steel in the north it would face the burden of high freight charges and the keen competition in bad times of rail mills striving to keep at work by turning to this business.

In spite of all the obvious disadvantages and possible snags, the desirability of working up some of its own pig iron—542,000 tons in 1897—made it essential for T.C.I. to enter steel-making. In 1898, a year in which Alabama made 1,033,000 tons of pig iron but only 9,692 tons of steel, the decision was taken to build a T.C.I. plant. Its new Ensley works contained ten basic open hearth furnaces, operated in conjunction with Bessemer converters from which the metal was transferred after the silicon had been removed. This first American duplex plant naturally cost more than a straight melting shop installation, and capital again proved difficult to raise. The T.C.I. executive committee provided one-third of it, and one third each came from local citizens and from the Louisville and Nashville Railroad; a little later the T.C.I. Directors subscribed more.[11] At the beginning of 1900, the first steel was shipped as slabs to Connecticut. That year Alabama made only 66,000 tons of steel. A new, local outlet for billets was obtained when the Chicago-financed Alabama Steel and Wire Company built a rod and wire mill at Ensley.

For a time Ensley seemed likely to go the way of its unsuccessful predecessors. In 1901 D.H. Bacon of the Minnesota Iron Company, Federal Steel's mineral subsidiary, became chairman of T.C.I. He later recalled that 'I found an empty treasury, and a property that needed millions for upbuilding. I also found the operations were greatly hampered, almost directed in fact, by labour organizations . . .' When he took control the rail mill was good but the open hearth department though new was very poor—it had indeed been built by men who were not really familiar with steel-making. 'There was scarcely any of the property that was right, if it was possible for it to be wrong.' the coalmines were mostly poorly equipped, only a few of the ore mines were cheaply operated, and until 1904 there was no modern blast furnace. The new furnace built that year could make iron for under $8 a ton, but the cost in the old ones was $2 to $3 a ton more. As a result of the accumulation of inadequacies, rail costs were $25 to $30 a ton.[12] In autumn 1901, Jeans, visiting Ensley, found that shortage of gas meant that not more than seven of the ten open hearth furnaces had yet been operated at one time. Lack of a mixer, of a ladle for making additions to the furnace, the generally cramped nature of the plant—not yet two years old—and other features 'combined to make it difficult to show the best that can be done in the way of producing cheap basic steel in the south'.[13]

In 1902 the Alabama Steel and Wire interests, perhaps dissatisfied with the contract—though Jeans believed its terms favoured them more than T.C.I.—, more probably unhappy with the material, for they had been accustomed to Bessemer steel in the north, built a plant of one blast furnace and four open hearth furnaces at Gadsden, over sixty miles from their Ensley mill.[14] Meanwhile, Bacon was striving to reduce costs. No dividends were paid between Autumn 1900 and 1905, all the small operating profits being invested in plant modernization and improvement of the mines.[15] In 1904 C.P. Perin, the steel engineer, examined southern conditions and two years later reckoned that $20 million would be needed to bring T.C.I.'s properties up to the standards of northern efficiency, while Bacon's own estimate at about the same time was $25 million.[16] Finance on this scale was not available in the South. By 1906 a group led by J.W. Gates and others from Republic Iron and Steel Company gained control. Their new managing director later recalled T.C.I. at this time, 'Its physical condition was very poor, and the property was run down at the heel.'[17] The steelworks and mills, only seven years old, were already badly behind the times and now had to be modernized. Plans for doubling steel capacity and for a second rail mill were drawn up. In the financial crisis of 1907 there occurred the event which, for better or worse, was to have such an important effect on future southern steel-making, the acquisition of T.C.I. by U.S. Steel.

T.C.I. AND THE UNITED STATES STEEL CORPORATION

By 1907 T.C.I. financing had so much ceased to be a local affair that a number of New York houses were deeply involved. One request to the Corporation to help check the 1907 crisis by acquiring T.C.I. was refused, but in November with Morgan as prime mover, and with express approval of President Roosevelt, the Alabama properties were brought into the group. Many believed that with such huge resources behind it at last Birmingham steel-making had endless possibilities of expansion—the *Birmingham News* reckoned the new owners would 'make the Birmingham district the largest steel-making centre in the universe'. Julian Kennedy, internationally renowned as a steelworks engineer, estimated that the T.C.I. properties in 1907 were worth $90 to $100 million. The Corporation acquired them for $35·3 million, mostly in its own bonds.[18] There was some dispute about the chief motive for the purchase; was it to eliminate a rival, above all in rails, or to control the ore and coal beds of the South?

Rails were not subject to Pittsburgh Plus pricing, mainly because railroads collected their own supplies from the works. Yet marketing advantages meant a great deal, buyers naturally seeking out the most accessible mills. By 1909 Birmingham was at least as well placed as Chicago or Pittsburgh for rail deliveries to 35 per cent of the area of the United States: its advantages were greatest in the west and south-west where new construction was concentrated.[19] (Table 62) Ensley made open hearth rails, and it was already becoming clear that they were in many respects superior to Bessemer rails. They commanded a higher price—$30 a ton as compared with $28—and it was realized that the railroads were about to switch to them on a larger scale. In 1907 the Steel Corporation was constructing Gary, a mmajor product of which was to be up to 900,000 tons of open hearth rails a year. In the east Bethlehem was building an open hearth rail mill, but the only operation of this kind already at work was T.C.I.'s. In spring 1907 Harriman placed an

Table 62.   *Freight rates on steel rails 1908*
                        (dollars per gross ton)

From	To: Mobile	New Orleans	San Francisco
Chicago	4·00	4·00	11·00
Pittsburgh	4·44	4·44	13·50
Birmingham	2·50	3·00	11·75
European mills (sea rate)	3·35	3·35	7·50

Source: Butler, op. cit., p.1697.

order for 157,000 tons of Ensley rails for the Union Pacific and Southern Pacific tracks. However, it later became clear that T.C.I. costs of production were so high that it was a much less formidable competitor than this spectacular sale led observers to believe. Some of the rails were returned as defective and for much of the order production costs exceeded prices. To outsiders at this time it seemed that, publicly spirited or otherwise, the Corporation had absorbed a competitor before it became more troublesome. With hindsight it seems that desire to control the mineral resources of the South was probably a much more powerful motive.

There was common agreement that the T.C.I. ore and coal properties were immense though of rather poor quality. Gate's group put ore reserves at 300 million, or perhaps even 500 to 700 million tons. Kennedy's estimate was in the 400 to 700 million ton range, and the coal properties he assessed as 1,000 million tons.[20] From the start U.S. Steel had tried to gain a predominant position in minerals, especially ore. The extremely high royalty paid for the Hill holdings in 1906 was expected to force up the cost of northern pig iron, and so improve the position of companies with their own mines, using imported ore, or in the south. As *Iron Age* remarked, a year before T.C.I. was acquired, 'the Hill deal suggested to men of large capital that the psychological moment had arrived for investment in Alabama ore and coal, and in the plants already existing to turn them into steel'. Berglund, writing in 1907, but some months before the merger, suggested the South would provide the Corporation with its most formidable future competition and that to contain this and to obtain a true monopoly of the country's production—something which in fact foregoing events had showed it never could—it must obtain an interest in the South. Earlier the Corporation had made approaches to a number of southern companies including T.C.I., the Southern Steel Company, and Sloss-Sheffield.[21]

After the acquisition, it was found that T.C.I. still lagged far behind the standards of northern efficiency. After a few years of work improving the situation, T.C.I.'s new President evaluated the property with a realism which contrasted with the flights of fancy which still looked to Birmingham as the steel centre of the world. 'The Birmingham district', he concluded, 'is one that cannot be made like the rosy pictures I have heard described, but it can develop into a reasonably good business proposition.'[22] When he arrived, apart from the rail mill, T.C.I. had a plate and shape capacity of 60,000 tons a year. In its works and mines there toiled not only the dubiously 'cheap' negro workers, but also 700 convicts. Generally, at the end of 1907, the mill cost of rails was $29 a ton, only $1 less than the nominal price for open hearth rails. The tonnage loss in manufacture from steel to finished rail reached the extraordinary figure of 35–40 per cent (by Spring 1909 it had been cut to 10 per cent). Three million dollars of T.C.I. debts had to be paid when the Corporation took over, and in 1908 $56,000 was spent on auxiliary equipment to finish the first export order.[23] After this, longer-term reconstruction work could begin. By 1913 $23·5 million had been spent on improvements. Steel capacity had been increased to 600,000 tons. New coal and limestone properties were opened, by-product ovens which cut coke costs by $1 a ton were built, new reservoir water supplies provided. Yet, employing the duplex process, Ensley was unable to use more than a 20 per cent scrap charge and as a result it still cost $2 a ton more to make steel than in Pittsburgh where a 40 per cent scrap rate was usual for the melting shops.[24] Operating well below capacity Ensley mill costs for heavy open hearth rails in 1910 were $19·24 per ton; at northern works (mainly Gary) the cost was only $17·53.[25]

As a step towards the improvement of the market situation, the Corporation diversified its plants. By 1912 at the new town of Corey, soon renamed Fairfield, a rod, wire, and wire products mill was built by American Steel and Wire to be supplied by Ensley billets. At this time Ensley could roll 45,000 tons of rails a month and soon the capacity of the older plant at Bessemer was 6,000 tons of plate, angles, and merchant bars. By the time of the Iron and Steel Institute's Birmingham meetings in 1914, U.S. Steel was widely praised for putting southern steel-making firmly on its feet at last.[26] There was undoubtedly much justice in this, for the capital resources and technical knowledge of the Corporation were both unrivalled, and T.C.I. had clearly been in bad need of both, but conversely the Corporation was discriminating against its southern properties in ways which inhibited their full development. The 'Birmingham Differential', an arbitary imposition of an extra $3 a ton on Birmingham mill prices above the Pittsburgh level for the same product, was imposed in 1909. This was in lieu of Pittsburgh Plus which would have been even more burdensome, but the differential lessened the advantages of buying from T.C.I. rather than northern concerns and so narrowed its competitive sphere. For wire, Pittsburgh Plus pricing was adopted.

All other operations in the South were small compared with those of U.S. Steel. Important iron concerns such as Sloss-Sheffield and Woodward survived, and in cast iron pipe the area was of great significance, the United States Cast Iron Pipe and Foundry Company having five of its fourteen plants in Alabama and Tennessee by 1906. Republic Iron and Steel owned the Thomas furnace and the Birmingham Rolling Mill plant, but eventually ceased to make steel in the South. The Southern Steel Company plant at Gadsden failed twice before, in 1912, it was reorganized as the Gulf States Steel Company, operations being rationalized by relocating the Ensley rod and wire mill next to the steelworks. By 1914 Bowron put its ingot capacity at 300,000 tons—probably too high a figure—, its mills rolled rods and bars and rails, and the works turned out barbed wire and woven fencing.[27] In 1901 a small rerolling concern began in Atlanta. It built a melting shop four years later and by 1922, as the Atlantic Steel Company, had a capacity of 100,000 tons. By this time 77 per cent the total ingot capacity east of the Mississippi and south of Kentucky, West Virginia, and Maryland was at U.S. Steel's Ensley plant.

GROWTH OF SOUTHERN STEEL TO 1930

The First World War boosted southern steel. (Table 63) In 1917 U.S. Steel built shipyards at Mobile, and with the nearby Chickasaw Shipbuilding Company these provided an important demand. As a result late in 1917

Table 63.   *Steel ingot capacity of main south-eastern*
steel plants 1907—1930
(thousand gross tons)

	1907	1913	1920	1930
Gadsden	?	300?	288	280
Ensley	243+	600	1,120	1,020
Fairfield[1]	—	—	—	750

[1] Built 1925
Based on Report of Commissioner of Corporations, *The Steel Industry*, 1911; J. Bowron *T.A.I.M.E.*, 71, 1925; *A.I.S.I.*

T.C.I. began to build structural and plate mills at Fairfield. The Mobile yard was dismantled after the war, but much of it was reassembled at Fairfield and turned to the production of rail cars using the same steel products. By 1924 the South's first sheet mill was built there, and the following year a melting shop was added. (Fig. 26)

In 1914 the South was still barely industrialized. The six-state area served by the new Federal Reserve Bank of Atlanta then had 12·1 per cent of the U.S. population but only 7·5 per cent of the manufacturing establishments, 6·6 per cent of the factory workers and 4·7 per cent of the value added by manufacturing in the country as a whole. In 1914 U.S. per capita steel consumption was about 650 lbs.; in the South only 150 lbs.[28]   After the war

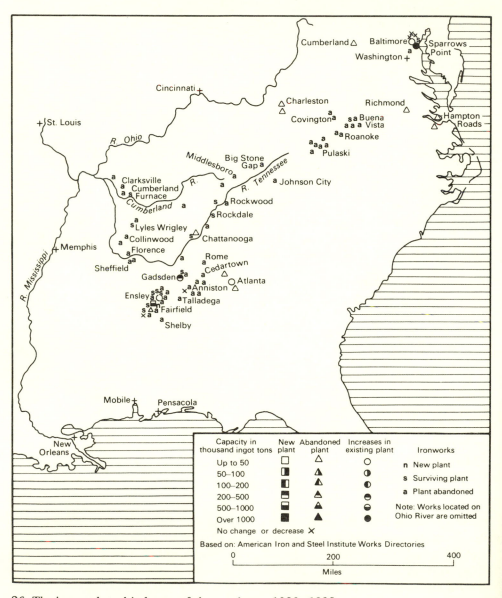

26. The iron and steel industry of the south east 1920–1938

the section still lagged far behind. (Table 64) By 1920 almost 68 per cent of the finished steel shipped by Illinois Steel Company was delivered within Illinois and Indiana; almost exactly the same proportion of T.C.I. shipments

Table 64.   *Value added in all manufacturing industries*
               U.S.A. and south-east 1899, 1909, 1919, 1929
                              (million $)

	U.S.A.	South Atlantic States[1]	Eastern Mississippi Valley[2]
1899	4,647	179	89
1909	8,160	368	182
1919	23,842	1,208	502
1929	30,591	1,662	688

[1] Virginia, North Carolina, South Carolina, Georgia, Florida
[2] Tennessee, Alabama, Mississippi.
Based on Bureau of the Census, *Census of Manufactures.*

went to the much larger area south of the Kentucky, West Virginia, and Maryland statelines and east of the Mississippi, though almost half the tonnage was used in Alabama itself.[29] (Table 65) Higher freight rates now

Table 65.   *Shipments of steel to south-eastern states[1] by*
               United States Steel subsidiaries 1920
                         (thousand gross tons)

	Total	Heavy rails	Bars	Heavy structurals	Plates
T.C.I.	280·7	122·1	50·7	29·8	61·3
Carnegie Steel	115·4	14·4	13·8	28·7	21·2
Illinois Steel	32·0	6·2	1·6	7·0	2·5

[1] Alabama, Florida, Georgia, North Carolina, South Carolina, Tennessee, Virginia, Mississippi.
Source: *I.A.* 19 Feb. 1925.

made penetration of distant markets more difficult. Writing of the slow development of southern steel in 1923, Woodbridge used slightly archaic terms but struck to the heart of the problem when he recognized consumption deficiency as still the main impediment. 'Lands devoted to cotton and negro labour', he wrote, 'will never build up another Pittsburgh, no matter how widespread they are.'[30]

There were already traces of the misallocation of Steel Corporation orders which proved such a drag on the competitiveness of the south-east in the thirties. In 1920 T.C.I. shipments to Texas were smaller than those of either of its northern heavy steel rival subsidiaries of the Steel Corporation; to the Pacific Coast it shipped only 2,926 tons whereas Carnegie supplied 42,579 and Illinois Steel 143,466 tons. T.C.I. was in fact excluded from the areas where, left to itself, it might have competed most effectively for business. Without Steel Corporation investment its process costs would have been much higher, and to this extent it would have been less competitive, but now, though efficient, marketing policy in turn slowed its expansion. Coal and ore mining costs had been higher than in the north. The best ore had been worked by the First World War, and later more costly mineral preparation was involved before smelting. In short the Birmingham area could still assemble materials more cheaply than anywhere else in the country, but in processing them its costs were as high as elsewhere. (Table 66) Costs of delivery to the consumer were generally higher than in the

Table 66.   *Alabama and northern assembly and total costs for merchant pig iron 1924—1926*
(dollars per ton)

	Assembly Costs	Total Costs
Alabama	2·65	14·46
Northern	6·45 to 8·05[1]	18·21

[1] Range from low of Buffalo to high in Valleys.
Based on R.H. Sweetser, *Blast Furnace Practice*, 1938 pp.326, 337.

North. By the second half of the thirties it was reckoned that in Republic Steel's southern operations, even after washing, southern coal had 30 per cent more ash than in the north, more furnace blowing power was needed to reduce the low-grade ore, and slag volumes were twice the common northern level. In the production of one ton of pig, one ton more ore and 600—700 lbs. more coke had to be handled.[31] On the other hand, by this time steel-making costs were lower in the south. Fairfield had an ordinary open hearth shop, and in spite of southern low standards of living and lack of steel-using industries scrap was now abundant in relation to demand, so that prices were low. By spring 1939 when No.1 heavy melting scrap was $15·50 to $16 a ton in Pittsburgh and Philadelphia or $13·50 to $14 in Chicago it was $12·50 to $14 in Birmingham.

The cheapness of southern labour had always been doubtful when real costs were taken into account. In 1906 *Iron Age* quoted the extraordinary situation of one southern iron company. Of 400 negro workers only one reported regularly for work six days a week. Another furnace manager at that time had three names on his payroll for each job.[32] Sample figures for merchant iron production for the mid-twenties show per ton labour costs of $1·42 at a northern works and $1·416 at a southern plant.[33] With new deal wage legislation any labour cost advantage probably almost completely disappeared, and James Hasler, Republic's chief engineer in the Birmingham district, concluded: 'Nowhere can there be found a more misleading statement than the old one that "iron can be manufactured cheaper in the south".'[34]

The number of southern iron-making plants declined rapidly, especially after 1920, the industry becoming more concentrated in the Birmingham area. The scattered brown orefields of the south-east were too small or their ore supply too unreliable to support a modern iron plant, and were certainly not big enough to justify large-scale steel-making. As a result small, outdated merchant ironworks lingered on there until the lower costs of Birmingham furnaces, the business recession of the twenties, and the great depression picked them off one by one (Table 67). Alabama ironworks had larger local

Table 67.   *Alabama and Roanoke, Virginia, costs of material assembly and pig iron delivery to New York 1924–1926*
($ per long ton)

	Iron ore	Fuel	Flux	Total assembly costs	Freight to New York	Total
Roanoke and vicinity	5·00	2·50	0·15	7·65	5·54	13·19
Alabama	1·25	1·10	0·30	2·65	9·24	11·89

Source: Sweetser, op. cit., p.326. © 1938. By permission of McGraw–Hill.

outlets, especially in the cast iron pipe works, and made up for their longer haul to northern foundries by their lower operating costs.[35] In the First World War there were still furnaces scattered along the Great Valley in Virginia and Tennessee, and to the west from Nashville to the neighbourhood of Sheffield on the Tennessee. They were notably more sensitive to the trade cycle than the Birmingham plants, and this was shown even more clearly in the 1921 depression. Between the end of 1921 and 1931 the number of blast furnaces in Tennessee fell from 17 to 6. Virginia has 16 iron-making plants in both 1920 and 1926, but from 17 furnaces in mid-1926 the number fell to 1 within 10 years. Some of the most hopeful projects of earlier generations now went under. The Middlesboro works, built in the Cumberland Gap in the early 1890s with liberal amounts of English capital, was a notable failure. The Lookout Mountain Coal and Iron

Company's plant at Battelle had seemed almost unassailable by virtue of its raw material endowment when iron production began in 1904, but by 1918 the furnace had been sold and shipped to the Tata Iron and Steel Company.[36]

NOTES

[1] J. Bowron, 'Steelmaking in Alabama', *T.A.I.M.E.*, 71, 1925, pp.398–403.

[2] *I.A.*, 17 Nov. 1898, p.10.

[3] H.H. Campbell, *The Manufacture and Properties of Structural Steel*, 1896, pp.164–8, ('Economic Conditions in the South').

[4] V.S. Clark, History of American Manufacture, 1929, vol.2, p.240; Bowron, op. cit., p.399.

[5] E. Armes, *The Story of Coal and Iron in Alabama*, 1910, p.409.

[6] Armes, op. cit.; F.R. Crockard, 'Progress in Steelmaking in Alabama', *I.A.*, 19 Dec. 1912, pp.1436–8; *Clark*, op. cit., vol.2, p.241.

[7] *I.A.*, 29 July 1897, p.18.

[8] Armes, op. cit., pp.463–4; Bowron, op. cit., p.400.

[9] Armes, op. cit., pp.461–2.

[10] *I.A.*, 6 Nov. 1890, p.807, 29 July 1897, p.18.

[11] Bowron, op. cit., p.400.

[12] Armes, op. cit., p.409, 507–8; Stanley Committee Hearings, *I.A.*, 18 Apr. 1912, pp.978–9.

[13] J.S. Jeans, *American Industrial Conditions and Competition*, 1902, pp.202–3.

[14] *I.A.*, 19 Nov. 1914, p.1186.

[15] Clark, op. cit., vol.3, p.51; Armes, op. cit., p.513.

[16] *I.A.*, 30 May 1912, p.1349, 18 April 1912, p.979.

[17] *I.A.*, 10 Aug. 1911, p.314.

[18] *I.A.*, 4 Apr. 1912, p.888.

[19] A.R. Burns, *The Decline of Competition*, 1936, p.304.

[20] *I.A.*, 1 June 1911, p.1310, 4 Apr. 1912, p.888.

[21] *I.A.*, 22 Nov. 1906, pp.1388–9, and A. Berglund, *The United States Steel Corporation*, 1907, p.161.

[22] G.G. Crawford, *I.A.*, 3 July 1913, p.23.

[23] J. Farell, *I.A.*, 22 May 1913, p.1240.

[24] *I.A.*, 3 July 1913, p.23.

[25] Bureau of Corporations, *Report on The Steel Industry*, part III, 1913, pp.465–8.

[26] *I.A.*, 26 Nov. 1914, p.1229, 5 Nov. 1914, p.1087.

[27] *I.A.*, 15 Jan. 1914, p.230a.

[28] *I.A.*, 29 Oct. 1914, pp.992–3; 'A Survey of 1914–1964', Federal Reserve Bank of Atlanta, *Monthly Review*, Nov. 1964.

[29] *I.A.*, 19 Feb. 1929.

[30] D.G. Woodbridge, *I.A.*, 17 May 1923, p.1418.

[31] J.M. Hasler, 'Offsetting increased Labour Cost in Southern Blast Furnace Operation', *T.A.I.M.M.E.*, Iron and Steel Division, 125, 1937, and W.E. Curran, 'Trend of Southern Pig Iron Business', Ibid., 1938.

[32] *I.A.*, 22 Nov. 1906, p.1389; H.H. Campbell, op. cit., 1907, p.483.

[33] R.H. Sweetser, *Blast Furnace Practice*, 1938, p.337.

[34] J.M. Hasler, op. cit., p.47.

[35] R.H. Ledbetter, 'Blast Furnace practice in the Birmingham District', *A.I.S.I. Yearbook*, 1924, pp.263–84.

[36] *I.C.T.R.*, 24 May 1918, p.587; *I.A.*, 4 Aug. 1904.

CHAPTER 12

The Interwar Years.
I. Pricing Policies and the Changing Balance of Locational Advantage

GEOGRAPHICAL PRICING SYSTEMS

The price of steel is a major item in the cost of relatively few finished products; steel transport cost is naturally much less important still. Even so, geographical pricing systems have been the centre of fierce controversy. Change in delivered costs may substantially affect the consumer's profit margins; to the producer, with a large capital investment, co-operative regulations of delivered prices may contribute to the stability which he so much desires. With a heavy product, and a large country, it is not surprising that uniform delivered prices have not been used for steel in the U.S.A. Basing point pricing, in which only a selected number of points determine the delivered prices which others must quote the customer, dominated before 1948, when all but two of the major fully integrated works of today already existed.

Even before the Revolution Philadelphia was the basing point for the works of eastern Pennsylvania. For about a century, as communications improved, it extended this role over a wider area. It was suitable as the chief market for a wide circle of ironworks, and as the point at which home works had to meet the prices of foreign material. Philadelphia price plus transport therefore indicated the competition which any firm in the region had to meet. A small producer in central Pennsylvania, connected for the first time with the east through the new Pennsylvania Railroad would be tempted to change from F.O.B. (Free on Board or Free on Truck) mill pricing to F.O.B. Philadelphia. Eastern mills operating on a much larger scale than their local markets justified would now provide him with effective competition.

With growth of the industry west of the Alleghenies, high freights gave the protection from eastern or British iron necessary to cover higher operating costs, including those for supplying a small and scattered market. Economic growth, new raw materials, and eventually the opening of Lake Superior ore and Connellsville coke, cut their costs, and, with new, independent pricing, it was found that they could win a larger share of the market from the eastern works. The key mill point of the west, Pittsburgh, became the chief western basing point, but, as other centres grew to a size at which they too could supply other than local markets, there occurred the process later described as 'natural basing point development'.[1] The trend to multiple basing points and to the logical conclusion of the development, a return to F.O.B. mill pricing, was complicated by the emergence of dominant companies within the industry, and from the early twentieth century by government attempts to maintain fair competition.

PITTSBURGH PLUS

Through to the 1870s, F.O.B. mill pricing was usual, but when a basing point was quoted it was commonly Philadelphia.[2] After this Pittsburgh Plus gradually emerged, now in one product, now in another, and for a long time not persistently in any. Not until after the 'Gary dinners' began in 1907 did it become a fully fledged system; though by no means sacrosanct, it persisted in this condition from 1909 to 1921, and three years later was abandoned. The first important instance was structural steel, the Pittsburgh price becoming the determining factor in the mid-1880s. In the Beam Association at that time the west was allocated to Carnegie, the east was divided between three eastern firms—at Passaic, Phoenixville, and Trenton—but prices for each zone were based on Pittsburgh mill prices plus an average freight rate, so that Carnegie was well placed to compete with them all. The zonal system lasted until 'inequalities' occurred as the number of mills increased.[3] Billets went on to the basing point system in 1896, and from 1900 Carnegie Steel priced plates in this way. In 1900 the new National Tube Company, whose largest and lowest-cost works was at McKeesport, chose Pittsburgh as a basing point. Bar makers followed in 1902, and plate and structural producers in December 1903. Wire nail makers had set up a zonal pricing system in 1898 but six years later changed to Pittsburgh Plus. At this time there seems to have been an assumption that such a system was inevitable. To the observation that 'in some cases' consumers were charged Pittsburgh Plus prices, Schwab replied in May 1901, 'If you will point out a method of avoiding that we shall be very glad.'[4]

Most of these examples could be explained as natural choices of a basing point depending on one or more of a group of criteria—cheapest point of production, main producing point for a particular category, or centre of capacity surplus to local needs (and therefore likely to provide competition for other mills in all parts of the country). The choice of Pittsburgh by American Sheet Steel and American Tin Plate was of a completely different order, for neither had Pittsburgh plants. In the early nineties the tinplate industry seems to have operated under Swansea Plus—that is South Wales prices plus transatlantic freight, import duty, and land carriage. This gave the high profit expectations which made the building of a home industry a worthwile speculation. Formed in 1898 as an amalgamation of 39 plants, American Tin Plate two years later opted for Pittsburgh Plus pricing.[5] The nearest tinplate works were to the east in the Kiskiminetas Valley, but the main capacity was scattered through the country to the west. The Indiana gas belt had a number of important plants, including the largest in the United States, but the company now announced that it would no longer sell

F.O.B. Indiana mills but on a Pittsburgh base, 300 miles away.[6] American Sheet Steel, an amalgamation of twenty different companies, also had interests in the Kiskiminetas Valley but not in Pittsburgh; in 1900 it began selling on a Pittsburgh Plus basis.

After the 1907 panic the steel industry, finding co-operation paid better than cut-throat competition, generally followed the price leadership of U.S. Steel. Pittsburgh Plus now reached its peak, in the years after 1914 some Steel Corporation subsidiaries even providing all their rivals with free copies of a booklet listing Pittsburgh prices and freight rates from there to chief points all over the country.[7]

The single basing point system enabled a Pittsburgh mill, or any nearby producer, as in the upper Ohio or the Valleys, to offer equal competition with a producer anywhere in the United States in terms of delivered price. However, at a distance the service provided by a Pittsburgh mill was inferior, except when market-located company warehouses were used in place of direct shipments. Quality of service was a weakness exposed by the increasing significance of steel deliveries to consumption goods industries. Already, after reaching a low in the late nineties rail freight rates were rising again, so that without Pittsburgh Plus, upper Ohio Valley mills would have been pushed out of western markets. With rails always sold F.O.B. mills, this happened early. (Table 68)

Table 68.   *Allegheny County production of steel rails and other finished steel as a percentage of U.S. total 1906–1914*

	Rails	All other rolled steel
1906	21·4	—
1907	21·2	29·9
1908	14·0	26·1
1913	11·4	26·4
1914	12·4	24·0

Based on *A.I.S.I.*

Distant producers gained substantially from phantom freight—freight charged the consumer as part of Pittsburgh Plus, but not in fact borne by the supplier near to a consumer. There was something to be said for the higher returns they enjoyed, for capital costs in new plants and mineral properties were rising, though whether this justified the whole pricing system was another mater. Judge Gary recognized that the system—Judge Gary's umbrella as it was sometimes called—fostered growth of new steel capacity at a distance from Pittsburgh. In October 1924 he spoke to the Iron and Steel

Institute on the recently abandoned Pittsburgh Plus system. '. . . without larger selling prices at Chicago, for instance, than Pittsburgh, furnaces and steel mills would not have been erected in the Chicago district, for investors could not afford to do so.'[8] But the new plant was more efficient than the older, and with fuel economy and efficient ore unloading, production cost trends also favoured them. Inland Steel in their brief to the Federal Trade Commission supported Pittsburgh Plus, pointing out that independent Chicago firms had paid more for their ore lands than the longer-established Pittsburgh firms—Inland's first mine was leased in 1906—but then went on, rather unconvincingly, to claim costs were also higher in the Chicago area.[9] In 1908 Gary suggested that Chicago's assembly cost advantage over Pittsburgh was $0·21 per ton of iron. In the same year No.1 heavy melting

Table 69.   *Deliveries by area from the heavy steel divisions of the United States Steel Corporation 1920 (thousand tons)*

	Carnegie Steel	Illinois Steel	Minnesota Steel	T.C.I.	Combined shipments
New England states	106·6	2·1	—	0·1	108·8
New York	164·2	3·8	—	0·5	168·5
New Jersey, Delaware, D.C., Maryland	271·2	6·2	—	0·1	277·5
Virginia and Kentucky	91·2	29·7	0·2	43·7	164·8
South-east states*	36·1	26·8	—	279·9	342·8
Pennsylvania and West Virginia	2469·1	17·4	—	1·1	2487·6
Ohio	698·5	97·2	1·7	6·4	803·8
Michigan	179·5	134·9	3·7	0·1	318·2
Indiana	22·0	532·3	19·9	4·3	578·5
Illinois	41·0	1062·0	30·4	4·4	1137·8
Wisconsin	6·0	209·9	12·5	—	228·4
Minnesota	2·1	115·7	23·0	0·1	140·9
Mississippi to Great Plains	39·5	356·6	5·5	104·9	506·5
Arizona, Idaho, Nevada, Utah	0·3	19·0		0·2	19·5
Pacific Coast	42·6	143·4		2·9	188·9
Total within U.S.A.	4169·9	2757·3	96·8	448·9	7472·9

*South of Virginia and Kentucky and east of the Mississippi.
Source: *I.A.*, 19 Feb. 1925.

scrap was $14·51 per ton in Pittsburgh but $12·45 in Chicago.[10] By-product
coking and rising land freight rates, added to the advantages of newer plant,
must have swung costs still more in Chicago's favour. By the early twenties
production costs for steel in U.S. Steel plants there were about 20 per cent
lower than in the Pittsburgh district, and the Steel Corporation's wide scatter
of plants put it in a uniquely favourable position for supplying regional
markets.[11] (Table 69) As costs fell, so rising freight charges gave Chicago
plants more phantom freight. From $3·30 in 1907—11, the charge on steel
Pittsburgh to Chicago had become $4·30 by late 1917' and in August 1920
$7·60 a ton. The Chicago market was not only large but unsatisfied by local
production. When the surveys for Gary were made production in the Chicago
district—whose limits were not indicated—was reckoned 2 million tons a year
short of consumption.[13] By 1919—20 the area was said to produce only half
of what it consumed. Local mills generally had short hauls to their markets,
and although in some reckonings Carnegie Steel could produce better figures
still for sales within Ohio and Pennsylvania, markets in those two states were
much more scattered. (Tables 70 & 71) Between 1914 and 1917 rolled steel

Table 70.   *Carnegie Steel Company and Illinois*
                    Steel Company sales 1920
                    (local sales as percentage of total U.S. sales)

	Carnegie Sales in Pennsylvania and Ohio	Illinois Sales in Illinois and Indiana
Plates	75·2	57·6
Bars	69·6	63·5
Heavy structurals	63·9	55·5

Source: Federal Trade Commission data quoted *I.A.* 19 Feb. 1925.

production went up by 65 per cent in Pennsylvania, 75 per cent in Ohio but
90 per cent in Illinois and 100 per cent in Indiana.[14] (Table 72)
    Beyond Chicago there was even more protection by distance and phantom
freight. Cold metal works located near big markets were especially favoured.
By 1911 the rate on finished steel Pittsburgh to St. Louis was $5·04 per ton,
and that year the Laclede Steel Company built a melting shop at Alton,
within the area of greater St. Louis, in order to sell 'for more than it cost on
account of the Pittsburgh Plus in existence at that time'.[15] Granite City
Steel on the eastern edge of the St. Louis industrial area had begun to make
steel for associated stamping works in 1895, and from 1908 supplied outside
customers as well. In the east low scrap prices, organization of foreign ore

supplies, rationalization of production, and market growth were favourable factors, but revival there was helped by the rising cost of transport and increased phantom freight. Eastern Pennsylvania's share of state output had fallen but now rose again (Table 73).

Table 71.   *Finished steel delivery by Inland Steel*
Company 1920
(percentage of total)

To:	Percentage
Chicago switching district and Illinois	40—50
East of Chicago switching district, mostly Indiana	5
Wisconsin	20
Iowa, Nebraska, Kansas	15
North-west	7·5
South-west	5
Scattered	0— 7·5

Source: *I.A.* 15 Feb. 1923, p.510.

Yet in spite of phantom freight to distant producers Pittsburgh Plus was sometimes an encumbrance rather than a help. In depression the distant company preferred to increase its share of the local market and operate at a higher level of capacity rather than continue to reap a dollar to two in

Table 72.   *Production of steel ingots and castings*
1912, 1920, 1925
(million gross tons)

	1912	*1920*	*1925*
U.S.A.	31·2	42·1	45·3
Pennsylvania	15·6	17·6	16·5
Allegheny County	7·8	8·5	7·7
Ohio	6·8	10·1	11·9
Illinois	2·8	3·6	3·6
Indiana	2·0	3·8	5·2

Based on *A.I.S.I.*

phantom freight at the cost of still having to meet competition there from Pittsburgh mills. In these circumstances some companies proved willing to break away and quote F.O.B. prices. Thus in 1909 U.S. Steel's Chicago area rivals cut their prices, but they were brought to heel by a threat that the price leader would do the same in order to keep its share of the market.[16] In 1911 too, steel was sold on a Chicago base, with plates and structurals selling at $1 to $3 a ton below the Pittsburgh Plus level, so that eastern producers were shut out. From 1910 levels there was a sharp fall in 1911 in Allegheny's

Table 73. *Allegheny County and Lehigh Valley production of rolled iron and steel as percentage of Pennsylvania total*

	Allegheny County	Lehigh Valley
1903	59·9	1·7[1]
1912	49·0	4·7
1920	48·5	6·2
1925	43·7	9·7

[1] Lehigh and Northampton County
Based on *A.I.S.I.* except for 1903 which is from H.H. Campbell, *The Manufacture and Properties of Iron and Steel*, 1907.

share of national production—in structurals from 41·9 to 37·5 per cent and in plate and sheet from 27 to 24·8 per cent again the crisis passed and the Pittsburgh price was restored.

Some firms spurned Pittsburgh Plus. One was Corrigan McKinney of Cleveland. Colorado Fuel and Iron was another which now sold on a Pueblo base, though earlier it had sold on a Chicago base with great profit.[17] As early as 1913 it was observed that Pittsburgh Plus did not hold as completely as previously because of the growth of producing centres distant from Pittsburgh. To steel consumers Pittsburgh Plus was anathema.

Steel prices were lowest at Pittsburgh. As early as 1911 a St. Paul fabricator, probably buying from Chicago, or perhaps Milwaukee, had to pay $6·40 a ton in nominal freight from Pittsburgh. As freight rates increased these burdens grew. For some more sophisticated new trades, notably the automobile industry, this was a relatively small affair, for steel price was a minute part of total costs. Pittsburgh signally failed to attract any significant share of this new industry, and soon lost that which was located there. Yet it was not only in the new consumer durables that Pittsburgh had a small part of consumption, but in other older lines.

By the Pittsburgh Plus hearings it was reckoned that the average profit to a fabricator of structural steel was $5 a ton, but in 1920 a Pittsburgh fabricator could obtain steel for $7·60 a ton less than Chicago rivals. He was thus able to win contracts in Chicago territory. In that year the mill costs of structurals at U.S. Steel Pittsburgh plants were $9·4 per gross ton more than in Chicago, but the Chicago price was the Pittsburgh mill price blus $7·50 freight. To the east of Chicago, though perhaps much nearer that district than to Pittsburgh, a fabricator could obtain steel cheaper to the extent that the freight rate from Pittsburgh was less than the Pittsburgh—Chicago rate. One example quoted at the Pittsburgh Plus hearings was a sale of Inland Steel in Detroit at $2·68 per cwt. at a time when Inland was selling the same grade of steel in Chicago for $2·72. U.S. Steel could allocate western orders to its Chicago mills, while its subsidiaries there could obtain steel at ex-works cost or, if they did pay Pittsburgh Plus prices, this was no more than a book-keeping item. Not surprisingly by 1924 American Bridge, established at Gary only thirteen years before, was said to be twenty times as big as its nearest competitor in the structural fabricating business.[19]

Consumer opposition to phantom freight was late in coming to a head. At one stage there were suggestions that western consumers were planning new works near Pittsburgh, but few could afford to do so. Cleveland Chamber of Commerce sent questionnaires to consumers of one-third of the steel used in the area in the second half of 1920. In that period they obtained 47,800 tons of steel from Cleveland plants, 64,200 from the Pittsburgh and Johnstown areas, 27,100 from the Valleys and 22,700 from elsewhere. Strict adherence to Pittsburgh Plus pricing would have cost these consumers $1·5 million a year in fictitious freight charges.

Chicago, further away, with much greater steel consumption, was where the irritation of the steel consumer was transformed into the crusade which changed the system. Beyond the industrialized Lake Michigan shore, in Wisconsin, the Manitowoc Shipbuilding Corporation always had to pay Pittsburgh Plus prices for 42,000 tons of plate and other steel over the years 1917—22, although 90 per cent of its material came from Chicago. It claimed that as a result it had lost business to Lake Erie yards. 150,000 to 200,000 tons of steel were fabricated a year in Milwaukee, and it was reckoned that firms there paid over $1 million a year in unearned freight as a result of the system.[20] At Moline, Deere and Company used 100,000 tons of steel annually, paying about $488,000 in phantom freight, which had to be passed on to the farmer in higher prices for equipment. A year later, it was claimed that the 825,545 farms in the states of Illinois, Iowa, Wisconsin and Minnesota had in effect to pay an average 'levy' of $10 a year each as a result of Pittsburgh Plus pricing.[21]

Industry boomed in the Middle West in the First World War—in Calumet alone the number of industrial plants increased by more than half between 1917 and 1919.[22] During the first ten months of control over steel prices, up to 1 July 1918, the War Industries Board made Chicago a basing point with mill prices equal to Pittsburgh's. To sell in Chicago Pittsburgh mills had to absorb $4·30 freight. In June 1918 the rail freight between the two was raised to $5·40, and a week later, on Gary's suggestion, Pittsburgh Plus pricing was restored, prices to Chicago consumers going up by the amount of the new freight rate, though some big consumers managed to obtain concessions.[23] The following January the Western Association of Rolled

Steel Consumers for the Abolition of Pittsburgh Plus was formed. According to Gary its early contacts with U.S. Steel were open and friendly. The association's aim was limited, the restoration of Chicago as a basing point.[24] After consideration of its complaint in July and August 1919 the Federal Trade Commission decided it could not assume jurisdiction. In September 1920 rail freight rates went up by 40 per cent and the opposition revived. Re-hearing of the case was begun that autumn. The following May the F.T.C. issued a formal complaint against the United States Steel Corporation and its subsidiaries. The Middle West was still the seat of most of the opposition, 75 per cent of the 1923 membership of the Western Association being in Illinois, Iowa, Minnesota, and Wisconsin. Even so, meetings were held through much of the west and it was clear that merely to add a Chicago basing point would not be enough. In July 1924, after extensive and pro-tracted hearings, the F.T.C. ordered U.S. Steel to cease and desist from the practice of selling everywhere on a Pittsburgh base; other companies fell into line. New basing points were announced in September. By the end of 1924 Chicago had been added as a base for sheet and tinplate, there were three basing points for pipe, and nine for wire.

It had been expected that the end of single basing point pricing would be followed by a greater concentration of activity in the Chicago area. The freight rate on pipe to certain south-western markets was $4 from Chicago, but $7 from the Valleys, and during the hearings Youngstown Sheet and Tube was said to have had a Chicago site under option.[25] In the following year it acquired the Steel and Tube Company of America works at Indiana Harbor. Between 1922 and 1925 there were persistent rumours that Jones and Laughlin would build at Hammond, but in the end it decided not to go ahead.[26] In August 1925 Bethlehem Steel formally denied newspaper reports that it had bought 2,000 acres near Michigan City.[27] There was no sudden influx of steel-using firms either. McClintic-Marshall, the biggest independent steel fabricator, largely Pittsburgh based, had long held a site at Indiana Harbor, but after the pricing system changed chose not to build on it. It did, however, quickly buy up two Chicago fabricators, the Kenwood Bridge Company and the Morava Construction Company.[28] Four years later, Malcolm Keir noted that an immediate result of the changed pricing system was a flood of orders for Chicago mills and a check to demand in Pittsburgh, but in fact the effect was small.[29] Rolled steel production in 1925 was 18·8 per cent greater than in 1924 in the whole country, and 18 per cent up in Allegheny County. By early 1926 *Iron Age's* Pittsburgh correspondent was writing, 'it is some time since there was as much steel plant extension and betterment work in progress and in sight as is true at present.'[30]

One reason why the change had less effect on Pittsburgh than had been expected was because elements of discrimination were carried over into the new multiple basing point system. Relatively few new basing points were established. The buyer of steel now paid the price for steel at the nearest basing point—the so called 'ruling' basing point—plus transport from there, but the mill price at most basing points was set well above the Pittsburgh level. (Table 74) The distant consumer was better off than under Pittsburgh

Table 74.   *Basing point differentials on wire, January 1925*
         (Dollars per ton above the Pittsburgh mill price)

Basing Point	Differential	Basing Point	Differential
Ironton (Ohio)	nil	Joliet	$2
Cleveland	nil	Duluth	$2
Anderson (Ind.)	$1	Worcester	$3
De Kalb (Ill.)	$2	Birmingham	$3

Source: *I A.* 1 Jan. 1925.

Plus, but still not so well placed as a consumer in Pittsburgh, and certainly not so favoured as if mill prices reflected production costs. With these basing point differentials, Pittsburgh remained competitive over a very wide area. On long hauls the tapering of rail freight rates sometimes gave Pittsburgh an advantage over nearer works with higher mill prices. The rail freight on pipe from Pittsburgh to California was only $3 more per ton than from Evanston or Indiana Harbor, the only other base points, whose mill price was $4 per ton higher. The new Chicago price for sheet—whose 1920 mill costs in U.S. Steel Chicago mills were $8 a ton less than in Pittsburgh mills—was $2 a ton more than at Pittsburgh, and an arbitrary $1 a ton delivery charge into Chicago from Calumet was added. Even on Chicago sales local mills had an advantage of only $3·80 a ton—a small advantage with such a high-value product.[31]

There was increased resort by inland firms to water shipment, especially along the Ohio. The advance in rail freights would, in any case, have encouraged this, but the new pricing system gave further incentive. Jones and Laughlin pioneered new long-distance steel shipments with deliveries in 1921 through to New Orleans. In 1924 and 1926 it built warehouses at Memphis and Cincinnati. By mid-1929 Jones and Laughlin was sending two tows a month, 20,000 tons of steel, down the river. Some years later it reckoned to save $2 to $3 a ton on western shipment made by water rather than by rail.[32]   U.S. Steel began barge shipment to New Orleans in 1922. Pittsburgh Steel acquired barges and warehouses at Houston, St. Louis, and Memphis, and by 1925 and 1926 sent about 40,000 tons of steel annually by water.[33]   Independent barge companies built up a considerable steel trade, of which smaller or non-riverside companies made use. Bethlehem's Cambria plant, for instance, shipped to Glassport on the Monongahela where material was transferred to barges.

Yet in spite of basing point differentials, and growth in water movements, the position of Pittsburgh and neighbouring interior districts worsened in the 1920s as shipments east and west became more difficult. Pittsburgh steel firms began to take a noticeably keener interest in fighting freight rate advances. Commenting on the 1927 Interstate Commerce Commission freight rate hearings *Iron Age* noted, 'although there was no special reference in the hearings to the Pittsburgh Plus method of operating, it is rather generally accepted as having been primarily responsible for the indifference on the part of Pittsburgh producers in former years to rate discrimination.'[34] U.S. Steel claimed then that over the seven years 1919—26 Carnegie Steel Company shipments of heavy shapes and bars to eastern points had fallen 160,000 tons. In the same period national production of these classes had gone up from 11·3 to 14·4 million tons.[35] Bethlehem, U.S. Steel pointed out, had, effectively, a tariff wall protecting its operations. Jones and Laughlin maintained that abolition of Pittsburgh Plus caused their shipments of steel to consuming points west of or adjacent to the Mississippi to fall from about 260,000 to 30,000 tons a year—though this claim is difficult to reconcile with their growing water shipments.[36]

Table 75.   *Production of finished rolled iron and steel*
by selected districts 1920, 1925, 1930
(Percentage of U.S. total)

	1920	1925	1930
Allegheny County	20·6	16·4	14·9
Shenango Valley	4·1	5·7	6·3
Mahoning Valley	8·7	9·8	8·6
Ohio River counties	1·7	1·8	2·0
Ohio Lake counties	5·8	5·5	5·0
Indiana and Illinois	18·5	19·6	20·5
Michigan, Minnesota, Wisconsin	1·5	1·4	2·7

Based on *A.I.S.I.*

The evidence from regional capacity expansion indicates an increasing preference for the Chicago area. In the Valleys, Youngstown Sheet and Tube capacity went up 172,000 tons or 7·9 per cent 1922—1930; at Indiana Harbor the increase was 210,000 tons or 30·4 per cent. In the same period the Steel Corporation extended its Pittsburgh and Valley works by 550,000 tons or 5·3 per cent. In the Chicago district its capacity increased by 2,158,000 tons or 35 per cent. In production, however, the picture was far more complicated. Indiana and Illinois increased their share of the national output, but some interior districts did very well too. This may be explained by differing regional specializations and varying degrees of company enterprise and modernity of plant. Yet overall the abolition of Pittsburgh Plus exposed the declining competitive strength of interior centres (Table 75 and Fig. 27).

MULTIPLE BASING POINT PRICING

Abolition of Pittsburgh Plus was neither followed by revolutionary changes in the geography of production nor by price chaos. The industry had realized the value of co-operative action and was too used to U.S. Steel's lead to break violently away. By producing a price list any mill could be made a basing point but the independents only slowly took the initiative. A distant producer could 'absorb' freight to compete for the business with the ruling basing point. There were few basing points, and the 'differentials' did not reflect production costs, so there was a good deal of cross-hauling. This was encouraged by a wish not to be too dependent on any one market.[37]

Under depression the multiple basing point system did not always stand the strain. The independents were tempted to 'shade' the market to a degree which U.S. Steel was not happy to tolerate. Between 1925 and 1929 Chicago base prices for soft steel bars had been 10 to 13 cents per 100 lbs. above the Pittsburgh price, but by 1931 the differential had been cut to 7 cents. In April 1930 National Tube abruptly reduced pipe prices by $4 a ton, the biggest change for ten years, to meet price cutting by rivals. In May 1931 James Farrell spoke in most forthright manner of the evils of price cutting, then rife.[38]

As early as 1927 Bethlehem Steel made Lackawanna, Coatesville, Bethlehem, and Sparrows Point basing points for shapes, plates, and bars, and by 1934, at the time of the National Recovery Act Steel Code, there were eight basing points for plate and seven for shapes. Merchant bars and concrete bars had eleven points, but sheets and hot rolled strip only four and three respectively. On sheet and strip the mills of western Ohio and Detroit enjoyed large phantom freight, which in the one area helped preserve old sheet mill locations and in the other helped to build up new ones (Table 76). The effects of the system on Pittsburgh, a surplus area, on Detroit, an otherwise favoured area, and in the south, may be considered from the heyday of the system.

In the early years of the multiple basing point system Pittsburgh was the ruling basing point for much of the central manufacturing belt and, for some products, for the east as well. Chicago on the other hand soon had a whole range of basing points, so that Pittsburgh producers had to make large freight absorptions to compete there. (Table 77) Detroit was a special case. It was a big market for Pittsburgh but was not a basing point. With lower mill prices than its rivals, Pittsburgh should have been able to compete. But from 1929 National Steel Corporation was expanding rapidly at Ecorse in the centre of

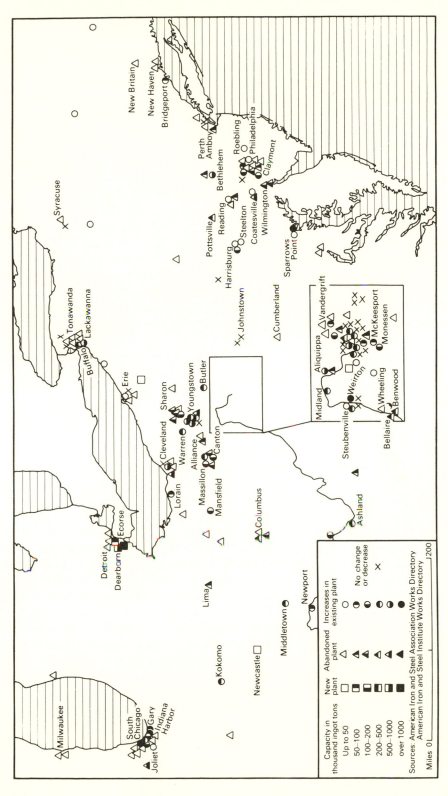

Capacity in
thousand ingot tons

	New plant	Abandoned plant	Increases in existing plant	No change or decrease
Up to 50	□	△	○	×
50–100	▬	◀	◑	
100–200	▬	◀	◑	
200–500	◼	◀	◑	
500–1000	◼	◀	◑	
over 1000	◼	◀	●	

Sources: American Iron and Steel Association Works Directory
American Iron and Steel Institute Works Directory

Miles 0 � 200

27. Changes in steel ingot capacity in the Manufacturing Belt 1920–1938

this market, and other lakeside producers were shipping in very large tonnages by water at well below the open mill price plus rail freight. By 1932 it was reckoned that 750,000 tons of steel came by water into Detroit,

Table 76.   *Capacity for bars, shapes, and sheets in relation to basing point location 1934*

Product	Number of basing points	Capacity		
		At a basing point	Within 50 miles	More than 50 miles
Merchant and concrete bars	11	29·1	41·0	29·9
Structural Shapes	7	26·1	65·6	8·3
Sheets	4	6·3	37·6	56·1

Source: C.R. Daugherty, M.G. de Chazeau and S.S. Stratton, *The Economics of the Iron and Steel Industry*, 1937. © 1937. By permission of McGraw–Hill.

and there were suggestions that U.S. Steel might consider relocating some capacity near to Detroit unless relief could be obtained.[39]   Ford was now building a strip mill, General Motors had been negotiating for a link with Corrigan-McKinney and there seemed a possibility that automobile firms

Table 77.   *Pittsburgh sales and freight absorption by areas summer 1934*

	Tonnage sold (thousand tons)	Average freight absorption per ton ($)
Pittsburgh consuming district	431	0·16
Metropolitan New York	85	0·62
Chicago	54	4·93
Detroit	129	0·40
Massachusetts	24	0·66

Source: National Reconstruction Act Report quoted Daugherty, de Chazeau and Stratton, op. cit., p.688. © 1937. By permission of McGraw–Hill.

would become less dependent on bought steel. Fearing this, in 1933 the steel firms adopted the so-called Resolution 21 whereby absorption of some of the freight was agreed, so that Detroit delivered prices for sheet fell $3 to $5 a ton below the base price plus freight. As Detroit was not an official basing point, apparent freight absorption by Pittsburgh producers was small. In spite of this Detroit steel-makers had advantage in giving good service, a mere rail switching or cross-city trucking operation as compared with the long haul from distant mills. In the first quarter of 1934 the steel operating rate (production as percentage of capacity) was 100 per cent in Detroit but only 50 per cent in Chicago and 30 per cent in Pittsburgh.[40]

The effects of pricing policies were sometimes masked by other factors, as for instance the gentleman's agreement between Bethlehem and U.S. Steel at the end of the twenties not to build in each others territory, an agreement which ended in 1930 with the Bethlehem attempt to take over Youngstown Sheet and Tube. The following year Bethlehem was said to be negotiating to take over Pittsburgh Steel of Monessen.[41] Conditions in the South were complex. Tennessee Coal Iron and Railroad Company's main product in 1907 was rails. From 1905 its other finished products were sold on Pittsburgh Plus. After its acquisition by U.S. Steel, some southern customers began to show their resentment by buying northern steel, paying full freight rather than the same amount in phantom freight to local mills. T.C.I.'s management pointed out to the Corporation's head office that the system penalized their operations. With straight F.O.B. pricing, freight rates on bars and plates were at least as favourable as from Pittsburgh or Chicago to 46 per cent of the United States. T.C.I. was given permission to replace Pittsburgh Plus pricing by an arbitrary 'Birmingham price' plus freight on sales of plate, bars, and shapes. The Birmingham price was fixed $3 a ton above the Pittsburgh mill price, an arbitrary figure but incorporating rather higher production costs as a consequence of smaller runs and inferior equipment.[42] In 1920, when rail rates were increased, the Birmingham differential was raised to $5. Wire was priced differently. American Steel and Wire had taken over the Ensley rod mill and the Birmingham wire works, but by 1908 had proved that if they sold on the standard Birmingham price southern wire would still have an advantage even in some northern markets, which, as the company's Vice-President observed 'of course would not be desirable'. American Steel and Wire therefore ensured that wire should be on Pittsburgh Plus, which raised the Birmingham price to about $15·30 a ton over that in Pittsburgh.[43]

On simple finished products using steel, the Birmingham differential sometimes amounted to from 4 to as much as 12 per cent of selling price, according to evidence given in the Pittsburgh Plus hearings, though these must have been unusual instances. Northern boilermakers, it was said, controlled two-thirds of the southern market, but for this their own economies of scale must have been much more important than the differential. The abolition of Pittsburgh Plus left the Birmingham differential intact, it now became just one of the multiple steel basing points. After 1933 competitors sold in the South at below the Birmingham base price, and to meet them U.S. Steel had to do the same. For big consumers the differential had been waived before, now it was in practice eliminated, so that by 1937/8 finished steel was selling there at below the Pittsburgh price.[44] However, it was at this point that other factors retarding the development of the southern steel industry were most clearly revealed.

Fairfield steelworks was a highly efficient plant which did well throughout the depression, but its product range was unbalanced. Built to roll plates and structurals, after 1926 it had old type sheet mills, and in 1929 production of cotton ties and hoops was started. But tinplate, for which the south was a big market, and, more understandably, high-quality sheet was not produced at all. Over the period 1935—8 the exhaustive survey of Steel Corporation structure made by the consultants Ford, Bacon and Davis showed to what a large extent orders which, on cost grounds, should have been allocated to Birmingham had been filled instead by northern mills.[45] On the west coast $1 million a year was lost by supplying from northern rather than from Alabama works. In 1937 all companies in the Pittsburgh—Johnstown area delivered over 34,000 tons of tinplate into the nine south-eastern states, but even on shipments to the Atlantic seaboard market Birmingham could produce and deliver for $2.48 a ton less than Pittsburgh. In spite of this U.S. Steel central planners seemed reluctant to build Alabama tinplate capacity, and not until 1938 was a 200,000 ton cold reduction plant for tinplate built in Birmingham.[46] When basing point pricing ended, the opportunity to make up lost ground was slipping, for Alabama was no longer a low-cost iron district. By 1950 pig iron production costs were even said to be lower in Pittsburgh, and although cheap scrap permitted low cost open hearth manufacture, in steel finishing Birmingham remained disadvantaged as compared with the larger, more specialized northern rolling mills.[47]

THE ABOLITION OF THE MAIN BASING POINT DIFFERENTIALS, JUNE 1938

Bolstered by the differentials, shipping some steel by water or rail and water even though rail freights were charged, Pittsburgh seemed to come out of the depression as well as most other centres. With the industry buoyant, early in 1937 the Pittsburgh operating rate was above both the national average and the Chicago rate. In autumn demand slipped, and by June 1938 the national operating rate was under 30 per cent, and that of Pittsburgh even lower. This was another dangerous time for maintenance of unanimity in regard to pricing policy. In addition, having just completed an expansion programme in Chicago and a smaller one in the South, the Steel Corporation needed to increase its share of the market. On 24 June, U.S. Steel's Carnegie-Illinois subsidiary reduced steel prices—by an average of $3 a ton in Pittsburgh, $4 to $5 a ton in Chicago, and $10 in Birmingham. Most of their basing point differentials disappeared. Other companies followed with new basing points. The effect was to limit still more the area in which Pittsburgh was competitive. This may be illustrated by reference to sheet steel.

Jones and Laughlin and U.S. Steel had both just built new Pittsburgh area strip mills. The latter's Irvin works was designed to serve eastern markets and outlets through Ohio to Cleveland and, in times of peak demand, Detroit, in all of which areas Pittsburgh was the ruling basing point until 1938. In Cleveland hot rolled sheet had sold at the Pittsburgh price plus $4 a ton freight, but now, bringing into production the biggest strip mill yet built, Republic Steel made Cleveland a basing point. Selling there, Irvin or the Jones and Laughlin Pittsburgh mill had to absorb freight and reap a reduced mill net. In the first half of 1938 the Cleveland operating rate was below that in Pittsburgh, in the second half it was almost as consistently above. Youngstown, Middletown, Sparrows Point and Lackawanna were now made into basing points by the major companies operating there. In each case they took little of the market territory of two of the three pre-existing northern basing points, Chicago or Granite City, but were mostly carved out of the area previously dominated by Pittsburgh.[48] Sparrows Point had a freight rate of $3·4 per net ton on sheet to Philadelphia and selling there on the Pittsburgh base price of $48 plus freight of $6·4 had netted $3 per net ton as phantom freight. With Sparrows Point as a basing point, Pittsburgh had to absorb freight heavily to keep a footing in the east. For some products freight absorption was impracticable, now that Sparrows Point was a basing point for plate, the freight absorption needed to sell Pittsburgh plate in Baltimore would reduce the mill net to below the cost of production. Yet within 75 miles of Pittsburgh 1935 capacity for all types of steel was reckoned to be three times the demand.[49]

END OF BASING POINT PRICING

Revival of business in 1939 eased the position of Pittsburgh mills, and after a decline in spring 1940 there followed the war boom. When the war was over the problem returned, accentuated by rapidly rising rail freight rates. Each freight increase penalized most distance shipments, shortening the radius within which sales could be made before freight absorption became necessary. In 1938 the Pittsburgh freight disadvantage as compared with Bethlehem in New York was $3·80 a ton, in 1946 $4·20 and by January 1947 $4·60.[50] The progressive subdivision of the market therefore continued.

In July 1948, a month after the Federal Trade Commission declared against a similar system in the cement industry, the steel industry abandoned basing point pricing. With an F.O.B. system and few surviving differentials, each mill had a natural tributary area. Freight absorption was still permitted. As in 1924 there were suggestions of new mills near major markets, and of consumers moving to steel surplus areas.[51] The high costs of new plant stopped radical change, and Fairless was the only new works. Until the end of 1953 markets were buoyant, and there was little need for freight absorption. In the 1954 recession keen competition led to the whittling away of the surviving Detroit differential. In autumn 1962 the much bigger west coast differential was removed by Kaiser Steel in an effort to eliminate eastern, western interior, and foreign competition. By the late 1950s Pittsburgh and Valley operating rates, and after 1960 the index of operations—based on 1957—9 averages—indicated how much these areas lagged behind other districts. One reaction was the further development of water transport, another, more truck transport for shorter hauls. In 1937 only 313,000 of U.S. Steel shipments of finished products went by truck; by 1955 railroads carried 4·9 million tons of sheet steel in eastern territory, trucks 13·4 million tons.[52] Belatedly, railroads introduced 'creative pricing' and also unit trains, which for big shippers promised some amelioration. Another development has been to ship semi-finished steel to be finished in company plants nearer the markets. Yet another resort has been specialization on quality, of which the Irvin-Vandergrift specialization on high-grade sheet, including electrical qualities, is an example.

Geographical pricing systems in iron and steel have gone almost full circle in the course of the last century. In the process they shaped, even distorted, the industrial map of America.[53] The industry has not been able to adjust its capacity distribution so easily, and major surplus capacity areas remain today, in part at least, as a memorial to the heyday of a spatially discriminating pricing system.

NOTES

[1] T.N.E.C., *Hearings*, Part 27, 1939—1940, p.14630.

[2] C.R. Daugherty, M.G. De Chazeau, ad S.S. Stratton, *The Economics of the Iron and Steel Industry.* 1937, p.533.

[3] Ibid.; also *I.A.*, 16 Nov. 1922, p.1287.

[4] *Report of Industrial Commission*, 1901, XIII, *Trusts and Industrial Combinations*, p.469.

[5] Bureau of Corporations, *Report on the Steel Industry*, part I, 1911, p.90; A. Berglund, *The United States Steel Corporation*, 1907, p.57.

[6] Federal Trade Commission Hearings, *In the matter of the United States Steel Corporation and Others*, 1924.

[7] *I.A.*, 5 May 1921, p.1176.

[8] *A.I.S.I. Yearbook*, 1924, p.241.

[9] *I.A.*, 4 Sept. 1919, p.643.

[10] *Tariff Hearings*, 1908—1909, p.1694 and *I.A.*

[11] Federal Trade Commission, 1924 hearings, p.1259.

[12] Ibid., p.1245.

[13] *I.A.*, 15 Feb. 1923, p.480.

[14] *I.A.*, 6 Mar. 1919, p.611.

[15] *I.A.*, 1 June 1911, p.1298, and evidence of Laclede Steel to *Senate Committee on Interstate Commerce*, 1936.

[16] A.R. Burns, *The Decline of Competition*, 1936, p.79.

[17] J. Farrell, *I.A.*, 22 May 1913; *Report of Industrial Commission*, 1901, IX, pp.852—3.

[18] *I.A.*, 18 Sept. 1919.

[19] Federal Trade Commission, 1924 hearings, p.1254.

[20] *I.A.*, 23 Feb. 1922, p.541.

[21] *I.A.*, 20 Apr. 1922, p.1070, 15 Mar. 1923, p.741.

[22] *I.A.*, 15 May 1919, p.1287.

[23] *I.A.*, 6 Mar. 1919, p.611.

[24] *A.I.S.I. Yearbook*, 1924, pp.244—6; *I.A.*, 6 Mar. 1919, p.611.

[25] *I.A.*, 1 June 1922, p.1569.

[26] *I.A.*, 12 Jan. 1922, p.162, 19 July 1923, p.140, 18 Sept. 1924, p.694, 12 Feb. 1925, p.515, 4 June 1925, p.1630; see also J.B. Appleton, *The Iron and Steel Industry of the Calumet District, 1925*, p.27.

[27] *I.A.*, 20 Aug. 1925, p.477.

[28] *I.A.*, 1 Jan. 1925, pp.56—7.

[29] M. Keir, *Manufacturing*, 1928, p.208.

[30] *I.A.*, 28 Jan. 1926, p.295.

[31] Quoted T.N.E.C., *Hearings*, part 27, 1939—1940, pp.14544—5; *I.A.*, 1 Jan. 1925, p.54.

[32] *I.T.R.*, 4 July 1929; Transport Investigation and Research Board, *Economics of Iron and Steel Transportation*, 1945, p.48.

[33] *B.F.S.P.*, Jan. 1928, p.73; *I.T.R.*, 9 May 1929, pp.1273—4.

[34] *I.A.*, 31 Mar. 1927, p.950.

[35] *I.A.*, 28 Apr. 1927, p.1232, and *A.I.S.I.*

[36] *I.T.R.*, 10 Jan. 1929, p.169.

[37] T.N.E.C., Monograph 42, *The Basing Point Problem*, 1939—1940; G.W. Stocking, *Basing Point Pricing and Regional Development*, 1954.

[38] Stocking, op. cit., p.54; Burns, op. cit., pp.80—90; J. Farrell, *A.I.S.I. Yearbook*, 1931, pp.40—5.

[39] *I.A.*, 19 Jan. 1933, p.138.

[40] *I.A.*

[41] Burns, op. cit., p.89.

[42] Federal Trade Commission 1924 hearings; Burns, op. cit., p.304; Stocking, op. cit., pp.63—4.

[43] Federal Trade Commission 1924 hearings, pp.1277, 1281.

[44] T.N.E.C., *Investigation of the Concentration of Economic Power*, part 19, Nov. 1939, section 'The Birmingham Differential', pp.10542—52.

[45] Stocking, op. cit., pp.106—8; also House of Representatives, *Study in Monopoly Power*, 1949, 40, *Steel.* (evidence of Stocking, Fairless and Levi).

[46] A.V. Wiebel, *Biography of a Business, T.C.I.*, 1960, p.39.

[47] Stocking, op. cit., p.149.

[48] See T.N.E.C., Monograph, 42, *The Basing Point Problem*, pp.83, 85 and *Pittsburgh Business Review*, 30 Sept. 1938, pp.1, 2.

[49] Burns, op. cit., p.345; *I.A.*, 7 July 1938, p.84D, 25 Aug. 1938, p.51.

[50] *I.A.*, Jan. 1947, p.101.

[51] *I.A.*, 1 July 1948, p.121.

[52] Transport Investigation and Research Board, *Economics of Iron and Steel Transportation*, p.137, and J.C. Nelson, *Railroad Transport and Public Policy*, 1959, p.49.

[53] R. Duncan, *Today*, 2 May 1936; F.A. Fetter, *Journal of Political Economy*, 45, 1937, p.601.

CHAPTER 13

The Interwar Years.
II. Sheet Steel and the Establishment of New Steel Centres

Sheet steel was for long a relatively minor mill product. In 1890 output of thin sheets was one-twentieth the total output of rolled iron and steel. In 1907 production was still exceeded by that of rails, plates, wire rods, structural shapes, skelp and merchant bars. From the First World War there was a very rapid advance. (Table 78) By 1917 sheet output was 7·1 per cent

Table 78. *Production of sheets of 13 gauge or lighter*
(thousand net tons)

1890—	300 (estimate)
1900—	700 (estimate)
1905—	983
1909—	1,249
1915—	2,520
1920—	3,232
1921—	1,693
1922—	3,312
1923—	3,926

Based on *I.A.*

of the total production of hot rolled iron and steel, in 1929 12·7 per cent and by 1940 24 per cent. The motivating force was a rapid growth in consumption and a change in the quality of sheet demanded. As early as 1902 Jeans remarked that the United States appeared to have 'a most devouring appetite' for steel sheet, and rapid plant expansion was already going on.[1] Much of this was for use in railway passenger cars and tram cars. By the twenties there was demand from the office furniture and domestic appliance industry, but far and away the major market, and the most exacting requirements, were those of the automobile industry. By 1926 American rolling mills shipped 4·9 million tons of steel to the motor industry of which 2·3 million tons were sheet or strip.[2] Methods of production of sheet were revolutionized. During the first twenty years of the century conventional sheet mills were greatly improved, but after 1926 the process of change was accelerated by introduction of continuous mills producing what was first called strip/sheet—a product of the gauge and width of sheet but in strip-like lengths. The wide strip mills, along with the cold reduction mills by which they were soon accompanied, were to be the most important technical innovation in the steel industry in the first half of the twentieth century. Between 1926 and 1940 twenty-eight strip mills were built; largely because of their competition the number of 'old-type' sheet mills fell in the same period from 1,264 to 750. Growth, concentration in new, large units, and the closing of small old-type mills were accompanied by substantial change in location.

At the end of the nineteenth century the location pattern of the small sheet industry differed considerably from that of older trades such as the manufacture of bars, plates, and rails. The sheet capacity of the Pittsburgh area as early as the 1870s was in the Kiskiminetas Valley, well away from the other heavier works of the city. Chicago had no sheet plants at all in 1890. By the early 1920s half the nation's sheet mills were in Ohio, even though for rolled steel generally Ohio's output was then only just over one-fifth of the national total. (Fig. 28) An important factor in this localization of sheet manufacture was the late development of the Ohio steel industry. Until the nineties much of the state's production, especially in the Valleys district, had been iron. With backward integration occuring elsewhere iron-makers entered the steel and rolled steel trades to preserve their business. After the McKinley Tariff Act tinplate, and less spectacularly, sheet manufacture, began to expand so that many of the new Valley works went into these lines. By 1922 the Ohio section of the Valleys district had a finished rolled product capacity of 2,851,000 tons—excluding U.S. Steel mills—of which 705,000 tons was sheet and light plate. Elsewhere in Ohio, sheet-making was frequently carried on by rerolling plants, or in small steel and mill plants, having no blast furnaces. Their viability is explained by the high value of their product. Sheet manufacture was a favourable activity for locations at which costs of iron or steel production were high, but where intensive labour inputs could compensate, and skill was much more important than with, for instance, rails or structurals. Corollaries of the location of sheet-making at high-cost producing points were the large number of companies, and still more of producing units, and conversely their small average size. There was a closer approach to conditions favouring free competition than in most branches of the steel industry.

Immediately after the First World War large extensions of existing sheet plants were begun, chiefly to meet motor industry demand. At this time a number of new companies were established, sometimes by successful managers from existing concerns. The entrepreneur with small capital but reputation and intimate knowledge of the capabilities of local labour would tend to build in that same area himself, especially when with a high-class product like autobody sheet, there seemed little advantage in seeking out a new low-cost location.[3] From 1922 to 1926 the number of sheet mills went up 10·4 per cent but capacity by 17 per cent. Much of the new, like the existing sheet capacity was at rerolling plants working up purchased billets or

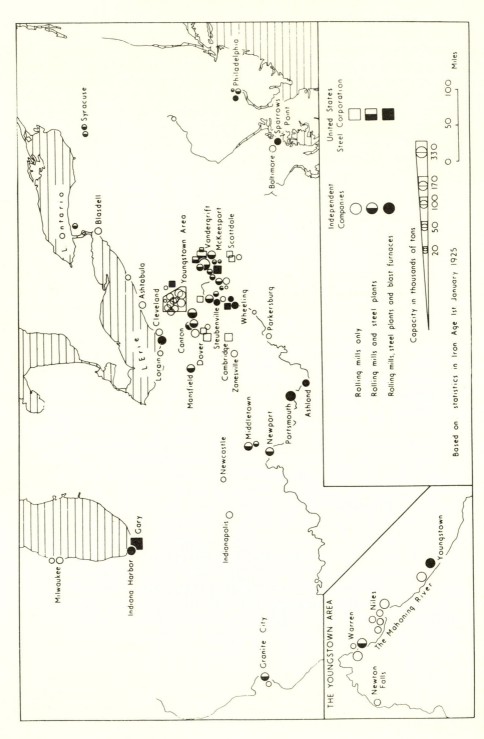

28. Steel sheet and light plate capacity 1922

sheet bars. Other mills, as at Canton, Mansfield, or the Steel Corporation's Vandergrift works, had cold metal steelworks; only very rarely was there full integration back to iron-making as at Youngstown Sheet and Tube, or Otis in Cleveland (Fig.28 and Table 79).

Table 79. *Ohio sheet steel works 1922*

Type of plant	No. of plants	Sheet and light plate capacity (thousand tons)
Ironworks, steelworks, and mills	5	477
Steelworks and mills	7	537
Rolling mills only	21	1,108

Based on *I.A.*, 1 Jan. 1925, pp.10, 15, 16.

Even the big sheet plants were little more than large agglomerations of small mill units. They had certain advantages including capital resources to provide for technical improvement, and the ability to obtain more favourable terms on purchases. There were incentives to backward integration to reduce the variables affecting the business, and gain control over steel quality, but only a big finishing plant could support steel capacity sufficient to use all the iron from one blast furnace. A small firm, later of great importance, integrating backwards at this time was Weirton Steel Company. In 1915 this company had twenty-six mills at Weirton and twelve at both Steubenville and Clarksburg. In 1919 a blast furnace, and the next year a melting shop and blooming mill was built at Weirton, in 1923 the first coke was made there; the same year another black and galvanized sheet plant was built.[4]

The strip mill emphasized the advantage of the big integrated concern. The strip mill had a long ancestry but commercial significance began with the completion of a 'continuous sheet mill' at Ashland and a continuous strip mill at Butler. By mid-1927 their annual capacities were approximately 500,000 and 400,000 tons respectively.[5] In 1926 Weirton Steel contracted for a strip mill and this began work in the summer of 1927. A little later work was begun on strip mills at Warren and Middletown, Ohio. As early as May 1927 *Iron Age* commented '. . . owners of old mills will need to do no small amount of planning if they desire still to be numbered among the purveyors of rolled steel.' New sheet mills, though of improved design, continued to be built, in summer of 1929 new mills being brought into production or decided on for Kokomo, Indiana, Monroe, Michigan, and at Tarentum on the Allegheny river,[6] but the strip mills had immense labour cost advantages—Butler by summer 1927 had a capacity well over twenty times that of an ordinary sheet mill but a crew of roughly the same size.[7] High productivity was however achieved only at very much higher capital investment, especially if the desirability of iron, steel, and primary mill capacity was taken into account.

Apart from the recession of 1927, a very mild one in sheet and strip, the first three years of the strip mill's commercial operations were ones of general prosperity. In the thirties the struggle between the strip mills and the older mills became acute. By the beginning of 1930 there were seven strip mills (including Ashland). Five more were built in the next five years, and from 1935 to 1940 sixteen were added—an almost inevitable overbuilding of capacity. As a result competition for business to cover capital costs was keen and big consumers, sensing their initiative, helped to push prices still further down. In 1923 the selling price of 20-gauge cold reduced sheet was $135 per net ton; by 1941 only $62. Meanwhile, to meet the quality competition of strip mills, and to cut operating costs, sheet mill firms merged into bigger groups, and committed large outlays of capital to modernize their mills only to find that the market still slipped from their grasp. In spite of large capitalization the strip mills held their trade much better than the hand mills (Table 80).

Table 80.   *The performance of continuous strip mills*
                   in the depression
                   (thousand net tons)

	Total Production of Sheet, Strip, and Black plate	Production from Continuous Strip Mills
1929	8,303	1,380
1930	6,595	1,403
1931	5,055	1,522
1932	3,440	1,572
1933	5,663	2,619
1934	5,530	3,168
1935	8,104	4,850

Source: D. Eppelsheimer, *J.I.S.I.* 1938.2.

Continued advance of the strip mill, in terms of output, quality, cost, and range of product caused large-scale abandonment of hand mills and then of improved old-type mills. (Table 81) Big firms built continuous mills and closed their old mills, though occasionally reopening them in times of boom. Smaller sheet firms, lacking resources big enough to finance a strip mill, struggled on, accepting small orders and small profits, sometimes trying to push into special grades where their high costs were not so much a disadvantage. Even there the closure rate was great. Investment in hot strip mills by 1939 was put at $500 million while in the same period between $100 million and $200 million of old-type plant had been discarded.[8]

The older steel firms, Bethlehem, Jones and Laughlin, and U.S. Steel, felt little incentive to build new strip/sheet mills before the depression for they were busy in their own trades. Seven strip mills had been brought into

Table 81.   *Sheet and strip production and capacity*
1930, 1935, 1940

	Sheet and Strip Production	*Continuous Strip/ Sheet Mill Production*[1]	*Continuous Strip/ Sheet Mill Capacity*[1]
1930	5,998	?	4,254
1935	8,762	4,147	7,059
1940	13,783	14,427	17,616

[1] Production and capacity of strip/sheet mills includes some light plate.
Based on *A.I.S.I.* and Geo. Armstrong & Co., *An Engineering Interpretation of the Economic and Financial Aspects of American Industry*, 1952, X, p.52.

production before the end of 1929. Three of them were Armco mills—at Butler, Ashland and Middletown—others belonged to the Trumbull, Weirton, and Wheeling Companies. American Steel and Tinplate had a mill for rolling tinplate stock at Gary. All the early mills, with the exception of Gary, were along the Ohio or its tributaries. In the long recession of the heavy trades in the thirties the major steel firms were forced to enter the sheet and strip business which, after a sharp downturn, fairly quickly revived to become the only conspicuously bright field in a generally dreary industrial landscape. In most cases these older, heavy firms also built at old locations, but some of them, like Bethlehem or the Corporation, choosing between their various existing plants, brought out the value of new locations for strip mill operation.

The steel Corporation abandoned a great deal of old plant in the late twenties and thirties, and almost one-third of all the capacity scrapped was in sheet, light plate, or black plate. In the Wheeling district six out of nine sheet plants were abandoned, in the Pittsburgh district six out of ten. Vandergrift, the chief old-type Pittsburgh area sheet mill, surpassed by the new Irvin strip mill, was gradually shifted to higher-quality sheets, especially electrical grades. By the late fifties it became wholly a reroller of material from Irvin.[9] Only one of the strip mills built before 1940 was on a greenfield site, the plant at Ecorse, Detroit, which was completely new except for the connection with the existing Hanna blast furnaces on Zug Island. By the late thirties some plants were being equipped with their second strip mill. Sometimes this was to meet demand for wider sheets. Older, narrower mills could be reconstructed for this purpose, as was done by Otis in 1937, but in other cases it was found better to build a new mill. The profitability of the various mills was the outcome of a complex of factors—location, enterprise, and the connections built up with consumers. Some firms stood out as much more enterprising than others, Inland and National being in very different ways conspicuous in this respect. But Youngstown Sheet and Tube's Campbell works, though disadvantaged by location, was second only to Inland in commercial results in the second half of the 1930s.

THE LOCATION OF STRIP MILLS AND THEIR MARKETS

Most sheet capacity at the time of the introduction of the strip mill was in the basin of the Ohio. The largest markets were already in the industrial centres along the shores of the Great Lakes. These too were the lowest-cost centres for steel production. Only slowly did the distribution of strip mill capacity begin to conform more closely to the optimum location suggested by these conditions. (Table 82 and Fig. 29). In 1922 sheets, light plate, and black plate for tinning were made in 105 separate locations, having an

Table 82.   *Sheet and Strip Mill Capacity 1922 and 1940*
(million tons)

	1922[1]	*1940*[2]
U.S. Capacity	6·55	16·0
On Lake Erie	0·68	4·0
On Lake Michigan	0·31	3·4

[1] Sheet, light plate and blackplate.
[2] Strip mills alone, excluding also the two continuous plate mills of Homestead and South Chicago.
Based on *I.A.* 1 Jan. 1925 and American Association of Iron and Steel Engineers, *The Modern Strip mill*, 1941.

average annual capacity of 62,400 tons. Only nine works were on the lake shores. By 1940 twenty-six strip mills had an average capacity of 610,000 tons; twelve of them were on the Great Lakes.

The largest single markets were Detroit and Chicago. Development in the first of these is of such especial interest that it justifies separate fuller treatment. When American Sheet and Tinplate built at Gary and Inland at Indiana Harbor much of their produce was used within the Chicago industrial area, though their locations were excellent also for access to the body shops of Indiana and Southern Michigan. Chicago was so attractive that within three months of its formation National Steel acquired a 1,000 acre

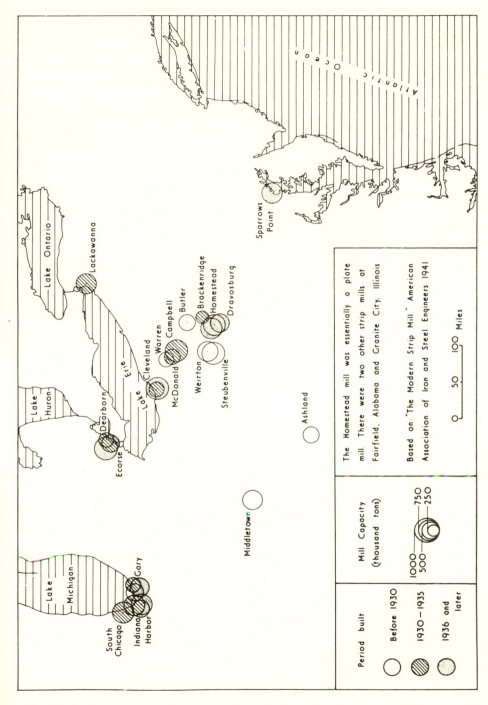

29. Continuous sheet and wide strip mills 1940

site six miles east of Gary. Here it planned to spend $40 to $50 million—half as much again as at Ecorse—on coke ovens, a blast furnace, an eight-furnace melting shop and finishing mills. The depression caused postponement and then war put off a plan to start with finishing mills for sheet, tinplate, and bars with backward integration to come afterwards. This scheme was taken up twenty years later.[10]

Cleveland provided substantial markets for sheet in local automobile and appliance firms and in its host of diverse metal industries. It was also well located for water shipment to Detroit. On the other hand, site conditions in the lower Cuyahoga Valley were by now extremely difficult, which increased both assembly and processing costs. In 1935 Republic took over Corrigan McKinney and two years later decided to build a Cleveland strip mill there. The river had to be diverted for about half a mile, and every part of the plant had to be mounted on piling. Even after this slabs for the hot mill had to be carried almost two miles from the old Corrigan McKinney steelworks.

Lackawanna had better site conditions and, in addition, a commanding position with respect to the eastern markets. New England scrap, supplementing local supplies, made scrap prices there especially low. In 1939 Buffalo shipped 348,000 tons of finished steel by lake, far more than any other port. By rail to Detroit it had a disadvantage as compared with Cleveland of 89 cents a ton, but by water of only 23 cents. The rail/water rate from Pittsburgh and Youngstown to Detroit was $3·97 and $5·54 per ton respectively but from Buffalo only $1·29. Not surprisingly Bethlehem Steel chose to build its first strip mill there in 1936. For a time Lackawanna supplied hot rolled coil to the Sparrows Point tinplate mills, but by the end of 1936 the decision was taken to build a strip mill of similar capacity there. The whole eastern seaboard was tributary to Sparrows Point, though for the centre of demand both for sheet and tinplate a location further north on the Delaware rather than on Chesapeake Bay would have been better. Greater New York was a major centre of sheet-using domestic appliance trades and New Jersey an important automobile state with six important divisions of General Motors alone by 1937.[11] Until June 1938 Pittsburgh was the nearest basing point so that on all eastern sheet sales Sparrows Point reaped substantial phantom freight. After this, as a basing point itself, the plant was sheltered from western competition by the high freights which shipments from across the Alleghenies had to absorb—$3 per net ton to Philadelphia and $2·4 to Harrison, New Jersey, from Pittsburgh.[12] By the end of the twenties it had been proved that cold reduced coil provided the ideal material for tinning, and in marketing tinplate Sparrows Point had advantages commensurate with those it enjoyed for sheet. Camden, the cannery centre of the New Jersey trucklands, was 110 miles away by rail, and in 1937 48·3 per cent of New Jersey's consumption was in the Baltimore district.[13] Moreover, as the only integrated tidewater works, Sparrows Point had an important position in the west coast tinplate market, U.S. Steel's subsidiary in Pittsburg, San Francisco, supplying only a very small part of west coast needs from old-type mills. However, the railroads allowed a very low 'tinplate rate' from Pittsburgh and the Mid-West to the Pacific coast, and this, and the importance of the market—almost one-fifth of the national consumption was in the three Pacific coast states in 1937—made producers west of the Appalachians willing to compete even with Sparrows Point water shipments on the Pacific coast. (Table 83).

Table 83.   *Tinplate shipments to three Pacific coast states*
1936, 1937, 1938
(thousand net tons)

	1936	1937	1938
Total received	345·6	382·1	209·9
From Chicago	70·3	89·7	56·0
Pittsburgh, Youngstown, and north Ohio River	198·2	222·4	87·3
Baltimore	40·0	32·2	16·8
West coast	36·6	37·2	28·3
Birmingham	—	—	21·1

Based on G.W. Stocking, *Basing Point Pricing and Regional Development*, pp.262–4.

South-eastern tinplate markets were shared largely between Sparrows Point and Ohio Valley plants, the latter retaining the larger share. They also continued to dominate the Gulf coast market. Not until 1938 was the first strip mill built in the South, a T.C.I. mill for tinplate at Fairfield. High-quality cold reduced sheet steel was not made there before 1946.

In the older mill towns of the manufacturing belt there was often physical difficulty in building large new strip mill units at existing works. This problem did not arise in the meander cut-off along which the Weirton plant expanded or in Armco's Middletown plant in the flats of the Miami Valley, but elsewhere it forced uneconomic plant layouts or increased construction costs. Youngstown Sheet and Tube's Campbell mill, built in 1934 in the middle of the Youngstown industrial district, exemplifies this, but the two Pittsburgh strip mills provide even more outstanding examples.

One of the chief problems facing the engineers of the Jones and Laughlin mill was to find a large enough site, and a great deal of demolition was involved before construction began.[14] When U.S. Steel decided to build strip mills, Gary was an obvious choice for early development, and by 1936 three of its four mills were there. The other was at the MacDonald works near Youngstown. For production of wide strip MacDonald was, however, both a small and high-cost producer. It obtained slabs from the Ohio steelworks ten miles away, thus adding extra freight charges to costly iron and steel production. By 1938 MacDonald capacity was 118,000 tons a month but only about 43,000 tons of this was wide strip, the rest being narrow strip and bars from smaller mills. For the eastern interior and still more for the eastern market a larger, cheaper producer than MacDonald, and nearer plant than Gary was needed. In 1936, the Corporation decided to build a new mill, the Irvin Works, at Dravosburg west of McKeesport. Edgar Thomson, which had lost most of its rail business, was equipped with a new forty-four inch slabbing mill to supply the new works. There were no suitable sites near the city on the meander flats of the river so that Irvin works was built, after very extensive levelling and filling operations, on the plateau top. Irvin's 5,000 labour force was of high quality, carefully picked from 25,000 candidates, many from the old American Sheet and Tinplate works which lay derelict in isolated and now blighted little mill towns.

Local deliveries of Irvin sheet in the Pittsburgh area were far exceeded by shipments to other areas (Table 84). The new mill caught the 1938 recession,

Table 84.   *Shipments of sheet and strip from the*
Pittsburgh-Johnstown district[1]
to north-eastern states 1937
(thousand tons)

	Hot Rolled Sheets	*Cold Rolled Sheets*	*Hot Rolled Strip*	*Cold Rolled Strip*
Pennsylvania	76·8	20·8	46·0	2:1
Ohio	74·1	26·9	56·9	11:5
New England	23·1	9·3	12·3	2:3
New York	18·0	13·4	13·7	9;0
New Jersey	9·3	5·9	6·8	4:2
Michigan	126·5	133·7	89·8	30·8

[1] Pittsburgh-Johnstown district includes Allegheny, Westmorland, Washington and Beaver Counties, Pennsylvania, Jefferson County, Ohio and Hancock, Marshall and Brooke Counties, West Virginia.
Source: Transport Investigation Research Board, *Economics of Iron and, Steel Transportation,* 1945, p.171

and in June 1938 was penalized by the removal of most of the mill differentials which had favoured Pittsburgh, and establishment of a number of new basing points. By 1939/40 the Pittsburgh rail/water rate to Detroit via Cleveland was $5·54 as compared with a Cleveland–Detroit water rate of $1·06 per long ton. The rail rate to Cleveland was $4·48 per net ton, so that shipping to a new Cleveland basing point equipped with the major new Republic Steel strip mill Irvin had to absorb freight of about that amount. New York had been expected to be a major market for Irvin tinplate, but by the early 1940s the rail rate on finished steel from Pittsburgh was $8·06 per long ton, whereas by water Sparrows Point could ship into New York Harbor for $1·68.[15] In short the Pittsburgh area and other interior centres like the Valleys or the Wheeling—Weirton district which had been forced more and more into the production of strip mill products because of the decay of their old staples, had inferior locations for the new lines of business.

THE GROWTH OF THE DETROIT STEEL INDUSTRY

Even before the twenties and the all-steel body car, Detroit had become one of the major steel markets. By 1919/20 Ford used about 1 million tons of steel a year; in 1920 Michigan made only 32,000 tons. By late 1922 Detroit area finished steel capacity was 65,000 tons—40,000 of bars, bands, and hoops and 25,000 tons of tubes—but there was no sheet capacity at all.[16] Growth occurred after this, but only on a comparatively small scale. (Fig.30) In December 1922 the Michigan Steel Corporation was established and began to build a 40,000 ton sheet plant. By 1929 its output approached 100,000 tons of sheet a year, much of which went to Ford.[17] Three months after the formation of Michigan Steel, the Detroit Steel Corporation was organized to roll cold reduced strip for the automobile industry. In 1929, Newton Steel, one of the largest and most rapidly expanding old-style sheet mill concerns of interior Ohio, decided to build in Detroit. This decision presumably in part reflected the diseconomies of further increase in the firm's wholly unintegrated operations 200 miles from its major markets, a problem accentuated by the prospect of keener competition from other old-type mills following mergers and rationalization, and also a reaction to strip mill competition. In March 1929 construction began on a 600 acre site east of Monroe on the Raisin River, two miles from deep water in Lake Erie but over thirty from the centre of Detroit. Initially Monroe was designed for about the same capacity as Michigan Steel, but plans were changed to envisage a 250,000 ton full-finished sheet plant. After an uncertain career the plant failed.

Michigan, Detroit, and Newton steel were all small ventures whose growing capacity failed to keep pace with the increase in demand in southern Michigan. By 1928 annual steel consumption by Michigan automobile firms was about 5·5 million tons made up of 2 million tons of sheet, 1 million tons of strip and 2·5 million tons of bars, plates, forgings, billets, and other categories. In 1928 rolled steel production there was under 400,000 tons. By 1929 Michigan ranked fourth among the states in steel consumption but seventeenth in production. Yet in addition to local consumption this area was excellently located for access to other lakeside markets, and for raw material assembly. Detroit was as well located as Chicago for rail shipments of West Virginia or Kentucky coal, and coal boats from Toledo or Sandusky could sail direct to plant sites there. All the 36·5 million tons of Upper Lake ore delivered to Lake Erie ports in 1928 passed through the Detroit River, each year the body shops, chassis, engine, and component plants of the motor industry, and other processing plants, turned out steel scrap, shipped

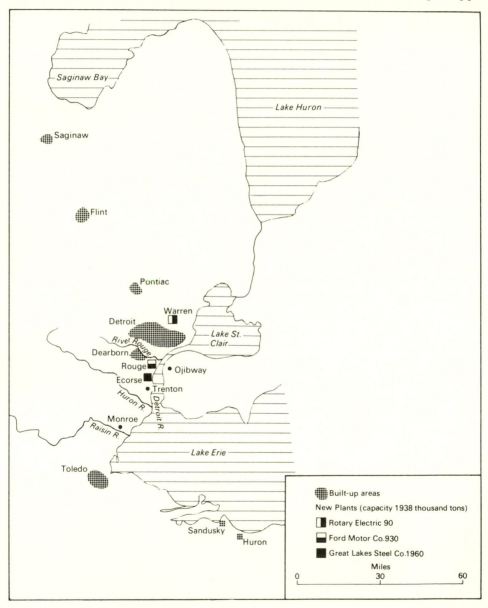

30. The Detroit district between the wars

out of the area to the extent of 700,000 tons a year.[18] One serious Detroit disadvantage was a lack of good sites for large heavy industrial plants. There was room enough for small steel concerns like those which built in the 1920s, but a fully integrated works needed both a larger acreage and a riverside location. The scarcity value of good sites near the centre of the district rocketed in the late 20s, the last considerable one near the river selling for a reported $35,000 an acre.[19] Even on a constricted site this could burden a plant with carrying charges for the site alone of about $2 per ton of finished product. There was more and cheaper land away from the Detroit River, but only along the Rouge was there a deepwater connection through to it, and there the land was already closely held in a few hands. For the rest it was necessary to go further afield perhaps like Newton to the Raisin, or perhaps to Saginaw Bay, whose approaches to the lake were only now being improved, or, alternatively, to tackle some of the swamplands, until this time considered unusable. Plant engineers from a number of companies had been sent to examine possible Detroit developments and they had commonly reported against construction in the area because of site difficulties.   Early interest shown by the big companies in the area evaporated or came to unprofitable ends.

As early as 1907 the local Great Lakes Engineering Company appears to have temporarily interested Jones and Laughlin in a proposal for a Detroit River integrated plant—presumably before that company had finally decided to build at Aliquippa.[21] Some six years later U.S. Steel began what was at first envisaged as a very large plant on the Canadian side of the river at Ojibway. There was ample undeveloped, cheap land in that area, but a tariff of $10 a ton on bars and $16·40 a ton on full-finished sheets made it a hopeless location to supply the nearby Detroit market. Sixteen years after it was begun Ojibway was a sadly incomplete plant—the shells of two blast furnaces, unfinished ore docks and bins and a small wire plant.[22] By the 1920s another factor, understandably not well documented, seems to have checked the building of Detroit plants by major companies—a so-called 'gentleman's agreement'. Whether or not this agreement ever existed, something certainly proved effective against all the pressures of commercial temptation. In 1929 it was said that U.S. Steel had been urged to build in Detroit, but instead it chose to expand in Chicago. In the following months the American Rolling Mill Company was rumoured to have an option on a 200 acre Wyandotte site, a mid-western firm to be planning finishing mills near Detroit, and eastern interests—presumably Bethlehem—to be thinking of a mill at Huron, between Toledo and Cleveland and nearer to Detroit than any existing major plant.[23] Perhaps some of these schemes had no stronger foundation than gossip, perhaps it was only the collapse of the bull market in the autumn of 1929 which prevented others being carried to completion, but certainly none of them was built. However, big consumers and small steel producers were not bound by the conventions surviving from the Gary era.

For the biggest motor firms control over steel production had substantial commercial attraction. A major element of costs could be controlled for the first time, and with their varied demand—for sheet, strip, bars, sections, forging billets—General Motors or Ford could support a large steel plant, and still leave sufficient to be supplied by purchase, so as to ensure high-level operations for their own mills even under the clouds of mild recession. 'Tapered integration' of this sort could also provide a measuring rod to assess the prices quoted by steel producers and might put the automobile firm in a better bargaining position. With his early passion for integration Ford naturally led the way. As early as 1915 he was talking of making steel in Detroit, and by 1916 the rise in material costs, especially of steel, provided the occasion for action. The lower course of the river Rouge was at this time only in its infancy as an industrial district, and Ford acquired large sites near its mouth. Interrupted by the war, construction of the Rouge metallurgical section was resumed afterwards. Priority was given to blast furnaces, coke ovens and the iron foundry, but Ford reckoned that they should produce at least half their annual consumption of about 1 million tons of steel. In November 1925 the first rolling mill began work on purchased billets, and seven months later the Rouge melting shop tapped its first heat. In 1925 Michigan made only 42,000 tons of ingots, the following year Ford alone made 321,000 tons. The Rouge Works was backed up by Upper Lake ore properties and Kentucky coalmines linked to the Rouge by the company's own Detroit, Toledo and Ironton Railroad.[24] By the late 1920s it was clear that Ford had over-extended itself in its search for more backwards integration, and this perhaps contributed to General Motor's decision not to make steel. It was credited in autumn 1928 with plans for furnaces and steelworks on the Saginaw River. A year later Fisher Body was said to be considering closer connection with Newton Steel's Monroe plant. Neither development occurred.

In 1928 Michigan Steel, in spite of good connections with the automobile firms, rolled only about 100,000 tons of sheet, or one-twentieth the southern Michigan consumption. At the beginning of 1929 it decided to form a new company, Great Lakes Steel, and to build a wholly new and much larger plant on a large swampy site on the Detroit River at Ecorse, three miles from the existing works and claimed to be the only large un-developed site in the Detroit switching area with rail and water transport. It was planned that much of the steel for a finished steel capacity of 400,000 tons of strip, light plate, shapes, and hoops a year should be made from local scrap.[25] However, in October 1929 Hanna, Great Lakes and the Weirton Steel Company merged as the National Steel Corporation. Great Lakes' contribution was the site and development programme; Hanna's Zug Island furnaces provided molten iron backed by its large ore properties, ten lake freighters and Pennsylvanian, Ohio, and Virginian coalmines. Weirton brought into National Steel a fully integrated works and strip mill, two smaller works and 7,000 more acres of good coal lands in West Virginia and Pennsylvania, and in return obtained a more secure footing in the Detroit market, absorbed the threat of a potentially major competitor, and ensured the large proportion of its own ore supplies which came from Hanna. Strip was first rolled on the Ecorse strip mill in September 1930, and by summer 1931 the first stages of development had been completed at a cost of $29 million.[26] The plant and National Steel as a whole performed well even through the depression. In 1931 U.S. Steel operated at only 38 per cent of capacity and in 1932 at 18·3 per cent, but National's rates were 61·9 and 39·4 per cent respectively. In 1932 Bethlehem lost $19 million, U.S. Steel lost $71 million, but National managed a profit of $1·6 million. Not surprisingly further growth at Ecorse was one of the outstanding features of the 1930s. Announced with an ingot capacity of 600,000 tons and a finished rolled steel capacity of 400,000 tons, by spring of 1936 it already had two strip mills and an ingot capacity of 1 million tons, and was pile driving for a 500,000 tons steelworks extension.[27] By 1938 ingot capacity there was 1·96 million tons.

There were other important though less spectacular Detroit developments. Early in 1934 Ford announced a $10 million 500,000 ton strip mill. In the same year two much smaller, but ultimately important growth points were established. The Rotary Electric Steel Company began a cold rolled strip plant at Warren, using electrical steel for material of high quality, and on South Livernois Avenue, in the centre of the industrial district, surrounded at a short distance by large motor firms, McLouth Steel Corporation installed a Steckel hot strip mill, a type of mill much smaller, cheaper, and with a lower output than the wide continuous mills.[28]

Detroit remained a steel deficit area. In 1932 two out of the three million tons of steel used there by the automobile industry came from outside, and further growth of Detroit steel production was matched by extension of consumption.[29] Ecorse and the Rouge were low-cost producers, and their marketing expenses were slight—even for the former no more than the Detroit area switching charge, reckoned in 1929 $4 a ton less than the rail freight from Pittsburgh.[30] Great Lakes Steel capitalized on its locational advantage by charging a higher mill price than its outside competitors, but its delivery prices were still below theirs. In recession it could be expected to hold its trade while their sales declined, a situation encouraged under those conditions by hand-to-mouth buying by the automobile companies. In fact, as 1938 showed, some Detroit firms became too dependent on local outlets and therefore performed very badly that year.

There were a number of ways in which distant companies could meet the challenge from new Michigan firms. One was to build in the area or to purchase an existing plant. Building was ruled out by difficult site conditions, by fear of further retaliatory building, and for some companies, such as Bethlehem or U.S. Steel, a pressing need to rationalize existing plant.

There were rumours of offers to Ford for its steel division in 1934, but there may have been little substance in them. An alternative was to reduce delivery costs either by all water or by rail/water shipment. Bethlehem pioneered direct water delivery in the late twenties from Lackawanna. Cleveland works, notably the Otis plant, followed, and in 1929 U.S. Steel began to ship structurals and bars from Gary to Detroit. Interior producers frequently shipped by rail to a Lake Erie dock and then by water to Detroit. They gained a saving over all-rail hauls but it was less than a comparison of line haul rates would suggest, largely because of increased handling charges. Resolution 21 of 1933 was another response. Believing that other motor companies might build their own plants, the steel industry made Detroit an arbitrary basing point, so that Chicago and Pittsburgh for the first time had to absorb freight to deliver there. One result was that steel delivered to Toledo from the Valleys or Pittsburgh, cost more than 60 miles further on in Detroit.

None of these devices could solve the problems of the distant suppliers. Detroit steelmakers retained an unassailable advantage in the local market. Other lakeside plants were well placed to make up much of the rest of Detroit area consumption. To the Ohio basin producers Detroit remained an important market, but, in spite of their early leadership in the sheet trade, and subsequently in strip mills, their share of the business now decreased.

NOTES

[1] Jeans, *American Industrial Conditions and Competition, 1902, p.305.*

[2] *Steel,* 16 Apr. 1945, pp.82–3.

[3] *B.F.S.P.,* Apr. 1919, p.201, May 1919, p.243, June 1919, p.310, July 1919, p.357, Aug. 1919, p.409, Dec. 1919, p.612.

[4] *B.F.S.P.,* Jan. 1927, p.55.

[5] *I.A.,* 16 June 1927, 19 May 1927, pp.1435–9; see also R.W. Shannon, *Sheet Steel and Tinplate,* 1930; E.S. Lawrence, *The Manufacture of Steel Sheets,* 1930.

[6] *I.A.,* 19 May 1927, p.1462; *I.T.R.,* 26 Sept. 1929, p.779, 18 July 1929, p.174; *I.A.,* 2 Jan. 1930.

[7] *I.A.,* 19 May 1927, pp.1435–9.

[8] T.N.E.C., *Investigation of the Concentration of Economic Power,* part 19, 1939–1940, pp.10689, 10691.

[9] *I.A.,* 7 Apr. 1938, pp.70A–70D; *Harpers Magazine,* Jan. 1940; *B.F.S.P.,* Feb. 1939, p.195, Oct. 1939, p.1021.

[10] *I.A.,* 16 Jan. 1930, p.245, 5 June 1930, 16 Mar. 1939, p.84.

[11] *I.A.,* 6 Jan. 1938, p.74.

[12] *I.A.,* 7 July 1938, p.84D.

[13] Stocking, p.263.

[14] *I.A.,* 16 Dec. 1937, p.49.

[15] Transport Investigation and Research Board, *Economics of Iron and Steel Transportation,* 1945, p.125.

[16] A. Nevins and F.E. Hill, *Ford,* 1954, p.290; *I.A.,* 1 Jan. 1925, p.9.

[17] *I.A.,* 3 Jan. 1924, p.50, 18 Apr. 1929, p.1066, 18 July 1929, p.173.

[18] *I.T.R.,* 18 Apr. 1929, p.1066.

[19] Ibid.

[20] *Fortune,* June 1932, p.35.

[21] *I.T.R.,* 14 Mar. 1929, p.757.

[22] *J.I.S.I.,* 1922, 2, p.299; *I.T.R.,* 18 Apr. 1929, p.1066.

[23] *I.T.R.*, 18 July 1929, pp.173—4, 19 Sept. 1929, p.734.
[24] Nevins and Hill, op. cit., pp.289, 290, 292; E.D. Kennedy, *The Automobile Industry,* 1941, pp.76,
    92; *B.F.S.P.,* May 1927, pp.212—15.
[25] *I.T.R.*, 28 Feb. 1929, 18 Apr. 1929; *Fortune,* June 1932, p.35; *I.S. Eng.,* Apr. 1955.
[26] *I.A.,* 13 Aug. 1931, p.449.
[27] *I.A.,* 6 Feb. 1936, p.89, 5 July 1934.
[28] *I.A.,* 5 July 1934.
[29] *I.A.,* 22 Dec. 1932, p.976.
[30] *I.A.,* 8 Aug. 1929, p.351.

CHAPTER 14

The Interwar Years
III. The West

Repeated failures in attempts to make steel west of the Rockies contrasted sharply with the dynamism of the western economy generally. Between 1920 and 1940 the national increase of population was 24·5 per cent. The eight mountain states almost exactly matched this rate but in the three Pacific coast states the increase was 74·8 per cent and in California just over 100 per cent. Already the westerner was wealthier than the average American and in the late 1920s motor vehicle ownership was already as high as 337 per 1000 of the population. As the Editor of the *Iron Trade Review* put it in 1928: 'In considering the industrial development of the Pacific Coast, Easterners should abandon all yard-sticks by which industry is measured in the East. The environments are so different and the entire economic structures have so little in common that it is futile to even attempt to make comparisons.'[1]

There were major western outlets for steel in primary industry—forestry and farming, the booming oilfields, and the canneries which served truck farming districts and fishing ports. Urban growth and road improvement led to a large demand for shapes and bars for structural purposes. Yet, in spite of prosperity, the western steel market was still small as compared with that of the north-east. Many manufacturing lines lagged behind its general growth, demand being supplied from the east. West coast demand for tinplate was put at 400,000 tons by 1927, 165,000 tons steel was needed for its oil wells, tanks and pipelines but structural shape consumption was no more than 100,000 tons. For this reason metal fabrication was poorly developed, so that steel consumption patterns were different from the nation's. Overall the whole area west of the Rockies consumed about 2 million tons finished steel a year, a little over 5 per cent of the national total. California dominated western consumptions. (Table 85) The pattern of its supply was changing. Ballast rate deliveries from Europe had long been an important factor, and considerable tonnages continued to come in over the tariff. Transcontinental railroads introduced concession rates to the Pacific to capture more of this trade and in the early twentieth century Chicago became an important source of supply. These concessionary rail rates survived in certain lines, as, for instance, in the very low 'tinplate rate' from the Middle West. The opening of Panama shifted the cost advantage in delivering to the Pacific coast. Increases in rail freight charges accentuated this, Chicago becoming still more disadvantaged in the west. By July 1920 the freight on steel to the Pacific coast was $0·70 per 100 lbs. from Sparrows Point, $1·05, from

Pittsburgh ($0·35 rail to Baltimore and $0·70 intracoastal) and $1·65 by rail from Chicago. In the depressed conditions of 1921 U.S. Steel started to supply the largest share of Pacific Coast demand from Pittsburgh plants.

Table 85.   *Pacific coast receipts of certain finished*
steel items 1919—1921
(thousand gross tons except for sheet)

		Receipts (average 1919—21)
Tubes:	California	153·1
	Washington and Oregon	20.2
Wire:	California	23·8
	Washington and Oregon	13·2
Tinplate:	California	83·7
	Washington and Oregon	32·6
*Sheets:**	California	2·3%
	Washington and Oregon	0·6%

*Percentage of national total.
Note: The Pacific Coast had no capacity for any of these grades.
Based on C.R. Daugherty, M.G. de Chazeau and S.S. Stratton, *The Economics of the Iron and Steel Industry*, pp.63—72.

Pittsburgh Plus pricing, and then the multiple basing point system after that, gave large 'phantom freight' to western producers. (Table 86) By the end of 1924 when Indiana Harbor and Evanston, Illinois were ruling base points for sales of pipe to California, their freight to that market was $20 a net ton.

Table 86.   *United States Steel Corporation and other deliveries*
of finished rolled steel[1] to
Pacific Coast states 1919 to 1921
(thousand tons)

	1919	*1920*	*1921*	*Average 1919—21*
Total receipts (all sources)	—	—	—	345·3
From Carnegie Steel	81·7	42·5	34·9	53·0
From Illinois Steel	194·8	114·5	11·4	106·9

[1] Excluding track accessories and rails.
Based on *I.A.* 19 Feb. 1925, and Daugherty, de Chazeau and Stratton, op. cit., p.62.

Yet in spite of this large price 'umbrella' and increases in demand, growth in western steel capacity was slow. By 1928 western works supplied about 500,000 tons of the rolled steel used west of the Rockies, foreign companies 120,000 tons while 1,500,000 to 2,000,000 tons came from eastern mills.

Firms east of the Mississippi enjoyed scale economies denied to the western producer, and, as over-capacity was one of the chief characteristics of the industry, they valued western consumption as an outlet especially in times of recession. In the west, there was not sufficient demand in any one line to fully engage a large, low-cost unit like a continuous mill for wire rod, billets or bars, let alone a wide strip mill. Overall the west had substantial consumption but for each product the market was 'thin'.[3] Further it was divided between three widely separated nodes of population and manu-facturing, Puget Sound, San Francisco Bay, and the Los Angeles basin. (Tables 87, 88) (Fig. 31) Proximity to the market was a considerable advantage for low-value products, so that the most characteristic line of west

Table 87.   *Steel ingot capacity in western states*
1920, 1930, 1938
(thousand gross tons)

	1920	1930	1938
California	374	760	758
Oregon	33	12	—
Washington	22	48	155
Utah	56	52	—

Based on A.I.S.I. *Works Directory.*

coast mills was bars. Supplying local engineering works and machine shops or delivering reinforcing bar direct to the local construction job the west coast firm had the advantage of speedy service as well as low cost. The big eastern firm could, however, obtain even this by warehousing steel in the main

Table 88.   *Population of main west coast centres 1920, 1940*
(thousands)

	1920	1940
Seattle, Tacoma, Portland	670	782
San Francisco, Oakland, Berkeley	779	1,022
Los Angeles and main suburban centres[1]	706	1,886

[1] Glendale, Santa Monica, Pasadena, Long Beach.
Based on census data.

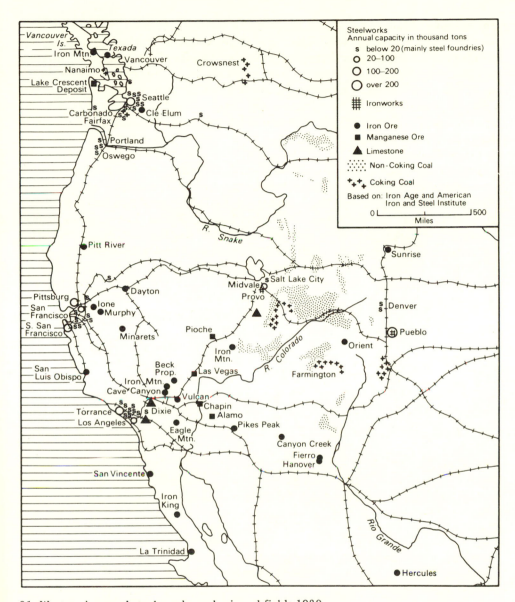

Steelworks
Annual capacity in thousand tons
- **s** below 20 (mainly steel foundries)
- ○ 20–100
- ○ 100–200
- ○ over 200
- ⌗ Ironworks
- ● Iron Ore
- ■ Manganese Ore
- ▲ Limestone
- ⋮⋮⋮ Non-Coking Coal
- ✛✛✛ Coking Coal

Based on: Iron Age and American
Iron and Steel Institute

0 ⊢————————————⊣ 500
Miles

31. Western iron and steel works and mineral fields 1930

centres of demand and in the mid-1920s Bethlehem set up warehouses in Seattle, Portland, San Francisco, and Los Angeles, and built several new freighters for the intracoastal trade at a total cost of $25 to $30 million.[4] On higher-value steels even high freights were little impediment to profitable sales by eastern mills. Tinplate, the chief finished product consumed in the west, illustrates this. In the three years 1919-21 average consumption in the three coastal states was 116,300 tons. By the end of the twenties western consumption as a whole was about 400,000 and by 1937 over 465,000 tons. Consumption in the three coastal states alone reached 382,000 tons in 1939.[5] Encouraged by growing demand, Columbia Steel put down black plate mills and tinning pots at Pittsburg, twenty-five miles north-east of Oakland. These had a tinplate capacity of 50,000 tons a year and benefited from the price umbrella of an eastern basing point. But tinplate stood transport well, and a little later cold reduced strip mill coil brought a better and cheaper product in the north-east. The 'thinness' in west coast consumption made a cold reduction mill inapplicable there, and the eastern producers gradually achieved complete control of the western market. In 1937 Pittsburg made only 37,300 tons of tinplate, less than one-tenth of the consumption in the three coastal states, and in the course of that year the mills were stopped, and later dismantled.

Partly because of small-scale operations, partly because of difficulties with respect to ore and coking coal all western steel-making was dependent on cold practice. Some Utah pig was used, especially at Pittsburg, but scrap was the main part of the charge. Scrap was cheap for supply far exceeded local demand—3·2 per cent of U.S. scrap was derived from Pacific Coast states in 1928 but they used only 0·9 per cent. Much of it was taken east in returning steel vessels, much was exported, largely to Japan. It was by now known that an acceptable coke could be made from Utah coal in the by-product coke oven, but carriage to the coast was costly. There were a number of iron ore deposits in the Sierra Nevada, in the southern part of the Great Basin and around Puget Sound, and others in western Canada, but many of them were inaccessible. The market, ore and coal situation suggested the possibility of a number of different types of iron-making location, but none were wholly satisfactory. Iron could be made with relatively short rail hauls in central Utah between the ore and coalfields of that state. If it was converted into steel at the same plant the problem of long haul to market had to be faced. Alternatively, Utah pig could be shipped to a plant on the coast. A tidewater plant would have heavy freight charges on minerals, and, because of the division of the market, high delivery charges to at least two of the three coastal centres of consumption. On the other hand, with cold metal plants, operations could be smaller, market and scrap orientated, and so successfully divided between all markets. In the thirties the largest and second-largest companies on the coast had plants in two and in all three centres respectively.

Two north-western furnace plants were projected in the early twenties. In 1920 the short-lived Western Rolling Mill Corporation announced plans for a works near the Bellingham hematite mines in the extreme north-west of Washington, using local and British Columbian ore and apparently intending to ship iron on to its Seattle steel plant. Four years later Pacific Coast Steel, having to buy east coast pig in spite of its large scrap charges, planned furnaces using foreign ores and Carbonado coking coal railed thirty miles to Seattle and thence by sea.[6] Although the company's melting capacity went up by 60,000 tons in the twenties it did not go ahead with this scheme. Perhaps the logistic difficulties of western smelting are shown best in the only successful project. In 1922 Columbia Steel, which had a 49,000 ton Pittsburg steel plant, acquired the Utah Coal and Coke Company and put up a blast furnace and small coking plant at Provo on Utah Lake. Ore was obtained from Iron Springs twelve miles west of Cedar City in the far south-west of the state and coal from Columbia mine, east of Price, in Carbon County, a rather roundabout journey of about 100 miles from the furnace. With a 1930 capacity of 175,000 tons of pig, Provo supplied the steelworks over 60 miles to the west. At the end of the twenties the biggest eastern firms first took a hand in western steel-making.

In autumn 1929 U.S. Steel acquired Columbia Steel's plants at Provo, Pittsburg, and at Torrance near Long Beach, the second largest Californian plant, which Columbia had bought in 1923. The arrival of the Corporation with its immense capital resources represented a long-term threat to other local firms and notably to Pacific Coast Steel which had a large plant in South San Francisco, another in Seattle, and control over the Southern California Iron and Steel Company of Huntington Park, Los Angeles—a total capacity of 380,000 tons. Bethlehem Steel's west coast outlets and its investment in new warehouses and extended steamer fleet was also threatened. A month after U.S. Steel Bethlehem too moved into western production by acquiring Pacific Coast Steel. For the next twelve years it seemed certain that the west as well as the south would be dominated by north-eastern firms.

By the end of the thirties overall western demand, though recovered from the depression, was still at very much the same level as ten years before. In some lines, notably merchant bars, western mills supplied over half the market, but in spite of large consumption in gas, oil, and domestic grades, no pipe or tube was made. Of 2·3 million tons of finished steel consumed in the seven western states local mills supplied only about 0·67 million tons. Of the 1·6 million tons from other districts an estimated 0·4 million was from Alabama, 0·3 million from Sparrows Point and 0·9 million from elsewhere, mostly the mid-west.[7] For a wider, eleven state area, including the large Pueblo plant of Colorado Fuel and Iron the position was as in Table 89.

The big eastern companies showed little readiness to think expansively after their arrival on the coast. The Corporation spend $20 million on

Columbia, and another $10 million of expenditure had been approved when war broke out, but ingot capacity at Torrance and Pittsburg increased only from 363,000 tons in 1930 to 441,000 in 1938.[8] Bethlehem did not expand

Table 89.   *Finished steel capacity and production in*
eleven western states 1937, 1938
(thousand net tons)

	Capacity 1938	Shipments by western mills 1937	Total shipments received 1937
Plates	25·2	28·8	215·3
Hot rolled sheet and strip	100·9	48·3	217·0
Cold rolled sheet and strip	—	—	28·3
Tinplate	38·9	37·3	465·6
Galvanised sheet	55·7	45·3	154·5
Heavy rails	336·0	186·6	265·6
Heavy shapes	99·3	49·1	164·1
Wire rods and bars	653·2	227·9	383·0
Pipes and tubes	—	—	431·9

Source: Senate Document 95, *Western Steel Plants and the Tinplate Industry* 1945, Appendix.

at all at Los Angeles or South San Francisco, while its Sparrows Point capacity went up 1·2 million tons or 69 per cent. U.S. Steel was now adopting a more realistic attitude to Birmingham marketing advantages so that its new Fairfield tinplate plant was allocated a large share of western orders. Even allowing for Republic's share of shipments from Alabama, U.S. Steel and Bethlehem appear to have shipped from the east quite as large a tonnage as all the plants in California rolled at this time. (Table 90) The general opinion was that western growth prospects were poor. In a paper to

Table 90.   *Estimated western steel supply 1940*
(thousand tons)

Western Production:	California	600
	Other states	300
Supply from East:	via rail	
	to 3 coastal states	686
	to rest of west	338
	via Panama	1,391
		3,315

Source: *I.A.* 21 Sept. 1944 p.67.

the Temporary National Economic Committee at the end of the period U.S. Steel representatives observed: 'Although it is an important steel consuming area, the West Coast cannot support more than limited steel-making capacity due to high assembly costs, particularly in the face of competition from Birmingham and Sparrows Point, both of which can serve this area on a more economical basis.' Early in 1940 Iron Age wrote, '. . . the iron ore situation on the Pacific Coast is not attractive and does not indicate the possibility of any large scale development.'[9] War changed the framework of thinking entirely.

NOTES

[1] 'From '49 to '29. A Review for the Western Metal Congress, Los Angeles, Jan. 1929', *I.T.R.*, 27 Dec. 1928.

[2] Ibid., p.1637.

[3] *Western Steel Plants and the Tinplate Industry*, Senate Document 95, 1945, p.3.

[4] *B.F.S.P.*, Oct. 1927.

[5] C.R. Daugherty, M.G. de Chazeau and S.S. Stratton. The Economics of the Iron and Steel Industry. 1937, p.65, *I.T.R.* 27 Dec. 1928, pp.1639—40. *Western Steel Plants and the Tinplate Industry*, p.29.

[6] *I.A.*, 1 July 1920, p.10, 27 Nov. 1924, p.1417.

[7] *I.A.*, 21 Mar. 1940, p.26., but see also *B.F.S.P.*, Jan. 1951, p.78.

[8] *I.A.*, 10 Aug. 1944, p.80.

[9] T.N.E.C., Monograph 42, *The Basing Point Problem*, 1939—1940. pp.111—12; *I.A.*, 7 Sept. 1944, p.65, 21 Mar. 1940, p.32.

CHAPTER 15

Second World War Steel Extensions

The Second World War was an interlude of great locational significance. There were important changes in raw materials and products; much more important, this was the period in which the west and south-west at last made the breakthrough into large-scale steel-making. As there was much excess steel capacity at the end of the thirties the industry was cautious in its attitudes to major defence expansion. Government planning for this began in summer 1940. By the following year there were suggestions of demand for 110 million tons of steel in 1942, over 25 million tons beyond current capacity. Between summer 1939 and spring 1941 steel capacity increased nearly 5 million tons. Between 1938 and 1945 the increase was 15·4 million tons or 19·2 per cent.

Table 91.   *Steel ingot capacity 1939, 1942, 1945 by process*
(million tons)

	Open Hearth	*Bessemer*	*Electrical*
1939	72·9	7·1	1·7
1942	78·1	6·7	3·7
1945	84·2	5·9	5·4

Based on *A.I.S.I.*

A number of conditions implied change in the geography of steel-making. One was strategic. In May 1941 President Roosevelt suggested that the possibility of an invasion of the eastern United States should be taken into account, and, to provide for this, some 5 million tons of steel capacity should be established in the far west. Much more important was the growth in consumption beyond the manufacturing belt. (Tables 92, 93, 94) In 1939 Pacific, Mountain, and West South Central States were responsible for only 10·3 per cent of the value added in American manufacturing; By 1947 12·4 per cent.

The shift to the west and south-west was much more important than that to the south-east. In spite of this most expansion was at existing north-eastern plants. The north-east had 93·2 per cent of U.S. steel ingot capacity at the beginning of 1940. It received only 74·3 per cent of the expansion over the next five years but this amounted to over 10 million tons, so that the area retained 90·5 per cent of U.S. capacity by 1945.

Table 92.   *Value added in Manufacturing divisions 1939, 1947*
(billion $)

	1939	1947
United States	24·5	74·3
New England	2·4	6·8
Mid-Atlantic	7·3	20·8
East north central	7·7	23·5
Manufacturing Belt	17·4	51·1
West north central	1·3	4·1
South Atlantic	2·2	6·9
East south central	0·8	2·9
South	3·0	9·8
West south central	0·8	3·0
Mountain	0·3	0·8
Pacific	1·5	5·5
West	2·6	9·3

Based on Bureau of the Census, *Census of Manufactures.*

Table 93.   *Estimated distribution of rail deliveries of*
iron and steel products 1935, 1940, 1945
(percentages)

	North-east and Middle West	West and South-west	South-east and South
1935	77·0	16·6	6·4
1940	76·9	14·7	8·4
1945	67·8	23·1	9·1

Source: Geo. S. Armstrong & Co. op. cit., (see above p.219), p.35.

Table 94.   *Estimated distribution of all finished steel*
consumption by states 1940, 1947
(percentage)

	1940	*1947*
Michigan	25·9	21·0
New York, Pennsylvania, Ohio	30·4	32·0
Illinois, Indiana	14·6	15.2
Texas	3·7	4·3
California	3·4	3·5

Based on *I.A.*, quoted by Geo. S. Armstrong & Co.

The north-east had a reserve of skilled labour and management, whereas elsewhere there was often a severe shortage of both. Expansion of existing plant was cheaper than building a new one—with services and ancillary plant already in being. Another point in favour of extending old plant was that, theoretically at any rate, it could be done more quickly. In practice it was not so simple. Some extensions involved a good deal of clearing before building, while new works were put up with unexpected speed. At Homestead works, Pittsburgh, strongly favoured by the Navy Department for armour plate extensions, demolition and clearance work began on 15 October 1941, but the first heat from the new open hearth furnaces was not tapped until 14 June 1943, and the slabbing and plate mills did not start work until early 1944. On the other hand, ground was broken at Fontana in April 1942, pig iron was first cast on 30 December, and steel was made the following May. A final factor favouring established centres was the reluctance of existing companies to expand well away from old customers; except when they were operating plant on behalf of the government the only example of this was Armco's construction of the Houston works, and even this was partly government owned. The general importance of existing plant in determining the location of expansion may be seen in relation to the rocketing demand for plate. In 1940 U.S. plate capacity was only 6 million tons, in part a reflection of the smallness of peace-time shipbuilding activities. In addition the strip mills had a light plate capacity of 1·3 million tons. It was estimated that wartime demand would be 12 million tons annually—a figure which increased still more later on. An expansion programme was drawn up which would contribute 3 million tons, of which just over half was at Homestead. The strip mills, which, except for Fairfield and Sparrows Point, were in the western half of the manufacturing belt, were converted to roll a higher percentage of plate, even of medium or heavy grade, and in a little over a year from May 1941 their output went up from 109,000 to 550,000 tons a month. Without this it was reckoned twelve to fifteen more new plate mills would have been needed. If located near to the coast, these would have had higher war-time production costs, and would have been either redundant or embarrassing new growth points in the post-war period. If placed in the Great Lakes area a peace-time use for this capacity would have been similarly difficult to find. The only two major east coast growth points were Bethlehem and Sparrows Point. A new plate mill considered at an early stage for Sparrows Point was instead later built at Geneva.[1] Alan Wood, and Lukens, though major plate-makers, had no steel capacity increases. A major project for government financing of additonal coke, iron, steel, and plate capacity at Alan Wood's Conshohocken works was cancelled.[2]

West of the Appalachians, the largest expansion was in the upper Ohio Valley, a development against the trend of the previous thirty years. (Table 95) Part explanation is to be found in the fact that over one-third of the Pittsburgh—Youngstown expansion, 2·2 million ingot tons, was at Homestead, Braddock, or Duquesne, all involved in the great Homestead plate mill development. Yet the increase in the Valleys, with its high pig costs and light steel emphasis, was greater even than in Allegheny County. In the Chicago area there were sizeable extensions at South works, Gary, and the two Indiana Harbor plants, but the biggest was the electrical steelworks, fully integrated with iron-making, which Republic Steel built for the Defense Plant Corporation at South Chicago. On the Erie shore there were important developments at Buffalo, but elsewhere a slight decline in rated capacity over the war period. The trend of expansion away from the Ohio Valley to the shores of the Great Lakes and to a smaller extent eastwards was temporarily halted. Outside the manufacturing belt existing plants were extended. At Ensley, Fairfield, and Gadsden, the three big Alabama works, capacity increased by 1 million tons. All the other important developments were in the west or south-west.

Table 95.   *Steel capacity by districts 1940, 1945*

	1940		1945	
	Capacity (million tons)	Percentage	Capacity (million tons)	Percentage
North Eastern—Buffalo	16·2	19·8	18·0	18·9
Pittsburgh—Youngstown	34·4	42·2	39·9	41·8
Cleveland—Detroit	7·9	9·7	7·7	8·1
Chicago—St. Louis—Duluth	17·5	21·5	20·7	21·7
Southern and Gulf	3·4	4·1	4·2	4·4
Western	2·2	2·7	4·9	5·1

Source: W.A. Hauck *Steel Expansion for War,* 1945, p.32.

New ore and coalfields were developed in the eastern half of the nation, but, more important still unprecedented demands were made on old sources. The national net increase in coke capacity was 13·4 million tons of which only 2·3 million were in the west and south-west. Pennsylvania, West Virginia, Virginia, and eastern Kentucky continued to supply most of the coking coal. In 1943 the surviving beehive ovens of the blighted Connellsville district, lit again, produced 7 million tons of coke—one-fifth of their record 1916 output and 12 per cent the total coke needs of the nation's blast furnaces. Extension of iron ore production east of the Appalachians was noteworthy. Republic Steel and Jones and Laughlin expanded in the Adirondacks, new mines were opened in the Clifton area of St. Lawrence County in 1942 and Alan Wood planned important developments of two mines until the Defense Plant Corporation withdrew support.[3] There was a built-in flexibility in Upper Lakes iron ore production, output being capable of rapid increase through the installation of more power shovels and railway facilities. As early as August 1941 a programme for sixteen extra ore carriers had been authorized. Because of this Mesabi bore the heaviest share of war demand, though Cuyuna, Menominee, and Marquette also reached record production levels in 1942–3. In addition to supplying demands in the Great Lakes area, Lake Superior had to meet a large part of eastern needs, for heavy losses to submarines caused the withdrawal of the Bethlehem ore fleet from the Chile run. (Table 96).

Table 96.   *Lake Range iron ore output 1937, 1942*
(million tons)

	1937 (Shipments)	1942 (Production)
Cuyuna	1·8	3·1(1943)
Marquette	5·6	6·6
Menominee	2·8	6·6
Mesabi	45·4	70·3

Based on American Iron Ore Association, *Iron Ore.*

WAR-TIME DEVELOPMENT IN THE WEST

United States shipbuilding both for merchant and naval needs was stimulated by the Merchant Marine Act of 1936; by 1940, with war in Europe, building rates accelerated. The west had important yards in each of the main population centres, with San Francisco Bay as the most important. Material came largely from the east, the eleven western states in 1937 shipping no more than 29,000 tons of plate while 186,000 tons came from outside. One of Bethlehem's four big shipyards was on the Bay and it now contemplated larger and more diversified coastal steel-making. Before the United States entered the war Bethlehem proposed a works of two blast furnaces and a steel plant at Los Angeles, where its steel capacity was little over half that of its South San Francisco plant. The War Department objected that tidewater operations were too vulnerable, and the government refused support. There were renewed considerations of Pacific north-west steel extension, but, though small, production there was nearer the level of consumption than in California, and, there does not seem to have been a dominating individual to push that region's case.[4]

In spring 1941 Henry Kaiser suggested construction of a number of small plants with combined ingot capacity of about 1·25 million tons and costing $150 million. Half of their finished steel would go to his own yards, already consuming almost 500,000 tons a year, the rest would be sold. The plants were to be in Utah, southern California, and the Bonneville Power Administration area of the north-west.[5] These would be cold metal plants which would therefore inflate west coast scrap prices. The lukewarmness of the industry to Kaiser's proposals appears to have been matched by that of the government. Yet a little later the government's consultant, W.A. Hauck, recognized that half the 1·1 million ton capacity expansion reckoned necessary on the Pacific Coast would have to be in new plant.[6] After Pearl Harbor activity rocketed, and with Germany also in the war the Gulf and Panama route for steel from the east soon proved insecure. Growth in western demand now required new western capacity, and the extent of that growth meant that it must be backed by iron-making. This raised once more questions of iron ore and coking coal supply, and a locational problem more complicated than that involved in the establishment of scrap-using melting shops.

The best coking coal was in central Utah, the biggest developed orefields those of Columbia Steel in south-west Utah. A nodal position in relation to the three major industrial agglomerations on the coast also suggested the economy of one major plant located there. This would be a tidier arrangement than Kaiser's three plants, two of which were market orientated. Utah too was safe from air attack whereas a coastal plant might not be. On the other hand, there was still little local consumption so that, in peace-time, a material-oriented plant would bear the heavy commercial disadvantage of a long haul to market. The earlier history of Utah steel-making provided no encouragement.

The Utah Steel Company had been formed in 1915 to build a steelworks at Midvale just south of Salt Lake City. By 1920 Midvale steel capacity was 50,000 tons, finished as bars. As with other western projects it had financial difficulties. Columbia Steel absorbed the mineral properties in 1922, but the steel plant passed to another concern and was abandoned in 1931. Columbia Steel appears to have contemplated steel-making as soon as it acquired interests there but was probably short of funds. In the mid-1930s U.S. Steel managers, now in control of Columbia, had strongly urged a Utah plant, but the Corporation would not entertain the idea, arguing that Birmingham could serve the west more cheaply.[7] Now U.S. Steel undertook to build and operate a Utah works for the government. Work began at Geneva on Utah Lake some ten miles north of Ironton in spring 1942. The following year the plant was at work. By the end of the war Geneva had cost $200 million and with Ironton and the Provo furnace was capable of 1·65 million tons of pig and 1·28 million tons of steel a year.

Kaiser meanwhile tried to secure government support for a tidewater works but, like Bethlehem, had this proposal vetoed on security grounds. In April 1942 the Kaiser interests broke ground at Fontana, San Bernardino County, forty-two miles from Los Angeles, backed with finance from the Reconstruction Finance Corporation. By 1945 Fontana was still a small plant as compared with Geneva, having a capacity of 380,000 tons of iron and 750,000 tons of steel. Their different balance of iron and steel tonnages witnessed both to the availability of shipyard scrap on the coast, and to the costliness of iron-making there. Meanwhile the older west coast cold metal plants expanded much less rapidly—by only 184,000 tons or 27·2 per cent in the case of the four Californian melting shops of the Corporation and of Bethlehem.

As the war approached its end, sober and generally unfavourable assessments were made of the commercial prospects of greatly expanded western capacity, and above all of the new plants. (Table 97) The general opinion was that Fontana would be less viable than Geneva. The next twenty years were to prove both assumptions incorrect.

Table 97.   *Finished steel demand and capacity in western*
states 1940 and 1945
(million tons)

		1940	*1945*
All finished Steel:	Demand	2·4	6·7
	Capacity	0·88	3·5
Ship plate:	Demand	0·20	3·0
	Capacity	0·00	1·025

Based on *I.A.* estimates.

THE SOUTH-WEST

By 1939 total estimated sales of finished steel in Texas were 1·46 million tons, but there were then only two south-western plants.[8] At Sand Springs near Tulsa, a small open hearth works was built in 1929, and in 1936 this was leased by the American Rolling Mill Company. At Fort Worth the small Texas Steel Company produced concrete reinforcing bars. Oklahoma and Texas together had a capacity of no more than 59,000 net tons. The rest of the supply came from outside via the Mississippi, by rail, or by sea. It seems likely that much of the steel sold in the south-west was 'dumped' thus providing little or no incentive to build up a new plant even though the market was large, and Texas each year produced about 600,000 tons of scrap.[9] Apart from tubular steel no one line of consumption was sufficiently large to support an integrated works (Table 98).

Table 98.   *Estimated consumption of finished steel in*
Texas, Oklahoma, Louisiana, and Arkansas 1939
(thousand tons)

Tubes and Pipes	650	Merchant Bars	60
Plate	125	Tinplate	100
Sheet	120	Wire Products	75
Structurals	100	Cotton ties	20
Reinforcing bar	50	Rails and fastenings	65

Source: *I.A.*, 28 June 1945, p.58.

The mineral endowment of the south-west is indifferent. In north-east Texas are large Eocene iron ore deposits of medium quality, reaching up to 42 to 44 per cent Fe. A coastal works near the biggest markets could have access to higher-grade imported ores, but would have a long haul for the coking coals of east central Oklahoma. A location near or on the east Texas orefield was better placed for material assembly. This area too was the centre of a large oil country goods demand, and well placed in relation to the much larger Gulf coast and mid-continent fields. In supplying mid-continent demand, such a location away from navigable water would be protected from the full force of competition from north-eastern or foreign suppliers of pipe.

In the First World War the Texas Steel Company drew up abortive plans for a plant at Beaumont, connected by ship canal with the Gulf of Mexico.[10] With the approach of the Second World War the coastal zone again received first attention, but by this time Houston had become the outstanding metropolitan area of the state. In January 1941, after heavy scrap exports from the Gulf Coast had stopped, the American Rolling Mill Company submitted plans for a new works on the Houston ship canal. Construction began in April 1941 and a year later the first steel was made. Houston made plate and structurals for war-time shipyards, bars, wire, and sheet. Production costs seem to have been high, for the works was small and its raw materials had to be railed long distances—coal, $3·40 a ton at the mine at McAlester and McCurtain, Oklahoma, cost $2·05 to carry to

Houston.[11] Assuming a 69 per cent coal/coke rate the assembly cost on coal per ton of pig iron would be about $2·97, a high figure for a plant which also had a long land haul on ore. On the other hand there was much local scrap so that Houston was equipped with only one blast furnace, its 1945 capacity being 274,000 tons of pig iron but 466,000 tons of steel. By 1948 most of the Houston works product was being sold either in Houston or along the Gulf lowlands, though 20 per cent went as far as Dallas and Fort Worth. By this time 25 per cent of its ore was Mexican, the rest was from north-east Texas. Coal came from Oklahoma but also from Arkansas, New Mexico, and Birmingham. All raw materials were still delivered by rail.

In April 1942 the Lone Star Steel Company was organized to build an integrated works for the government at Daingerfield on the orefields 150 miles east of Dallas. The ore was only six miles from the furnaces and was beneficiated on the way; freight from the Oklahoma coalmines was $1·20 a ton, 85 cents less than to Houston. The area was deficient in scrap and Daingerfield was equipped with a much larger furnace capacity—400,000 net tons a year. Capacity for 240,000 tons of Electric Resistance Weld (E.R.W.) pipe, 140,000 tons of plate and 40,000 tons of forging blooms was planned, but the scheme proved too ambitious, an illustration of the fallacy of thinking in assembly cost terms and neglecting a thorough study of marketing realities. Bigger northern plants could balance higher distribution by lower processing costs. Though shipping by water, they could secure a low delivery cost to coastal Texas as well. By the end of the war the Daingerfield blast furnace had not yet worked, the coke ovens operated 11 months then closed, and the steelworks and mills had not been built. Iron was first produced in 1947. The following year Lone Star Steel bought the plant and the ore and coalmines from the government. In June 1948 malleable iron production was begun, and in 1950 an 80,000 tons a year cast iron pipe foundry started work. Daingerfield had clearly been a less successful locational choice than Houston.

NOTES

[1] W.A. Hauck, *Steel Expansion for War*, 1945, p.23.

[2] H.R. Wood, *Alan Wood. A Century and a Half of Steel-making*, published by the Newcomen Society in North America, 1962, p.18.

[3] *B.F.S.P.*, Jan. 1946, pp.70–4.

[4] *I.A.*, 7 Sept. 1944, p.60; *B.F.S.P.*, Oct. 1936, p.911, Sept. 1941, p.1004; G.A. Kingston and F.B. Fulkerson, *The Pacific North West Steel Industry*, Bureau of Mines Information Circular, 8073, 1962, p.3.

[5] *I.A.*, 24 Apr. 1941, 1 May 1941, p.66.

[6] *I.A.*, 5 June 1941, p.81.

[7] *B.F.S.P.*, Aug. 1955, p.871; *I.A.*, 18 Jan. 1948, p.149.

[8] *I.A.*, 28 June 1945, p.58.

[9] Ibid.

[10] *B.F.S.P.*, Aug. 1918, p.347.

[11] *I.A.*, 28 June 1945, p.60.

CHAPTER 16

The Postwar Steel Industry
I. Materials, Technology, and Markets

Since the Second World War only two wholly new integrated steelworks have been built, and few big plants have been closed. Even so the framework within which the industry operates, its raw materials, processes, and markets have very largely changed. In response locational change has taken the less spectacular but still very important form of differential rates of growth between various districts and plants, new patterns of raw material assembly, and alterations in the source or destination of finished products. Technological change is conditioned by material availability and will not be undertaken unless market prospects are right; in short, materials, technology, and marketing are not independent variables. They must, however, be separately and systematically evaluated.

MATERIAL SUPPLY CHANGES

There has been relatively little change in the pattern of coking coal procurement for the iron industry. The northern part of the Appalachian plateau remains the chief source with Alabama, Utah, and Colorado as secondary suppliers. Until the mix-sixties Illinois coal was little used but after that became more significant and potentially of great importance. By the late sixties, with foreign demand taking large and increasing tonnages of the best Appalachian coking coals there were fears of long term shortages at home, and as supplies tightened a number of major U.S. companies began to sink new coking coalmines.

In tonnage terms U.S. 1971 shipments of iron ore were almost identical with those of 1954—76·9 million gross tons. However, the deliveries made in 1971 were of much more highly processed material with increased iron content and a superior physical form for blast furnace operations. Between the two dates imports of ore rose from 15·7 to 40·1 million tons. Although early postwar fears of an irresible flood of foreign ore, accompanied by a shift of steel-making to the seaboard, proved wide of the mark, imports have had very considerable significance. (Table 99) By 1968 a new study of world iron ore supplies was suggesting that 48 per cent of U.S. ore will be derived from overseas by 1980.[1]

Ore production in the south-eastern states has shrunk sharply since the mid-fifties and imports have increased—the Mobile customs district handled 1·4 million tons of foreign ore in 1962 but 4·8 million tons in 1970. In the Gulf south-west imports into the Galveston district grew to 400,000 tons by 1963 and doubled again to 1970. With major sales from California and to some extent from Nevada the west has been a net ore exporter, but by the early seventies the developing ore supply situation in the Pacific basin and particularly the growth of Australian production of high-grade ore seemed to point to future west coast imports. The ore situation in the Great Lakes and Ohio basin is complex.

Table 99.   *Sources of United States ore supply 1956,*
1971, 1975
(percentage)

	U.S.	Canada	South America	Other
1956	77	10	10	3
1971				
(actual)[1]	65·2	17·3	14·5	2·7
1975				
(forecast)	59	17·5	19·5	4

[1] Receipts at United States steel plants. These figures are not exactly comparable with those preceding.
Based on *I.A.*, 29 Nov. 1956, American Iron Ore Association, *Iron Ore* 1971.

The decline of direct shipping ores which had been the standard product of the Lake ranges for a century was compensated for by a very large investment in facilities to process the low-grade and hard 'taconite' ore. In the case of the twenty U.S. pellet plants the outlay amounted to over $2,000 million by 1971. A new inflexibility has been introduced into ore supply, for the capital outlay and therefore the standing charges per ton of annual capacity are so high that every effort must be made to operate the plant continuously. One 1955 estimate suggested that one ton of pellet capacity cost 8 to 10 times as much as 1 ton of open pit capacity and some 3 times that of an underground mine.[2] Within the sphere of movement of Great Lakes ore this expenditure has largely checked the influx of foreign ore. By the late sixties also steel companies were closing Upper Lake mines as their reserves of inferior grade non-pelletized ore gave out. (Fig. 32)

The Seaway introduced a new competitive factor into the ore supply situation west of the Appalachians, but its effectiveness was reduced by the relative smallness of the ore vessels used, by congestion on the Welland Ship Canal and by tolls. Even though foreign ore supplies to the Chicago district have increased greatly, when Canadian Great Lakes ore deliveries have been deducted these amount to only 10·4 per cent of the total and almost all of this is from Labrador. (Table 100) Rail transport of ore from the Upper Lakes is still relatively unimportant but has increased as large capacity unit trains have cut costs. By 1965 the Milwaukee Road was using 100 freight car

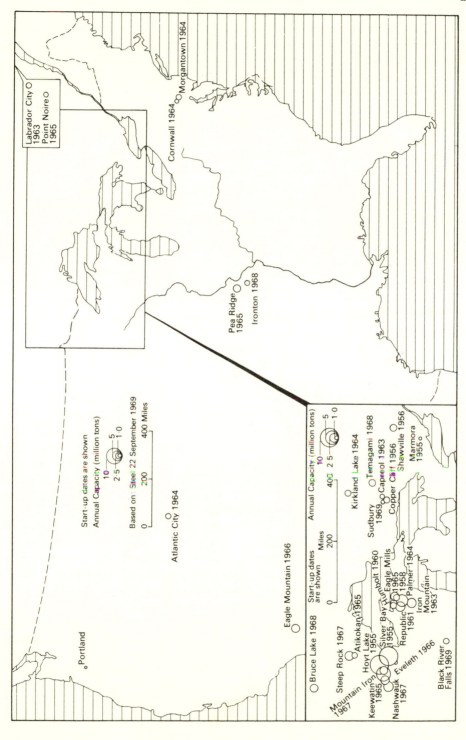

32. North American iron ore pellet capacity 1968

units to deliver ore from Republic, Michigan to Aliquippa on the upper Ohio, a rail journey of over 750 miles. More recently, and over a shorter distance, all Inland Steel's deliveries of ore from the Black River Falls operations in Wisconsin are being delivered 270 miles to Indiana Harbor by rail. The cost of operating a pipeline to carry 10 million tons of pellets annually from Mesabi to Chicago has been put at $3·97 a ton as compared with $4·08 by rail and water and $5·60 for all rail carriage. An advantage of such an installation is that not only would it provide for a one-third growth in Chicago area iron ore demand, but it would also permit all-year ore delivery so avoiding the need for the characteristic Mid-West steelworks stockpile of approximately six months' supply of ore by the beginning of December.[3] However, after many years of inactivity there has recently been new lake ore carrier construction so that costs of all-water shipment will also fall. If this fall in costs is substantial the advantages of the lake-shore centres will be increased.

Table 100.   *The ore supply for Illinois and Indiana*
1963, 1965, 1971
(consumption millions gross tons)

	1963	1965	1971
Total:	22·2	26·0	25·8
U.S. Great Lakes	19·3	20·4	19·7
Southern and Western[1]	0·2	0·2	1·7
Canadian Great Lakes	1·7	1·9	1·6
North-east Canada	0·4	2·9	2·6
Other foreign ore	0·6	0·6	0·1

[1] Includes Missouri.
Based on American Iron Ore Association, *Iron Ore*, 1963, 1965, 1971.

Generally the growth of ore imports and the reduced cost of this ore has increased the relative attractiveness of coastal locations, and has there focused advantage at very large deepwater sites. There is no sign that this situation will change. It is true that new methods of shipping ore as slurry, easing loading and unloading, are reported capable of cutting costs of handling by as much as 90 per cent, but this seems most likely to benefit smaller coastal works, leaving the giant operations dependent on more conventional ore carrier and handling methods.

Changes in ore supply have above all involved shipment of higher iron content material. This helps to stabilize the present pattern of iron- and steel-making by reducing the need for new iron-making plant—beneficiation leading to increased furnace output was reckoned to have saved Inland Steel alone some $60 million between 1947 and 1962 which would otherwise have had to go on new blast furnaces[4] —and by producing a higher-value, more transportable material. On the other hand, within this broad framework fuel economy has fostered the relatively greater growth of capacity at ore transhipment points than near to the older, coal-based iron centres.

IRON AND STEEL SCRAP

In the late fifties investment in pellet plants, and, in the sixties, expansion of oxygen steel capacity, emphasizing high-percentage molten iron charges, reduced the consumption of purchased scrap. By 1964 the industry was using 15·1 per cent more iron than in 1957 but bought 6·2 per cent less scrap. When they made 115 million tons of steel in 1956 the steel firms bought some 39 million tons of scrap; their 1969 output was 141 million tons of steel but their scrap purchases were only 37 million tons. The result was a near collapse in scrap prices for a number of years, a collapse most pronounced in the former high cost scrap areas. This low price was of especial value to smaller, unintegrated concerns with no capital tied up in ore mines, ore preparation plants, coke ovens and blast furnaces, and the reduction of price in the Pittsburgh area helped counteract some of the saving in fuel costs which had favoured the lake shore centres.

Scrap supply will remain at a high level—by the late sixties some 6 million automobiles alone were abandoned every year. Demand is likely to increase again as electric steel-making expands rapidly and as techniques for using a higher percentage of scrap in converter practice are adopted. Purchases one-quarter more than those of 1970 have been projected for 1980.[5]

TECHNICAL CHANGE. IRON ORE PREPARATION AND IRON-MAKING

Pellet plants, using electricity as their source of energy, and sinter plants employing coke breeze, powdered coal and gas have very much improved the grade of material supplied to blast furnaces. Direct shipping ores have decreased in importance and there has been a decline in the apparent consumption of 'ore' per ton of iron. (Table 101) In the fifteen years to 1971 average consumption of iron ore per ton of iron fell by 10·5 per cent limestone consumption by 36 per cent and the coke rate 28 per cent. In short, in spite of the higher-grade shipments of ferrous material, fuel economy has been even greater and 'ore orientation' has increased. This is yet another factor strongly favouring iron-making operations on the seaboard or lake shore as against those in interior locations. Changes in carbonization technology have so far been less significant but blending techniques have made possible a larger use of inferior coking coals and therefore again lessened the locational advantages of proximity to the prime coking coals of the mid Appalachians.

In early postwar years it seemed that major technical change would occur in iron-making. Many furnaces were equipped to operate with so-called high top pressure and some to use an oxygen enriched blast. Output was increased and there was fuel economy. The weakening of coalfield orientation was less than that which resulted either from the more highly prepared furnace burdens discussed above or from fuel injections. The injection of liquid, gaseous, or powdered hydrocarbons into the lower part of the blast furnace to supplement the work of coke passed from the experimental to the commercial stage in the late fifties. No other hydrocarbon could replace the essential mechanical function of tough, fibrous metallurgical coke in supporting the high column of the blast furnace burden, but the coke rate was cut as they took over some of its chemical function as a reducing agent.

Table 101.   *Great Lakes ore shipments 1959, 1965, 1971*
(million gross tons)

	1959	1965	1971
Direct shipping ore	11·9	11·8	2·0
Screened ore	11·5	3·7	2·3
Concentrates	11·7	22·5	14·1
Agglomerates (including pellets)	9·3	26·6	43·3
Total	44·4	64·7	61·8

Based on American Iron Ore Association, *Iron Ore.*

By the end of 1963 sixty-seven of the 134 active blast furnaces in the United States and Canada were equipped for some form of fuel injection. It was reckoned that the coke rate at Armco's Ashland furnace was reduced by 28 per cent as a result. The general decline in the coke rate was the product of a complex of factors—better furnace burdens, fuel injection, and the operation of fewer, bigger, and more efficient blast furnaces. In 1955 0·7 gross tons of coke was used on average for every net ton of pig iron, in 1962 0·62 tons and in 1971 0·56 tons.

Another development, potentially rather than so far of great importance, has been the introduction of direct reduction processes. In these a steadily moving body or agitated mass of iron ore is reduced in a hydrogen or carbon monoxide atmosphere to produce a crude iron which is suitable either for further processing in a blast furnace—with very great increase of productivity consequent on the decline in the amount of work which has to be done—or more important, for direct use in steel-making. By the late fifties it was reckoned that $10 to $20 million was being spent annually on research or pilot plants for direct reduction, and at Conshohocken a commercial scale plant was already at work, though it did not prove successful. It was suggested at this time that direct reduction employing natural gas might eventually bring the smelting industry to the mid-continent and Gulf natural gas fields, though the existence of very large established capacity in the north, pipeline transport of gas to that area, and concentration of consumption there very much moderated even if it did not wholly cancel out the force of this argument.

In the sixties development work on 'pre-reduced' or 'metallized' pellets of 90 to 95 per cent Fe content went on rapidly. Again there will be no sudden massive change in the structure of the industry but the prospects of locations near the gas fields, orefields, or at any rate well away from prime coking coals are being improved. There have been installations in the Pacific north-west, in the Gulf south-west, and even more significant a big installation in the Orinoco ore district, effecitvely a radical relocation of primary metallurgical activities outside the national borders. It has been estimated that direct reduction plant and electric furnaces to convert the metallized pellets into steel can be built at no more than 40 per cent of the cost of equivalent blast furnace and oxygen steel-making capacity—though this capital economy declines with increasing scale of operation. Not only is it cheaper but it is more suitable for continuous operation, removes the price uncertainties of the scrap market and, of increasing importance, does away with the pollution associated with coke ovens. By 1980 the United States may be using as much as 10 million tons of metallized pellets, or a tonnage more than one-tenth its maximum annual pig iron output to date.[6]

STEEL-MAKING

Changes in steel technology have been more fundamental and important than those in iron-making. They have been associated above all with the introduction and very large-scale adoption of oxygen steel-making processes. The expansion of electrical steel production has also been of great importance. In 1957 101·6 million out of 112·7 million tons of steel made in the U.S.A. was produced in open hearth furnaces. In 1964, when steel production again passed the 1957 level, only 98·1 million out of 127 million tons was made in open hearth furnaces. Electrical steel production in the period went up by 58 per cent and output of basic oxygen converters from 0·61 million to 15·4 million tons. By 1970 open hearth steel production was down to 48 million tons, oxygen steel had increased to 63·3 million, and 20·2 million tons was produced in electric furnaces. In the first half of 1972 basic oxygen converters made 55·6 per cent of U.S. steel, open hearth furnaces 26·5 per cent, and electric furnaces 17·9 per cent.

The new basic oxygen processes originated in Europe as the outcome of searches for ways of tackling raw materials for which conventional processes were unsuitable. The key to commercial success lay in a sharp reduction in the cost of large supplies of oxygen—tonnage oxygen as it is called. In the mid and late 1950s it was gradually realized that oxygen steel-making made possible large economy in capital outlay, increased productivity, and improved quality of product as compared with the open hearth furnace. Steel scrap usage was, however, little more than that arising in the associated rolling mills, so that the new process led to a decline in scrap consumption as considered above. By the early sixties it was reckoned that conversion costs were $8 to $9 a ton as compared with melting and refining costs of $10 to $15 in most American melting shops. Not surprisingly there took place a wholesale restructuring of steel capacity.

The locational effects of the switch to oxygen steel-making are complex. On the one hand, by lowering capital costs, the new process provides a convenient break-in point for new producers and areas. Two examples of this involved the backward integration of Acme Steel of Chicago and McLouth Steel of Detroit. Even more important, Bethlehem Steel installed oxygen furnaces from the start at Burns Harbor. The smaller new steel-makers usually make their steel in electric furnaces. Much more important, oxygen steel-making has provided the enterprising, established company with an opportunity to cut its costs. In making its major expansion at Sparrows Point in the late fifties Bethlehem Steel judged that oxygen steel processes had not yet fully proved themselves. The open hearth capacity which it then installed gave it both higher standing charges and higher process costs than with oxygen steel-making. Conversely Jones and Laughlin was enterprising in deciding early to install oxygen furnaces at its Aliquippa and Cleveland works, where by early 1962 it had 36 per cent of the nation's oxygen steel capacity. Not only were operating costs cut—the investment at Aliquippa was reckoned to be only 37 per cent that for open hearth capacity—but problems of space in the crowded Cuyahoga valley were reduced as well, for converter shops make possible a great space economy as compared with melting shops.

For half a century electric arc furnaces have dominated quality steel production. Since the mid-fifties a big increase in electric furnace size has made it economic to turn out large tonnages of carbon steels as well. Electrical steel output was only 5 per cent of the national total in 1947 but 15·3 per cent in 1970, and is rapidly increasing. Expansion of oxygen steel-making, with its reduced scrap consumption, has made the electric furnace an ideal companion process. By the late 1960s a few major companies, including McLouth and Granite City, were making all their steel in either oxygen or electric furnaces. In other cases old open hearth furnaces have been replaced by electrical furnaces in plants indifferently located for assembly of iron-making materials, as at Steelton or Butler. The development of metallized pellet production has made available a new material which may be charged directly into the electric furnace, and plants to use these have already been built in the Columbia basin, at New Orleans, and in the south Atlantic states.

Another development of great potential importance is continuous casting of steel for the finishing mills, a process which cuts out the ingot-soaking pit and blooming mill operations and again economizes in both space and capital outlay. In operation it reduces costs for heating and intraworks transfer, increases yields by reducing the scrap rate, and improves steel quality by speeding up process times and by avoiding defects on a large number of ingots. Continuous casting provides a cheap way for small- or medium-scale plants to enter production. Combined with electrical furnaces, and therefore scrap dependent, it can be largely market oriented, theoretically at any rate a great advantage as industrialization spreads and costs of supplying distant markets from existing works increase. Some existing small works have been converted to continuous casting, a considerable number of other units have been built at new works. This has fostered the growth of so-called 'mini mills', small—averaging some 100,000 tons—,serving local markets for relatively unsophisticated products, highly flexible in their operations, and giving prompt service. Between 1945 and 1967 about 20 mini-mills were built. A number of them have been in locations well away from the established steel centres but even some within the manufacturing belt have been notably successful. Roblin Steel of Tonawanda near Buffalo began production in 1959 and by 1975 expects to operate 400,000 tons capacity. North Star Steel of Minneapolis, which serves upper mid-western markets not only with merchant and reinforcing bars, but also with higher-grade products, has recently doubled its capacity to 300,000 tons. Continuous casting has also been installed in very large units at large integrated works since the mid-1960s.

Together oxygen steel-making and continuous casting have substantially cut operating costs. The converters at Bethlehem produced steel for $4 to $5 a ton less than the open hearth furnaces they replaced, and the large-capacity continuous caster at Gary has been reckoned to cut costs by $12 to $15 a ton.[7] Although in this fashion these developments perpetuate old locations, generally the technical advances of the last few years seem likely to lessen the structure of location factors and to favour a wider range of plant types and a wider scatter of locations. The situation is however also affected by changes in markets and in the capital cost of plant.

MARKETING DEVELOPMENTS

Over the postwar years the automotive industries have continued to provide both far and away the chief single market for steel and the one in which demand has grown most. Although there has been some de-centralization this has meant that the centre of demand has remained firmly in the western half of the manufacturing belt. (Fig. 33) General Motors alone consumed over 12 million tons of steel annually by the late 1960s, and, although it has built stamping plants from Los Angeles to Massachusetts to supply regional assembly plants, by far the largest proportion of its consumption is in Ohio, Michigan, Indiana, and Illinois. The prospect is for continuing growth in this industrial sector and in this area of the country.

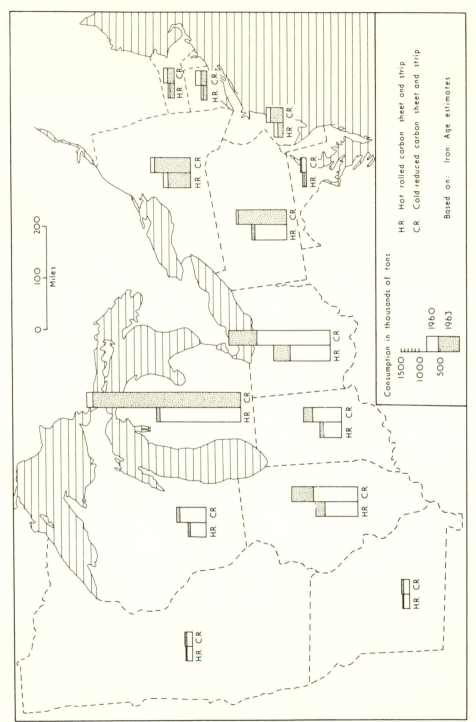

33. Estimated consumption of sheet and strip in the Manufacturing Belt 1960 and 1963

Half U.S. deliveries of cold reduced sheet are made to the automobile industry and it is estimated that national demand for this product will go up from 16 to 25 million tons in the 1970s.[8] Within the western manufacturing belt Michigan remains the leading single state in steel use but its share of the total has fallen. Especially noticeable has been the growth in demand in the districts within the tributary area of Chicago mills.

Marketing routes have changed. The most spectacular single development has been the opening of the St. Lawrence Seaway and thereby of direct foreign competition into the chief American steel markets. This penetration began even before the Seaway was opened with low quality imports, notably wire products, but subsequently the movement greatly increased and moved into higher grades of steel.

In national marketing the adoption of F.O.B. pricing in 1948 emphasized the advantages of nearness to markets. Certain regional price differentials remained to distort this picture. There was a slight east coast differential above the prices of most of the manufacturing belt and similar ones in St. Louis, but far more important was the west coast differential of $12 to $14 a ton which gave a fillip to western mill earnings until it was removed in 1962, a price reduction which however failed to shut out eastern or foreign deliveries.

Rising rail freight rates and the related switch of more and more traffic to truck transport has similarly emphasized the advantages of plants with regional markets, though the meaning of 'regional' must be widely interpreted, for studies suggest that, within the manufacturing belt at least, truck transport is more important than rail transport for distances in excess of 500 miles. It is however *selectively* more important, being especially suited for delivery of flat rolled steels for the consumption goods industries where reliable and speedy delivery, care in handling, yet economy in packaging are all important. Even as early as 1953 73·2 per cent of deliveries of sheet and strip in Eastern Territory were made by truck, but for plate, pipes and tubes, and structural shapes and piling the proportion was only 20·3 per cent.[9]

After losing steel traffic rapidly in early postwar years railroads have subsequently adopted more imaginative policies both in technical and organizational arrangements and in pricing policies. High capacity, specially designed rail cars for particular types of steel are an instance of the first, unit train operation of the second. These and new pricing policies have gained some notable successes such as contracts for slab delivery from Lackawanna to Burns Harbor on Lake Michigan or coil from Cleveland to Hennepin on the Illinois river in the mid-sixties. Railroad transport as always provides a means of breaking out of regional into national marketing. An example with a highly original twist is the 1972 Kaiser contract to supply General Motors with 300,000 tons of coil annually. This is to be shipped from Fontana to Hennepin in unit trains at a freight charge which reputedly will be $19 a ton as compared with the normal Chicago to West Coast freight of $58·40.[10]

PLANT COST AND LOCATIONAL CHANGE

In any past period changes in raw material supply and technologies on the scale of the postwar period would have been accompanied by major locational shifts in steel production. Such movements it is true become less easy with the emergence of massive units, necessarily more difficult to abandon than the small furnaces and mills of the nineteenth century, but this does not in itself prevent the building of wholly new plants. New technologies have assisted existing works to obtain increased output from already installed equipment at low additional cost. For instance, pellets were first used by Inland Steel in 1964, but by 1970 provided three-quarters of total ore supply. This was the chief factor increasing iron production at Inland's Indiana Harbor works by 43 per cent between 1964 and 1969. The largest Inland Steel blast furnace was designed for 1,000 tons of molten iron a day; by 1972 it was capable of 2,500 to 3,000 tons daily.[11] The overwhelming incentive to push such extensions of existing equipment to the limit has been provided by an unprecedented increase in the capital cost of plant and particularly a widening of the differences between the cost of extending existing works and that of building a wholly new one.

By 1947 it was estimated that the 80 million tons of prewar capacity had probably cost of the order of $59 a ton, that the 15 million tons added in the war cost $166 a ton and that the current cost was $250 a ton. Nine years later it was reckoned that a new integrated works on a greenfield site would generally cost $300 a ton, or three times the prewar level, but that extensions to plant could be obtained for as little as $100, or, if major extensions to finishing mills or melting shops were needed, the cost might creep up to $200 a ton. As a specific example Republic Steel reckoned in 1956 that its 1·8 million tons expansion programme would have cost $325 a ton if wholly new plants and raw material supplies had to be developed, but that as a rounding out programme the capital outlay could be cut to $85 a ton—but a further stage of rounding out might cost $200.[12] In spite of such cost-reducing innovations as oxygen converters in place of open hearth furnaces or continuous casting units for primary mills, investment costs have continued to rise and especially rapidly for new plants. By 1970 Bethlehem Steel's Burns Harbor works, only the second wholly new integrated works built since the Second World War, had cost $1,000 million.

Sometimes the problem of high capital cost as an impediment to expansion has been eased by loans from major consumers. This has benefited some firms supplying autobody sheet, and at a critical stage Luken's steel plate extensions were helped by General Electric loans.[13] Government approval for accelerated depreciation has also been of assistance in some cases, as with the Trenton steelworks built by U.S. Steel in the national emergency of the Korean War, but refusal of fast write-off concessions was a major factor causing the failure of two other schemes for new integrated works in the mid-fifties, one on the Delaware estuary, the other at Houston.[14]

Another way for a company to achieve a change in location without bearing the exorbitant cost of a wholly new works has involved mergers and new plant interlinkages. Examples of the first are the acquisitions which marked a Colorado Fuel and Iron attempt to penetrate eastern markets in the late forties and the fifties. On the whole the attempt failed. More important was Bethlehem Steel's determination to gain access to the Chicago area by acquiring Youngstown Sheet and Tube, an effort which like the first, made in 1930, was defeated. New company associations and semi-product movements have improved material supply to Acme Steel in Chicago and Detroit Steel from large operations acquired by them along the middle Ohio. Bigger links with bulk rail movement between plants have facilitated the penetration of the Chicago area by three outsiders since 1959.

The rigidity of the steel location patterns since World War Two may eventually soften. It should however be remembered that since the mid-fifties the industry has reinforced the strength of inertia represented by its existing works capacity by immense investments in ore mining and preparation facilities. Locational change will continue but it will not be precipitate. Postwar changes in the various steel districts and current developments alike indicate that, beneath the level of broad developments in raw materials, technologies, and marketing, a multitude of locational influences is at work.

NOTES

[1] *I.A.*, 18 Apr. 1968, p.82.
[2] *I.S.Eng.*, Jan. 1955, pp.117−18.
[3] Ibid., Jan. 1972, p.D-47.
[4] *B.F.S.P.*, Feb. 1962, p.145.
[5] *Industry Week*, 2 Feb. 1970.
[6] Ibid.; *I.S.Eng.*, Jan. 1972, pp.D-28, 29.
[7] *I.A.*, 3 July 1969, p.9; *Steel*, 28 Apr. 1969.
[8] *I.S.Eng.*, Jan. 1972, pp.80−2.
[9] J.C. Nelson, *Railroad Transportation and Public Policy*, 1959, p.49; E.E. Jones, 'Economics of Alternative Methods of Distributing Flat Rolled Steel', *I.S.Eng.*, Jan. 1964.
[10] *A.M.M.*, 1 Mar. 1972.
[11] *A.M.M.*, 14 Aug. 1969; Inland Steel, *First Quarter Report*, 1972, p.9.
[12] *Time*, 18 Aug. 1947, p.36; T. Diamond, 'This New Round of Steel Expansion,' *Harvard Business Review*, May−June 1956, pp.89−90; *New York Times*, 23 May, 1956, p.39.
[13] *New York Times*, 30 Dec. 1956.
[14] Ibid., 18 Dec. 1956, 29 Aug. 1956.

CHAPTER 17

The Postwar Steel Industry
II. Districts beyond the Manufacturing Belt

THE UPPER LAKES

The drift of technology in postwar years was in the main to the detriment of
the Upper Lakes, with beneficiation of ore, higher freight rates, and a switch
to road transport of steel. Even fuel economy, an absolute gain, lessened the
relative advantage of a low lake rate on coal to Duluth as compared with
points on the lower lakes. Use of large electric furnaces and continuous
casting plant has made theoretically possible a solution of the two main
weaknesses of the district, a long haul on coal and a market too scattered to
justify the large central operations which integrated iron- and steel-making
entails. Not surprisingly, the biggest market in the area, that of the
Minneapolis—St. Paul metropolitan area, has been chosen as the site of the
main new 'mini-mill', that of the North Star Steel Company.[1]

Interlake Iron Company had lakeside furnaces and coke ovens at Erie,
Toledo, Chicago, and Duluth. Even before the recession of the late 1950s it
was clear from allocation of new investments that it assessed the prospects of
the Duluth plant as poorer than those of all the others and especially of
Chicago. In the summer of 1960 its Duluth plant was closed with the
explanation that high production costs were there combined with an
inadequate local market. Two years later the plant was abandoned.

In the early postwar period of steel shortages there were extensions at
U.S. Steel's Duluth works but it remained small. Though it continued to
make wire and reinforcing bars it largely became an operating appendage of
Chicago area plants. By 1947 half the billets rolled into wire rod at American
Steel and Wire's Joliet mill were brought from Duluth, and already by the
mid-fifties half its output went down the lake as ingots or semi-finished
steel.[2] Duluth interests enthusiastically favoured the St. Lawrence Seaway,
but, even before the deepwater route was opened, foreign competition was
especially keen in wire and nails, and the importation of these and other
products into the Great Lakes basin subsequently went up still more. By the
early 1970s only 20 per cent of Duluth's steel went to the finishing mills
there, the rest was sent out for finishing. As steel shortage gave way to
surplus, and semi-finished steel became available from bigger, lower-cost, and
nearer works this became unacceptable. By 1971 Duluth Works had a steel
capacity of 800,000 tons, but it was '. . . one of the oldest and in many
respects the most marginal of the Corporation's existing plants'. In spring
1972 closure of the Duluth coke ovens, ironworks, steel plant, and blooming

and billet mills was announced. The rod, bar, and wire mills will continue, working up semi-finished steel brought from other U.S. Steel plants.[3] As a result of the liberal tax policies which Minnesota voters endorsed in the Taconite Amendment vote of November 1964 there followed a very large extension investment in ore beneficiation plants. Here would seem to be the employment, the long-term assurance of continued iron ore production, and the status which must now satisfy the state.

In spite of its great mineral wealth and unrivalled water route to the heart of the manufacturing belt, the Upper Lakes region retains most of the deficiencies which caused the first failures and bankruptcies in its iron trade 100 years ago. It is not the place to make iron or steel. The ore and coal movements on the Lakes which helped raise the great metal complexes of the Ohio Valley and on the southern edge of lakes Erie and Michigan have left only the cold hearths of old charcoal furnaces, the overgrown wrecks of more ambitious projects and disappointed hopes along the shores of Lake Superior.

THE POST WAR WEST

Early in 1945 the economic journalist C.H. Grattan, visiting the Pacific Coast, summed up his impressions on western steel. 'The Geneva plant, with additional equipment, can provide most of the needed steel for the western industry, and Fontana will provide a vital if supplementary role . . . It is slyly suggested, as it is so often in the East, that Kaiser is a creature of government spending, and that it is yet to be proved that he can make out on a big scale in a normal market.' A 'zealous and unprejudiced reporter' from *Iron Age* at the same time had yet to find '. . . a single substantial consumer of mill products, other than the Kaiser organisation itself, which seems to have present confidence that Fontana can or will produce and sell its products postwar to compete with intracoastal shipments from eastern mills or with older tidewater coast mills producing from low-priced scrap.'[4] There seemed justice for these opinions. They were, however, to prove wrong. In 1945 Colorado, Utah, and California made 3 per cent of U.S. Steel; in 1970 5·5 per cent.

A formidable reconstruction programme lay ahead of the western works when the war ended. Together in 1945 Fontana and Geneva had a plate capacity almost five times the total consumption of the eleven western states in 1937, and shipbuilding demand fell very rapidly as soon as war ended.[5] (Table 102) Investment had been high and costs of conversion to peace-time production would push outlay still higher—to an estimated $175 per ton of annual capacity for Geneva. (Table 103) Competition from other areas threatened the high level operations which under these circumstances alone could make western integrated works profitable. Towards the end of the war Pacific Coast mill prices were usually eastern prices plus the $11 a ton water freight from the Atlantic seaboard. Until 1944 the rail freight from Geneva to the Pacific was $12 a ton. (Table 104) Fontana was better placed in this respect, but had high rail freights on ore and coal, so that 1944 pig iron production costs there were $24·50 a ton as compared with $17·42 at

Table 102.   *Mill capacity of Geneva and Fontana works 1945*
(thousand tons)

	Geneva	*Fontana*
Plates	700	300
Structural shapes	250	210
Merchant bars	—	180

Based on A.I.S.I. *Works Directory.*

Table 103.   *Investment in plant per ingot ton 1943–1944*
(Dollars)

Geneva	148·44
Fontana	125·00
United States Steel Corporation	47·6
Bethlehem Steel Corporation	50·76
Republic Steel Corporation	40·48

The Company figures are for 1943, the figures for Fontana and Geneva for summer 1944.
Source: *I.A.* 7 Sept. 1944, p.62.

Geneva. With high-percentage pig iron charges steel-making at Fontana seemed likely to be uncompetitive; with high scrap usage costs could be brought more in line. (Table 105) There was, however, no assurance that coast scrap prices would remain advantageous. One by one the problems of plant conversion, of high capital charges, of marketing and assembly costs were tackled. (Tables 106, 107)

In February 1945 Benjamin Fairless of U.S. Steel informed a Salt Lake City meeting that his company would bid for Geneva and also for Fontana if the Reconstruction Finance Corporation no longer wanted it. In April R.F.C. engaged consultants to study west coast conditions, on the basis of which the government would decide a reasonable selling price.[6] A year later bids were invited for Geneva. Most of the seven received were from non-steel-making firms, and all but one asked for large government loans for plant conversion. Colorado Fuel and Iron proposed to lease the works, paying a $2 a ton fee on every ton of steel made and with an option to buy if the tenancy proved the plant was profitable. U.S. Steel made an outright offer of $47 million and committed itself to spend $43·5 million on new plant, $18·6 million of it at Geneva and the rest on a cold reduction plant at

Table 104. *Estimated reasonable freight charges on tinplate to Pacific coast markets 1944 (dollars per ton)*

From:	To: San Francisco	Los Angeles	Seattle	Portland
Chicago[1]	15·40	15·40	15·40	15·40
Pittsburgh[2]	14·60	14·60	14·60	14·60
Birmingham[3]	12·00	12·00	12·00	12·00
Sparrows Point[3]	8·60	8·60	8·60	8·60
Geneva[1]	6·00	6·00	8·00	8·00
Fontana	4·40[1]	1·10[1]	8·00[2]	8·00[2]

[1] all rail
[2] rail and water
[3] all water
Note: At this time neither Geneva nor Fontana made tinplate.
Source: Senate Document 95, *Western Steel Plants and the Tinplate Industry*, 1945.

Table 105. *Averages prices of No.1 heavy melting scrap 1939, 1944, 1945 ($ per gross ton)*

	Los Angeles	Pittsburgh	Chicago	Eastern Pennsylvania
1939	13·66	15·00	13·38	15·44
1944	16·45	20·00	18·75	18·75
1945 (Mar.)	14·50—15·50	20·00	18·75	18·75

Los Angeles figures are annual. For 1939 and 1944 the others are June figures.
Based on *I.A.* figures.

Pittsburg to finish 386,000 tons of coil a year shipped from Utah. Though it was below a quarter of construction costs the U.S. Steel bid was accepted. Annual interest and depreciation charges per ton of capacity were thereby cut from perhaps $15 per ton to about $6—$7.[7]

The Corporation improved Geneva's marketing position by adding strip mill stands to the plate mill. With 200,000 tons of Geneva structural capacity, and also a still large plate output to dispose of the Steel

Table 106.   *Estimated material costs of pig iron at*
Geneva, Fontana, and eastern plants 1944
(dollars per ton)

	Iron ore	Coking coal	Flux	Total
Geneva:				
At mine	1·85	3·52	0·45	10·76
Freight	2·78	2·16		
Fontana:				
At mine	1·85	3·52	0·45	15·83
Freight	3·05	6·96		
Birmingham:				
At mine	4·45	3·76	0·30	11·40
Freight	1·25	1·64		
Sparrows Point:				
At mine	2·22	2·51	0·45	14·19
Freight	5·26	3·75		
Pittsburgh:				
At mine	4·80	2·77	0·45	14·21
Freight	5·88	0·31		
Gary:				
At mine	4·80	2:80	0·40	15·56
Freight	3·52	3·99		

Source: J. R. Mahoney 'The Western Steel Industry', summarized *I.A.* 21 Sept. 1944.

Corporation felt the lack of west coast fabricating yards. In 1948 it bought the large structural and plate fabricating business of the Consolidated Western Steel Corporation.

The chief remaining Geneva problem was to secure reduced freight rates to the coast. (Table 108) Some success was achieved, but it is on this point that the wisdom of the Geneva purchase eventually proved most questionable. The local market was insignificant; in 1946 less than 0·2 per cent of national rolled steel consumption was in Utah. Eight years later the figure was about 0·29 per cent. The figures for the three coastal states were 4·9 and 4·7 per cent respectively.[8] To the east Pueblo dominated the central Great Plains market. In 1944 two western railroads agreed to reduce the

Geneva to Pacific rate from $12 a ton to $8 for the rest of the war and for six months thereafter. Until March 1947 the formal rate to Los Angeles and San Francisco was $15 but this was never charged on large shipments, and

Table 107.   *Estimated raw material costs of ingot steel 1945*
(dollars per ton)

| | | Type of furnace charge | |
	All pig	*50% scrap*	*75% scrap*
Geneva	10·76	12·97	14·08
Fontana	15·83	14·62	14·01
Birmingham	11·40	13·29	14·24
Sparrows Point	14·19	15·47	16·11
Pittsburgh	14·21	16·04	16·95
Gary	15·56	16·15	16·45

Note: Scrap prices are assumed to be $17 per gross ton at Geneva and $15 at Fontana.
Source: Senate Document 95, *Western Steel Plants and the Tinplate Industry*, 1945, p.19.

Table 108.   *Rail freights on steel to Los Angeles and*
San Francisco 1952, 1953
($ per ton)

From:	*Los Angeles*	*San Francisco*
Geneva	14·49	14·49
Fontana	1·82	7·00
Chicago	34·04	34·04
Pittsburgh	38·41	38·41

Source: Federal Reserve Bank of San Francisco *Monthly Review*, Sept. 1953, p.113.

that month the rate was cut to $9·60.[9] The situation thereafter remained one in which Geneva could succeed so long as western demand was booming, and while Fontana costs and prices were high. The initiative in cost reduction lay with Kaiser. The opportunities there were seized with such readiness that growth outmatched Geneva's. (Table 109, 110)

In 1947 Fontana carried much heavier overheads than Geneva, and was owned by a company with no steel-making experience. Through 1946 and 1947 Kaiser tried and failed to induce the Reconstruction Finance

Table 109.   *Ingot capacity of main western plants 1945, 1953, 1960*
(thousand net tons)

	1945	1953	1960
Geneva	1,283	1,675	2,300
Fontana	750	1,536	2,933
Bethlehem Steel	662	900	1,000
U.S. Steel cold metal plants	603	605	617
Pueblo	1,272	1,485	1,800

Based on *A.I.S.I.*

Table 110.   *Blooming and slabbing mill capacity, Geneva and Fontana 1954, 1957, 1964, 1970*
(thousand tons)

	1954	1957	1964	1970
Geneva	1,778	1,856	2,194	2,194
Fontana	1,300	1,423	3,931	4,368

Based on A.I.S.I. *Works Directories*

Corporation to write off part of its $123·3 million loan as representing the inflated cost of war construction. However it was allowed to use war-time shipbuilding profits, which otherwise would have been heavily taxed, to help pay R.F.C. Fontana market advantages were never in doubt. Los Angeles had become one of the biggest steel-consuming areas of the United States, by 1954 taking 77 per cent more steel than the San Francisco–Oakland industrial area. In 1956 some 65 per cent of Fontana steel was sold within sixty miles or so of the plant.[10] High production costs were now tackled along two main fronts: substitution of richer or nearer materials for poorer, distant ones, and installation of better plant.

The planners of Fontana had considered use of Washington or British Columbian coking coal but decided ash content was too high, and that the fields were too far away. Instead Kaiser leased the Sunnyside mines in Carbon County, Utah, less than five miles north of the Columbia mine which had supplied Ironton for thirty years. In order to make good metallurgical coke, Oklahoma low-volatile coal, carried 1,500 miles, had to be mixed with Sunnyside coal.[11] In the late forties two-thirds of the delivered price of Oklahoma coal at Fontana represented freight charges; even in the case of Utah coal the cost at the mine of about $4 a ton was increased by a further $5·09 a ton because of the long haul. Kaiser thoroughly modernized the Sunnyside mines. Crushing, blending, and other beneficiation processes reduced ash and sulphur levels.[12] At Raton, New Mexico, in the same coal basin from which Pueblo obtains its coking coal, a 500,000 acre coal property was acquired in 1955. Although superior to Sunnyside in most respects apart from ash, Raton is one-third further away from Fontana, 920 miles by the Santa Fe. Fuel economy through better burden preparation and more injection of hydrocarbons subsequently reduced still more the disadvantages of remoteness of coal supplies. By 1969 delivery costs were being cut as the western railroads introduced unit trains with 100 ton capacity wagons carrying 8,400 tons on each trip from Sunnyside to Fontana. Kaiser's Crow's Nest, British Columbia, coal operations, though mainly designed to supply Japan, will also make deliveries to Fontana.

Reduction of assembly costs for raw materials was most successful with iron ore. The Vulcan mine near Kelso on the Union Pacific Railroad in the northern Mojave desert had been the main wartime source, but large tonnages of Utah ore were also used. One hundred miles further south, on the ill-defined divide between the Colorado and Mojave deserts was the fault block ore body of Eagle Mountain, held, undeveloped, until 1943 by the Southern Pacific. In 1943 Kaiser bought title to it, but he was not able to overcome the restraint of another lease until 1946.[13] Eagle Mountain ore was little over 50 per cent iron. By the end of 1948 the costly fifty-three mile truck transport from the pit to railhead on the Southern Pacific at Ferrum on Salton Sea had been replaced by a company-owned spur railroad, and the superintendent of raw materials was already claiming delivered ore costs less than half those at Lake Erie ports.[14] Later costs were cut further by sintering at the works and ore beneficiation at the mine, and then in 1958 transport economies were pushed further when the Southern Pacific began to replace its 65 ton ore cars with 600 new ones of 100 tons capacity.[15] By the early 1960s Kaiser was shipping large tonnages of ore to Japan and the enlarged mine operations which this involved justified installation of a pellet plant at Eagle Mountain.

West coast scrap prices have risen so that Kaiser was wise not to follow the advice of those who suggested high scrap charges. (Table 111) The scrap rate by the early sixties was below the national rate, largely because of Kaiser's early adoption of oxygen steel-making. By 1964 whereas only 12·2 per cent of national output was from oxygen converters for Kaiser the proportion was 43·3 per cent.

Rolling mill operations continue to suffer from the relative smallness of western consumption in any one line, and from a rather unusual balance of demand. Mill runs are still too small to give unit costs comparable with those of the more specialized eastern mills; '. . . with the size of orders we get, we roll 100 tons of steel, and then have to change the rolls, while the eastern

Table 111.   *Average prices of No.1 heavy melting scrap*
1945, 1955, 1961
($ per ton)

	1945 (March)	1955 (December)	1961 (December)
Los Angeles	15·00	42·00	33·50
Pittsburgh	20·00	53·50	37·00
Chicago	18·75	51·00	36·00
Philadelphia	18·75	55·00	34·00

Based on *I.A.* figures.

companies roll 1,000 tons before they have to change theirs' was how an executive Vice-President of Kaiser Steel put it. Manufacturing consumption of steel in the west is relatively much less important, and use in construction more important than in the east. By 1963 40 per cent of the steel made at Fontana was ultimately used in construction. The structural mill, threatened by the loss of its shipbuilding market at the end of the war, has benefited. Plate shares the same outlet and also supplies large-diameter pipe works. Until 1958 the plate mill also served as the roughing plant for a hot strip mill, whose product was finished by cold rolled strip, sheet, and tinplate mills, but by 1958 the western market had sufficiently matured for a completely separate 86 inch hot strip mill to be built. Even so Californian steel consumption is disproportionately small in thin flat-rolled steel. (Table 112) Tinplate bulks much larger than in the east, and with an annual consumption of over 1·1 million tons of tinplate the seven western states took well over one-fifth the national total in the mid-sixties. Even so large tonnages of tinplate still come from the east. Consumption is concentrated in northern California, and the Pittsburg plant of the Steel Corporation has a better location to supply it than Fontana. The sheet and strip market on the other hand is concentrated in southern California, but in this field total western consumption is small, in relation to complex products. Eastern motor firms find it advantageous to assemble vehicles in the west, but most of their material comes from the east as crated knockdowns, and therefore is made from eastern steel. In 1964 it was estimated that only 6 per cent of the 900,000 tons of steel in the west coast's annual assembly of automobiles was from western mills, and Kaiser reckoned then that combined consumption of automobile and machinery makers of strip mill steel was only about 60,000 tons a year.[16]

Table 112.   *Estimated consumption of hot and cold rolled plate, bars, sheet, and strip 1963 (thousand net tons)*

	Plates	Hot rolled bars	Hot rolled sheet and strip	Cold rolled sheet and strip
California	380	218	400	607
Connecticut and Massachusetts	234	196	246	356
New York and New Jersey	555	375	828	1,456
Pennsylvania	819	368	834	1,240
Ohio	619	469	1,484	2,541
Michigan	369	895	2,086	3,685
Illinois and Indiana	828	920	1,637	2,654

Based on *I.A.* 26 Dec. 1963.

The general growth in the southern Californian market has been much more rapid than that of any other section of the west, and Kaiser's position there is unassailable. Between 1959 and 1970 steel mill product consumption there rose from 42 per cent of the consumption in the seven western states to 45 per cent.[17] (Tables, 113, 114)

Perhaps partly because of the dominance of Fontana in the south, the Steel Corporation concentrated its Pacific Coast expansion at Pittsburg. (Fig. 34) In 1948—9 a site for sheet mills served by semi-finished steel from Geneva was chosen at Torrance Works in the Los Angeles area. By 1955 it had capacity for 55,000 tons of sheet, but by 1957 the sheet plant had been discontinued, and the nominal capacity of the other mills there was written down slightly over the period 1954—64, while the finishing mills, though not the primary ones, at Pittsburg were largely increased. On the other hand, Bethlehem, having no western integrated works, chose throughout the postwar period to concentrate expansion in ingot and primary mill capacity

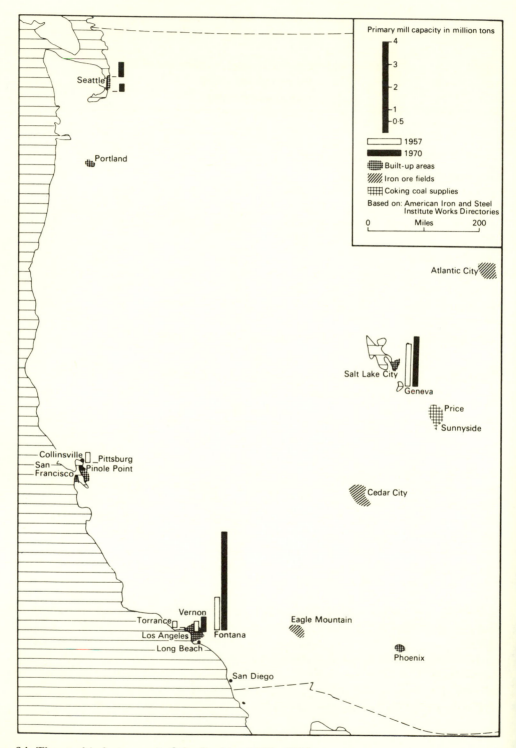

34. The steel industry west of the Rockies 1957–1970

in Los Angeles. In 1958 it replaced its Seattle open hearth furnaces with two large electric furnaces, and in the spring of 1962 decided to close the

Table 113.   *Western markets for tinplate, sheet, and strip*
1959, 1964, 1970
(thousand tons)

	Total seven western states	Southern California	Northern California	Oregon and Washington	Arizona, Utah, Nevada, Idaho
Tin mill products					
1959	1,120	280	660	160	20
1964	1,169	311	756	85	17
1970	1,394	380	869	135	10
Hot and cold rolled sheet and strip					
1959	957	607	222	95	33
1964	1,128	757	242	100	29
1970	1,462	978	283	142	59

Based on Kaiser Steel, *The Western Steel Market* (annual).

Table 114.   *Consumption of all finished rolled steel in*
western markets 1952, 1959, 1964, 1970
(thousand tons)

	1952	*1959*	*1964*	*1970*
Total seven western states	5,415	6,420	7,500	8,728
Southern California	2,227	2,642	3,405	3,940
Northern California	2,106	2,097	2,350	2,649
Oregon and Washington	794	1,031	1,218	1,449
Arizona, Utah, Nevada, Idaho	288	650	527	690

Based on Kaiser Steel, *The Western Steel Market* (annual).

276,000 ton melting shop and billet mill at South San Francisco and to supply the finishing and fabricating plant there from Los Angeles. (Table 115)

In 1958, at a time when it was investing $214 million in new plant, Kaiser Steel forecast a west coast steel consumption of 8 million tons in 1960 and almost 10 million by 1965.[18] There was in fact a sharp decline, and the highest level reached in the sixties was the 9·5 million tons of 1968. After

large losses in 1959 and a smaller one in 1962 Kaiser had already seen the
need for action to increase its share of the western market. The situation was
aggravated by a number of factors. One was the continuing importance of

Table 115.   *Mill capacity of Bethlehem Steel Corporation*
Pacific coast mills 1957, 1970
(combined annual rolling capacity,
thousand net tons)

	South San Francisco	Los Angeles	Seattle
Billets:			
1957	252	430	216
1970	–	475	–
Structurals, bars,			
rods, and plate:			
1957	234	392	216
1970	423	430	340

Based on A.I.S.I. *Works Directories* 1957, 1970.

eastern steel in the western market. In 1959 eastern mills supplied 33 per
cent of the steel market in the seven western states and this was a year in
which Kaiser lost $7·4 million. Even more troublesome was the growth of
imports of steel, especially from Japan. In the first half of 1961 imports into
Los Angeles were 133,000 tons, but in the whole of 1962 over 550,000,
while over these two years Kaiser sales fell by $32 million. In October 1962
Kaiser attempted to improve its competitive position by reducing prices an
average of $12 a ton. This more or less cancelled out the west coast
differential which was equal to the cost of carriage from eastern ports. On
Kaiser's 1961 sales of 1·4 million tons this differential represented a bonus
of $16·8 millions at the mill, or $4 million more than net earnings. By the
reduction Kaiser hoped to exclude the eastern and foreign mills and to
stimulate metal-working in the west so as to recoup the loss of this bonus.
The net loss of 1962 was turned into a large surplus in 1963 and a greater
one still in 1964, but this seems to have been due more to a general trade
revival than to a permanent improvement of Kaiser's position. Imports and
eastern shipments remained large, and competitors were stimulated to
retaliatory action.

Imports of steel mill products at Pacific Coast ports rose from 0·94
million net tons in 1962 to 1·32 million in 1964, and in the case of Los
Angeles, where the competitive advantage to be gained should have been
greatest, went up even more rapidly from 550,000 to 830,000. In 1964

Bethlehem Steel replaced the ten Liberty ships which it had used in intra-coastal trade with six new vessels to provide a greatly improved service, carrying steel and general cargo westwards and west coast lumber eastwards. In spite of this, eastern producers have not retained their share of the Pacific states market, foreign steel having taken over their higher percentage of deliveries in the course of the sixties. (Table 116)

Table 116.   *Steel consumption and supply in seven*
western states 1959, 1963, 1970
(million tons)

| | | Tonnage Supplied: | | |
	Tonnage Consumed	From Western Mills	From Eastern Mills	By Imports
1959	6·2	3·4(55%)	2·0(33%)	0·8(12%)
1963	7·0	4·0(57%)	2·0(27%)	1·0(16%)
1970	8·7	4·8(55%)	1·5(17%)	2·4(28%)

Based on Kaiser Steel, *The Western Steel Market* (annual).

Like Kaiser, the Steel Corporation has had to strive to improve plant efficiency and in this case high process costs—in part due to continuing dependence on open hearth furnaces—and long hauls to market increased the difficulties.[19] A coal preparation plant was installed near Wellington, Utah, in 1959–60, and in 1962 the first iron ore pellets were shipped from a new mine at Atlantic City on the extreme northern edge of the Wyoming basin. The ore is 30 per cent iron, but beneficiated to 67 per cent can well stand the 355 miles rail journey; by the summer of 1964 it already amounted to 45 per cent of Geneva's ore. New cold reduction mills, tinplate, and galvanizing units were built at Pittsburg, and in 1964 U.S. Steel undertook mill improvements at Geneva. By 1972 it was decided to modernise the 132 inch sheet and plate mill which serves regional markets for hot rolled products and the Pittsburg cold reduction mill.[20] These developments commit the Corporation to long-term operations at Geneva, but eventually solution to the problems posed by Kaiser by increases in west coast business, and by Japanese steel may involve new integrated operations at tidewater. Oregon and Washington provide a relatively small market—only consuming 19·3 per cent as much steel as California in 1960 and 21·9 per cent in 1970. Kaiser has a dominating position in the south, so that the San Francisco Bay area seems the most suitable location for new developments. There were rumours of a Steel Corporation works of at least one blast furnace and of 2 million tons ingot capacity at Pittsburg in the boom of the mid-1950s. The Corporation was then acquiring more land near to the existing works and until 1960 disputing Kaiser's claim to ore bodies on the fringe of Eagle Mountain.[21] In the sixties U.S. Steel systematically surveyed for ore in west-central Nevada and proved considerable tonnages in Lyon County. If a Pittsburg integrated works is built Geneva will presumably have to force its way into the Great Plains market, from which Chicago area plants would withdraw.

At the end of 1963 Bethlehem acquired 800 acres of tideland and over 1,000 acres of higher ground at Pinole Point, Richmond, on the northern part of San Francisco Bay, about 25 miles west of Pittsburgh and well located for service by the Santa Fe and the Southern Pacific, and for access by vessels. In 1964 it built structural steel fabricating shops, and in January 1965 announced a $20 million continuous galvanizing line to be supplied with coils from Sparrows Point.[22] Eventual full backward integration at Pinole Point was expected, though, as Bethlehem lacked western coal and ore reserves, it seemed probable that the Steel Corporation would have integrated operations at Pittsburg first. However, by 1972 the keenness of competition from imported steel led Bethlehem to the decision not to build iron or steel works at Pinole Point but instead to sell off all the site except those parts with already operating plant. In 1967 National Steel acquired a large site at Collinsville near the mouth of the Sacramento River and with a thirty foot channel to the Pacific fifty miles away. Early construction on this site is not expected.

Japanese steel is now a major factor on the west coast, its share of that market being twice its proportion of United States steel consumption generally. Until the early seventies the west was an important supplier of minerals to Japanese works, California and Nevada shipping 3 to 4 million tons iron ore annually. However, falling iron ore prices gradually made this business less attractive, and in December 1971 Kaiser cancelled the export contracts on Eagle Mountain ore and diverted it to its own use. Further developments in mineral traffic flows in the Pacific basin open new perspectives of raw material supply and of plant development.

The U.S. Steel Corporation has shown keen interest in Australian mineral supplies which might obviously be relevant in eventual backward integration at Pittsburg. Kaiser already has major investment in the huge iron ore deposits of Pilbara in the north of western Australia, and in British Columbian coal—indeed by 1969–70 Australian and Canadian mineral operations provided half Kaiser revenues. By 1970 Kaiser engineers completed a long-term feasibility study for a second Pacific tidewater plant. A significant pointer to the long-distance source of minerals was that water deep enough for vessels of at least 100,000 d.w.t. was reckoned essential. Puget Sound and San Francisco Bay were considered.[23] The former is well located for British Columbian coal, but the consumption of steel in the whole of Washington and Oregon is only half that of northern Carlifornia and one-third that of southern California. It seems probable that Kaiser, twenty-five years ago expected to succumb in postwar steel competition, will eventually build the first integrated steel plant on San Francisco Bay.

THE POSTWAR SOUTH

Since the Second World War the economy of the South has noticeably matured. Between 1947 and 1962 average per capita personal income in the states of Alabama, Florida, Georgia, Louisiana, Mississippi, and Tennessee rose from 66·7 to 73·3 per cent of the U.S. average. By 1967 incomes in these states ranged from 60·1 per cent of the U.S. average in Mississippi to 95·5 per cent in Florida. A wider group of southern states* had 15·1 per cent of the employees in U.S. manufacturing industry in 1947 but 20·3 per cent by 1967. For the same years their share of value added in U.S. Manufacturing went up from 13·8 to 18·6 per cent. In still more recent years the movement of industry southwards seems to have accelerated as firms seek out more amenable labour. However, much of this growth has not been of steel-using trades, as, for instance, in the spectacular expansion of textile and apparel manufacture in the Appalachian Piedmont. In some lines of manufacture which elsewhere use large tonnages of steel the South is generally poorly represented, as with automobiles or shipbuilding, but on the other hand it has the largest market in the nation for oil and gas country goods.

In 1946, consumption of steel by metal-working plants in Tennessee, North Carolina, Georgia, Florida, and Alabama together seems to have been only about 76·6 per cent that in Connecticut. In the first half of 1952 consumption of carbon steels in Alabama, Florida, the Carolinas, Georgia, Mississippi, Tennessee, and Arkansas amounted to about 3·56 per cent the national total.[24] The 1954 Census of Manufactures listed data for rolled steel consumption for forty-seven state economic areas: only three southern areas had consumption of over 200,000 tons. Houston, Birmingham, and New Orleans together then consumed 1·3 million tons of steel. By 1963 only Texas and Florida were among the 16 leading steel-using states.

A dominant theme since the Second World War has been the spread of southern steel-making to the south-west (Table 117) Figures of primary mill capacity, though a good deal less precise than those for crude steel capacity indicate an increase of about 11 per cent in the south-east between 1957 and 1970, but in the south-west a rise of over 60 per cent. This movement is largely a response to the growth of Gulf south-west demand, but it also reflects changing raw material supply conditions.

Demand for steel in the Texas Gulf coast area grew especially rapidly in the war and early postwar years—in the 1947—54 period growth in

---

*Virginia, North Carolina, South Carolina, Georgia, Florida, Tennessee, Alabama, Mississippi, Arkansas, Louisiana, Oklahoma, and Texas.

consumption in Texas was about twice that in the rest of the South, but thereafter little greater. This area had inferior mineral resources for steel-making, but its big coastal urban and industrial zone was well located for

Table 117.   *Southern steel ingot capacity 1938, 1945, 1960*
(million tons)

	1938	1945	1960
Alabama	2·5	3·5	5·5
Tennessee, Georgia, Florida, Virginia, Mississippi	0·1	0·2	0·5
Texas and Oklahoma	0·05	0·6	2·7

Based on *A.I.S.I.*

access to foreign raw materials. In the south-east, with a slower-growing, more widely spread demand, and greater existing capacity, the mineral supply situation was deteriorating.

The ore used by U.S. Steel's T.C.I. division in the mid-fifties was of about 37 per cent iron content. As a result furnace productivity was low—the average for Alabama in 1955 being 79 per cent of the national figure, though this level was depressed in part by the existence of a number of foundry ironworks. Subsequently northern furnaces turned to taconite pellets or high-grade imported ore and the less efficient furnaces there were closed. By 1964 Alabama furnace output averaged no more than 63 per cent of the national level. The richer Alabama ores have been worked out, upgrading and pelletizing of the lower-grade material has proved impracticable and the ore is phosphoric at a time when the emphasis on basic oxygen steel-making puts a premium on low phosphorus ores. Alabama ironworks have, necessarily, had to switch to a larger use of imported ores.

As late as 1957 the Woodward Iron Company made its foundry iron almost wholly from Alabama red or brown ores; by 1967 it operated only two Alabama mines and was also using ore from Missouri, Venezuela, and even India. Republic Steel produced 800,000 tons of Alabama ore in 1957 but by the end of the sixties had no active southern mines. U.S. Steel supplied its southern works from seven active Red Mountain mines with an annual ore-producing capacity of 5·6 million tons in the mid-fifties. By 1964 it decided to abandon these mines and become wholly dependent on imported ore.

Although the coke rate remains high by national standards, fuel economy, taken along with a new dependence on foreign ore, and a spread of steel demand, suggests that the ideal location for large-scale south-eastern steel-making is on The Gulf coast, probably near Mobile. In 1972 U.S. Steel's big

Orinoco metallized pellet plant came into operation, and this represented, in effect, a major relocation of primary capacity from the south-east, though initially a major market will be the electric furnaces of the Steel Corporation's new Texas works. The Birmingham area will remain a major centre of production but its growth will be noticeably slower than that of steel nationally. Its once superb resource endowment is no longer an asset.

The extension of electrical steel-making, low scrap prices, the arrival of continuous casting, and growth of demand for such unsophisticated mill products as reinforcing bar have encouraged the development of a number of much smaller outlying steel centres in the south-east. There were some of these already. Atlantic Steel began rerolling Pittsburgh bars into hoops and ties at Atlanta as early as 1901, and four years later had its own melting shop. By the mid-fifties its melting shop charge was scrap to the extent of 75 per cent and by the mid-sixties the 300,000 tons capacity there was wholly electrical. When the conversion from open hearth furnaces to electrical furnaces was decided it was reckoned that ingot costs would be cut from $66·5 to $60·8 a ton. The south-east had only one electric steelworks in 1938, but by 1970 nine unintegrated steel plants were at work south and east of Kentucky (Fig. 35).

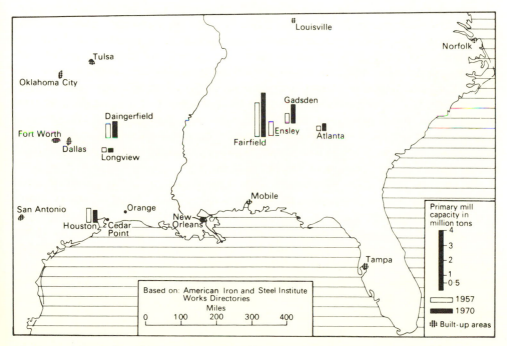

35. The steel industry of the south 1957—1970

THE SOUTH-WEST

In the 1940s the situation of the steel industry in the south-west was changed by the development of a local pipe and tube industry. Demand for tubular products had grown rapidly in the interwar years, and national capacity went up from 3·6 million tons 1915 to 8·3 million by 1935. The centre of consumption shifted south-westwards, but production remained heavily concentrated in the north-east. Demand for so-called standard pipe, used for plumbing, naturally broadly reflects the pattern of population. Consumption of mechanical tubing, used in fabrication, and of boiler and pressure vessel tubing is concentrated in the manufacturing belt. By the late 1940s these classes constituted half the demand; the rest was from the oil and gas industries. Line pipe is widely consumed, but the biggest demand is in the south-west where the pipeline network is closest. Demand for oil country goods—drilling and casing pipe—is wholly localized in oil- and gas-producing areas. By 1946—7 70 per cent of oil country goods were consumed in Texas, Oklahoma, Kansas, Louisiana, Arkansas, and New Mexico.[25] At that time the south-west produced no pipe. With the abolition of basing point pricing and rising rail freights, supplies sometimes proved difficult in that region. In 1949 Lone Star Steel of Daingerfield was awarded a Reconstruction Finance Corporation loan to build a steelworks and mills to produce line and casing pipe. In reaction to these conditions some north-eastern firms began to make larger use of water shipment, but the Lone Star project also sparked off south-western developments by northern pipe manufacturers.

U.S. Steel followed Lone Star in 1949 by announcing a 100,000 ton pipe mill for a deepwater site at Orange, Texas, to be supplied with steel from its northern mills. By 1954 Orange capacity was 350,000 tons, and Birmingham later joined Pittsburgh and Chicago as its supplier of semi-finished steel. From 1927 to 1949 the A.O. Smith Corporation, the pioneer of large-diameter electric weld pipes, supplied the south-west from its Milwaukee works. Jointly with Armco it now built a works at Houston to make E.R.W. pipe from Armco plate. In 1956 Jones and Laughlin applied for fast write-off concessions on a proposed $250 million integrated works on the Houston ship canal whose finished product would be tubes. The concession was not granted and Jones and Laughlin did not build the mill. However by 1957 Texas had an E.R.W. tube capacity of 1·3 million tons. Another, later tube plant was built at Houston by Tex Tube which was acquired in 1964 by Detroit Steel, whose Portsmouth, Ohio, works had supplied its semi-finished steel by water.

Pipe and tube mill demand for plate and strip, and general Gulf coast economic growth encouraged extension of steel capacity and entry to new lines of production. In the mid-sixties Armco committed itself to build cold reduced sheet and galvanized sheet mills there with, later, the construction of the south-west's first strip mill. By 1971 it had installed the area's first wide beam mill. It acquired a site twice as big as that of the existing works for expansion. In 1966 U.S. Steel began to develop a green field site at Cedar Point 35 miles east of Houston. By 1972 scrap-using electric furnaces and continuous casting units there supplied 60 per cent of the slabs used for a wide plate mill, the rest being brought in from the Corporation's other works. Cedar Point will take over the supply of plate to the Orange pipe mills sixty miles further east. When fully integrated Cedar Point will cover 4,000 acres but another 10,000 are earmarked for possible development by customers.[26]

For raw material assembly the coastal area in the neighbourhood of Houston is becoming steadily more attractive. Port facilities have been substantially improved, and the area is well located to handle Venezuelan or African ores. U.S. Steel's Orinoco metallized pellet plant is the ideal partner for the Cedar Point electric furnaces. Armco uses east Texas ores as well as imported, richer ores and by 1972 was building its own metallized pellet plant at Houston. Lone Star Steel, with a local market and sub-regional mineral supplies, is now seen to be in a less suitable location. As with Birmingham the value of its mineral resource base is depreciating. By 1970 it was still almost wholly dependent on the low-grade local ores but a few years after rationalization of its Oklahoma coal mines in 1963 it decided to abandon mining operations and purchase its coal.

SOUTHERN MARKETS, THE ECONOMIES OF SCALE, AND FOREIGN SUPPLIES

In spite of the growth of southern consumption, steel demand in that section is not sufficiently large to give optimum operating conditions for much of today's high-capacity plant. By 1963 consumption of steel by metal fabricators in the whole of the South was only a little over one-sixth that of the Manufacturing Belt. By 1964 the blooming mill capacity of Houston and Lone Star was no more than 8·1 per cent of that within seventy-five miles of Chicago. This has implied a wide spread of production lines and sometimes operation of technically inferior plant. Both the Fairfield and Gadsden strip mills were for a number of years operated as finishing stands to their plate mills. Even when market growth justified a separate strip mill it was significant that Republic installed a reversing rather than a continuous unit. By 1970 the capacity of the strip mill at Fairfield was only just over half that of mills at the Irvin, Fairless, and Gary works, and in the case of Republic the contrast with northern mills was even greater. The general superintendent of Fairfield has recently summed up the situation very well, though, rather surprisingly, he also claimed highly successful operations. 'The unique part of our mill is the wide variety of products from one location. We really have a collection of small mills for making everything from cotton bale ties to tin plate and prepainted sheet. It is among the leaders of our company in productivity and ranks well in profits.'[27]

The Gulf coast in particular has long been a major consumer of foreign steel. In 1962 Atlantic ports of the South imported 0·48 million tons steel, the Gulf took 1·09 million or 22·3 per cent of the national total. By 1971 the southern Atlantic seaboard ports handled 1·14 million tons of foreign steel, but the Gulf coast 4·37 million tons, again 22·3 per cent of the national imports, but a tonnage much greater than the output of Texas mills. If they could win back this trade operating conditions would be much more satisfactory.

NOTES

[1] *I.A.*, 6 Dec. 1965, p.51, 25 Nov. 1965, p.30.

[2] *B.F.S.P.*, June 1947, p.687, and *The Effect on Minnesota of a Liberalization of U.S. Foreign Trade Policy*, Report of Business Executives Research Committee, 1956, pp.92–3.

[3] *A.M.M.*, 13 Sept. 1971, p.1, and *'33 Magazine*, Apr. 1972.

[4] C.H. Grattan, 'The Future of the Pacific Coast', *Harpers Magazine*, Mar. 1945, pp.305–6; *I.A.*, 15 Feb. 1945, p.88; *Harpers Monthly Magazine*, 195, 1947, p.149.

[5] A.G. Roach, 'Steel for the Expanding Industry of California', *B.F.S.P.*, Jan. 1951, p.78.

[6] *I.A.*, 15 Feb. 1945, p.88; *Steel*, 30 Apr. 1945, p.79.

[7] U.S. *v.* Columbia Steel Co. *et al.*, Supreme Court Hearings, reprinted in *Government Regulation of Business*. Cases from the National Reporter System selected by R.R. Bowie, 1949, pp.1727–30.

[8] *I.A.*, and *Steel*, Estimates.

[9] *Steel*, 8 Jan. 1945, pp.77–9; 'Twelfth District Steel: A "War Baby" Grows Up', *Monthly Review of the Federal Bank of San Francisco*, Sept. 1953, pp.112–13.

[10] *I.A.*, 4 Oct. 1956, p.56.

[11] *B.F.S.P.*, Apr. 1946, p.475; *I.A.*, 18 Nov. 1948, p.134.

[12] *I.S.Eng.*, Feb. 1961, p.K-4.

[13] *B.F.S.P.*, Jan. 1947, p.82.

[14] *I.A.*, 18 Nov. 1948.

[15] Kaiser Steel, *Kaiser Steel—Built to Serve the Growing West*, 1957–8, p.6.

[16] *I.A.*, 10 Sept. 1964 p.120, Kaiser Steel, *Annual Report*, 1964, p.11.

[17] Kaiser Steel, *The Western Steel Market (annual)*

[18] *I.A.*, 10 Apr. 1958, p.79.

[19] *A.M.M.*, 13 Oct. 1966, p.1.

[20] *U.S. Steel Quarterly*, Aug. 1960, p.6, Nov. 1962, pp.3–4; *Steel*, 27 July 1964; *U.S. Steel Quarterly*, Nov. 1962, p.4.

[21] *I.A.*, 4 Nov. 1956, p.56.

[22] *I.A.*, 12 Sept. 1963, p.144; *B.F.S.P.*, Jan. 1965, p.59.

[23] *I.S.Eng.*, Feb. 1970, p.95.

[24] *I.A.*, 17 Jan. 1955, pp.125–6.

[25] *B.F.S.P.*, June 1947, pp.691–2.

[26] *I.A.*, 29 Apr. 1971, p.49.

[27] *Industry Week*, 10 Jan. 1972, p.29.

CHAPTER 18

The Postwar Steel Industry
III. The Manufacturing Belt:
General Considerations and the Ohio Basin

In spite of the impressive growth of steel capacity in the west and Gulf south-west the greatest growth in consumption, capacity and production continues to be centred in the Manufacturing Belt. In 1947 metal fabricating plants in fifteen states of this belt consumed 83·8 per cent of the nation's steel. By 1963 the share of the same states had fallen to 78·2 per cent but in tonnage terms they registered a 12·3 million ton increase as compared with 6·4 million tons in the rest of the nation.[1]

However, there has been a centrifugal movement within the area. (Fig. 36) In part this has resulted from increasing costs of distribution over long distances, while truck transport, which has become so important since 1945, is most suited to shorter hauls. Within the region the changes in the standing of sub-regions have been made easier by mineral supply developments. The anticipated need to move to dependence on foreign ore sources was moderated by large investments in ore beneficiation in the Upper Lakes region. None the less the increasing attractiveness of imported ore has improved eastern iron-making prospects. At the same time fuel economy has acted to the especial benefit of the districts away from the Appalachian coalfields. It too has improved eastern prospects but also those of the Great Lake shore and the western third of the Manufacturing Belt. The last area has gained particularly by changes in the distribution of demand. (Table 118) 1963 steel consumption in the United States was 18·7 million tons greater than in 1947. In Pennsylvania there was a decline of 300,000, in New England, New Jersey, and New York an increase of 1·4 million tons, in Michigan a rise of 3·1 million, and in Illinois, Indiana and Wisconsin the increase was 4·6 million tons. The postwar developments in the geography of steel-making in the manufacturing belt can be considered from two perspectives—an east-west contrast, that is one between the Atlantic sea-board, central, and western districts, or the different expansion patterns in the old Ohio basin steel towns and the Great Lakes shore. A compromise approach is adopted here. The manufacturing belt contains a large number of steel-making locations and companies so that the broad lines of locational change are modified for different product lines and degrees of integration, by variations in management skills, judgement, and policies. For these reasons study of locational change in this section of the country provides a useful corrective to a too facile determinism in relation to locational values.

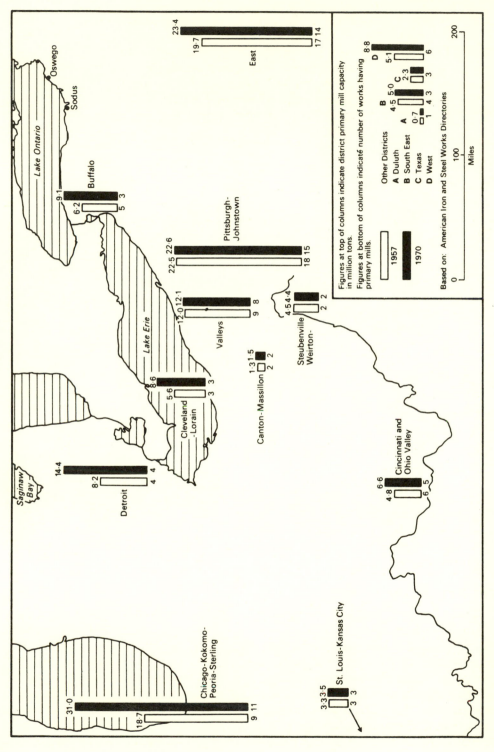

36. Primary mill capacity by districts of the Manufacturing Belt 1957–1970

Table 118.   *Consumption of steel mill products by metal*
fabricators in the Manufacturing
belt 1947, 1954, 1963
(percentage of national total)

		1947	1954	1963
North-east		11·8	12·6	10·5
of which	New England	3·5	3·0	2·7
	New York	5·4	6·3	4·9
	New Jersey	2·9	3·3	2·9
Central		27·1	25·2	23·5
of which	Pennsylvania	14·2	11·8	9·2
	Ohio	12·9	13·4	14·3
Western		44·9	43·1	44·2
of which	Illinois	12·7	11·7	11·6
	Indiana	5·7	5·4	6·3
	Michigan	17·2	17·3	17·0
	Wisconsin	4·7	3·7	4·1

Based on Federal Reserve Bank of Cleveland *Economic Review,* Oct. 1969, p.10.

THE OHIO BASIN

In the war, for speed, and afterwards, for cheapness, 'rounding out' of existing works was preferred to the construction of new ones. The interior centres of production benefited then and still profit from this policy. Even so the importance of the Ohio basin has declined sharply since the Second World War. In 1947 its widespread steelworks produced 42·7 per cent of U.S. steel as compared with 33·6 per cent for the centres along the shores of Lake Erie and Lake Michigan. By 1968 their shares were 32·1 and 38·8 per cent respectively. Trends in the various parts of the Ohio basin have been by no means uniform.

THE PITTSBURGH DISTRICT

There has been marked relative decline in the Pittsburgh district since the Second World War. As defined by the American Iron and Steel Institute the district includes Wheeling and other upper Ohio valley centres. Its output increased from 22·3 million ingot tons in 1947 to 25·3 million in 1968, a period in which the national increase was 46·2 million tons. The only other of the nation's eleven steel districts to register a decline in share of national output over this period was Youngstown, where the fall was much greater even than in Pittsburgh. In spite of this it is well to remember that Pittsburgh remains one of the worlds eight biggest steel-producing districts. Progressive fuel economy, growth of distant markets, and generally increasing difficulty in competing in these markets have been the chief factors in its worsening situation. As seen above, these problems are by no means new but they have been especially prominent since the early postwar years. (Fig. 37)

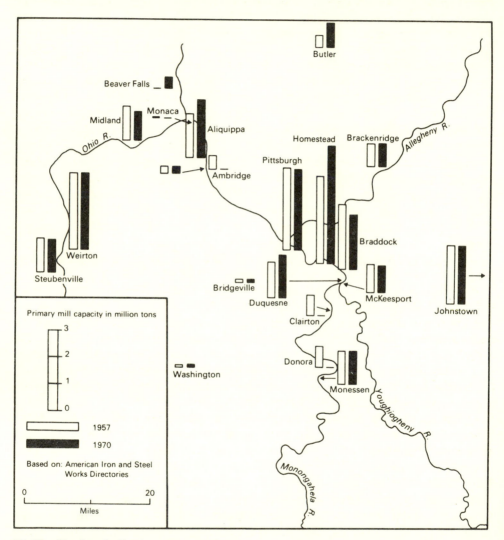

Primary mill capacity in million tons

3

2

1

0

☐ 1957

■ 1970

Based on: American Iron and Steel Works Directories

0   20

Miles

37. The Pittsburgh district 1957–1970

The President of Carnegie–Illinois observed in mid-1946 that 'the density of consumption has been moving away from Pittsburgh' and he anticipated that replacement facilities for that district's generally rather run-down plant would be built elsewhere.[2] A little later an ex-president of National Tube noted that several merchant pipe mills had already been moved from McKeesport to Lorain, and went on to suggest that over the next three years all McKeesport operations would be removed, many of them to Gary.[3] Yet in fact McKeesport was maintained and extended, though mainly in finishing rather than steel operations, in which expansion throughout the fifties was slower than at Lorain.

There were several causes for the failure of the predicted Pittsburgh decline to materialize. Cheapness of expanding existing plant and the localization of labour skills were undoubtedly important. Extension of truck shipments of light flat-rolled steels, and water shipments, notably of tubular products, checked the worsening situation in distribution. Buoyancy of demand was also relevant. In the 1949 recession the Pittsburgh operating rate fell below the national rate, but after that compared favourably with the national rate until the 1954 recession. In 1954 Pittsburgh performed noticeably less well than Chicago, and since then its operating rate has been almost always below the national average. It was in the fifties that Pittsburgh decline was most pronounced. Chicago district mills produced 19·1 million ingot tons in 1950 and 20·7 million tons in 1960; Pittsburgh mills 24·2 and 20·1 million. By 1968 their respective outputs were 26·7 and 25·3 million tons.

Between 1945 and 1959 U.S. Steel extended the steel capacity of its two Chicago works by 3·04 million tons and in the east built Trenton works which had a 1959 capacity of 2·7 million tons. Over the same period nominal capacity in its five works in Allegheny County fell slightly. Beyond the county line works were closed or sold. Mingo Junction works on the Ohio and the Farrell plant in the Shenango valley were sold in 1945—6. The 250,000 ton melting shop at Vandergrift was dismantled in 1955. Donora was extended in the fifties, but in 1962 the blast furnaces, steel plant, and blooming mill there, and all except the structural mills at Clairton works were abandoned (Table 119).

Other Pittsburgh district plants have been saved by means of new inter-linkages. Duquesne, threatened with closure after the war, has since been thoroughly modernized to become one of the Steel Corporation's most successful operations, and now supplies most of the steel for the McKeesport tube mills. Melting shop extensions at Duquesne, Edgar Thomson, and Homestead in the fifties were designed to support an extension of plate, structural, and forging output at Homestead.[4] At the end of the thirties the Irvin strip mill project had involved reconstruction of open hearth furnaces and a new slabbing mill at Edgar Thomson, and when a new E.R.W. tube mill was put up at McKeesport in 1963 it was supplied by strip deliveries from Irvin, in turn dependent on increased slab tonnages from Edgar Thomson. Even so, by the mid-sixties, Edgar Thomson was regarded as a semi-marginal producer and had only just over half its medium-sized open hearth furnaces at work. As demand for rails proved sluggish, and was concentrated away from the region, U.S. Steel supplied more of its share of that market from its Fairfield and Gary rail mills. In 1966 it finally took Edgar Thomson out of the trade in which its past glories had been so great. However, by 1971 a 3 million ton oxygen steel plant had been built at Edgar Thomson largely to support a doubling of the capacity for cold reduced sheet at Irvin works. Other producers have developed less complex interlinkages. Jones and Laughlin use the Ohio to link their Pittsburgh and Aliquippa works, and when in 1950, Pittsburgh Steel decided to build a hot strip mill at Allenport it supplied the necessary 800,000 tons of slab by barge from its chronically congested main works at Monessen.

Table 119.   *Primary mill capacity of U.S. Steel and Jones and Laughlin Pittsburgh and lake-shore works 1957, 1970 (million tons)*

		1957	1970
U.S. Steel			
	Homestead	3·48	4·67
	Edgar Thomson	2·56	2·16
	Duquesne	1·63	1·69
	McKeesport	1·14	1·06
	Donora	0·84	abandoned
	Clairton	0·81	abandoned
Total	Pittsburgh area	10·26	9·58
	Gary	3·12	5·16
	South works	3·81	4·65
	Lorain	2·25	1·95
Total	lake-shore	9·18	11·76
Jones and Laughlin			
	Pittsburgh	3·25	3·20
	Aliquippa	1·75	2·30
Total Pittsburgh area		5·00	5·50
	Cleveland	1·20	2·79

Based on *A.I.S.I.*

These developments indicate that massed capacity makes possible a good deal of flexibility and constitutes one of the continuing strengths of the Pittsburgh district. Even so its slimming down still goes on. McKeesport retains nominal steel-making capacity but its melting shop of small open hearths is closed while Duquesne supplies its needs. Costs for pollution control in a new, amenity-conscious generation add to the assembly, process, and marketing troubles of Jones and Laughlin's Pittsburgh works, the only major one within the city limits. It is reckoned that $100 million investment is needed to make this works fully competitive, but it is subject to such severe environmental constraints that it is doubtful if this expenditure can be justified. In recent years more and more Jones and Laughlin effort has gone into Aliquippa and Cleveland works, and labour forces in Pittsburgh are being run down.

The inadequacy of demand remains the most severe problem facing Pittsburgh steel-makers, for in times of recession they cannot retain distant customers located nearer to competing mills. The problem has existed for generations but all attempts to remedy it have failed. In 1948, at a favourable time of steel shortages, and helped further by the abolition of basing point pricing, the Pittsburgh Chamber of Commerce actively began to promote removal of fabricating firms to Pittsburgh. Considerable play was made with early successes, but in fact these were relatively small and failed to match either the extensions in local steel-making capacity or the growth in steel consumption elsewhere.[5] The 1954 Census of Manufactures indicated that in the Pittsburgh economic area metal fabricators purchased 2 million tons of steel annually as compared with 3·8 million in Chicago and 5 million in Detroit. By 1956–7 the annual value of steel and rolling mill products from the six counties near Pittsburgh (Allegheny, Beaver, Washington, Westmoreland, Butler, and Armstrong) was about $2,435 million, but consumption only $347 million.[6] Employment figures show that, although the situation is improving, there remains an extraordinary imbalance between steel-making and steel-using trades. (Table 120) By 1970 primary metals made up 6·8 per cent of U.S. manufacturing employment. In

Table 120.   *Employment by industrial category,*
Pittsburgh, Cleveland, and Cincinnati 1950, 1960
(thousands)

	Pittsburgh 1950	Pittsburgh 1960	Cleveland 1950	Cleveland 1960	Cincinnati 1950	Cincinnati 1960
Primary metals	148·3	134·9	40·6	39·8	7·4	5·3
Fabricated metal products	21·9	27·6	28·6	31·8	9·3	13·0
Machinery	44·6	51·9	61·7	70·4	27·5	30·3
Motor vehicles and equipment	2·3	3·2	26·8	37·5	9·0	12·3

Federal Reserve Bank of Cleveland *Economic Review,* Nov. 1967.

the Cleveland Federal Reserve district, spanning the so-called 'American Ruhr' from the Ohio to Lake Erie, the proportion was 16·9 per cent but in the four-county Pittsburgh metropolitan area which is part of this it was 40—45 per cent.[7]

As a result of this deficiency in demand Pittsburgh mills ship over longer distances than most of their rivals. Combined with higher raw material assembly costs and higher process costs than some other districts, this implies lower mill net earnings and therefore a lower rate of growth of capacity. (Table 121)

Table 121.   *Primary metal shipments from Ohio valley and Lake Erie shore economic areas 1963*

| Economic area | Primary metals as percentage of total shipments | Percentage distribution by distance of shipment | | | | | Mean distance of shipments |
		Under 100 miles	100— 199 miles	200— 299 miles	300— 499 miles	Over 500 miles	
Cleveland, Canton, Valleys	34·7	30·7	26·1	14·3	18·4	10·5	274
Pittsburgh, Weirton, Wheeling	65·5	24·0	24·4	16·9	19·5	15·2	335
Detroit, Toledo	15·6	59·3	16·5	16·3	5·8	2·1	138
Cincinnati, Middletown	23·2	31·5	19·4	29·4	4·9	14·8	267

Source: Federal Reserve Bank of Cleveland *Economic Review,* Nov. 1968.

The Pittsburgh area, like the rest of the Ohio—Lake Erie belt, contains a host of smaller firms. Some of them have steelworks, like Allegheny—Ludlum at Brackenridge, or even coke ovens and blast furnaces as with Crucible Steel's Midland works on the Ohio, though both of these also supply other works of the group with semi-finished material. Both, as with other, much smaller firms, mostly using electrical furnaces, buy scrap and to some extent cold pig iron. Other firms again are pure rerollers working up slabs or billets. They are untroubled by any locational disadvantages represented by changes in ore and coal requirements, for as was remarked about one of them, 'tonnage companies typically quote prices in dollars per ton. Crucible quotes in dollars per pound.' Even in marketing any problems stem not from freight costs but from the difficulty of giving prompt service to a distant consumer. These mills provide a small but growing market for the bigger, heavy steel firms. They are not, however, likely to bulk very large in the latter's order books. In these big plants, the units which in the past have given the district its character, modernization and extension will go on, and having such large capacity Pittsburgh will coast through into the twenty-first century as one of the world's leading primary metal districts. The greatest growth in capacity, and the construction of wholly new plant will pass it by.

THE OHIO VALLEY

Strung out along the Ohio below Pittsburgh are six integrated steelworks. Two, at Aliquippa and Midland, are closely integrated with Pittsburgh operations and may be regarded as part of that district. The others are in a different situation. Steubenville, Ohio, and Weirton, in West Virginia's panhandle, are well located in relation to the good coking coals of the Connellsville district, Ashland dates from the turn of the century as a steel-making location, but Portsmouth at the mouth of the Scioto has had a longer history. Elsewhere along the river there are steelworks and mills but not fully integrated works, as at Huntington and Newport. Other locations have been recently abandoned—a blast furnace at Martins Ferry, the small steelworks and mills at Toronto, and the Parkersburg sheet mills (Fig. 38).

It is easy to regard these plants simply as survivors of times when the Ohio was a great avenue of movement, but this ignores the twentieth-century establishment and notable success of Weirton, which by 1970 was a 3·5 million ton plant, and the changing circumstances which, whatever their origins, have made some of the others good points for expansion in the mid-twentieth century. The improvement of the Ohio has been of great importance. It has not only eased access to markets but has also enabled interworks transfer along the river, a process which not only built up riverside groupings but after the Second World War encouraged growth of outside interest in these plants. Changes in raw material supply conditions have helped too; Ashland, a high-cost point for assembly of lake ore, was significantly the first plant in the country equipped with a furnace especially designed to operate on a 100 per cent pellet burden. Each of the four integrated works has a strip mill. In marketing sheet it may be reckoned that they are at a considerable disadvantage as compared even with Valleys mills, for the country immediately to the south of the Ohio provides only a small market, and sheet and strip is not an ideal product for water handling. Two of the bigger works, Wheeling and Portsmouth, and the largest of the semi-integrated plants, Newport, make wire and pipe, and for these products water shipment to the West is a great advantage. From its formation in 1920 Wheeling Steel was involved in a good deal of river-borne interworks transfer. In 1946 it sold the Portsmouth works, and steel-making has finished at Wheeling, whose mills are now supplied from Steubenville. Within a distance of twenty miles along both banks of the Ohio, Wheeling Steel has operated seven plants, the rationalized survivors of the small independent units of the last century. This is still by no means an ideal pattern when movement costs and heat losses, quality control, interworks accounting, and the provision of

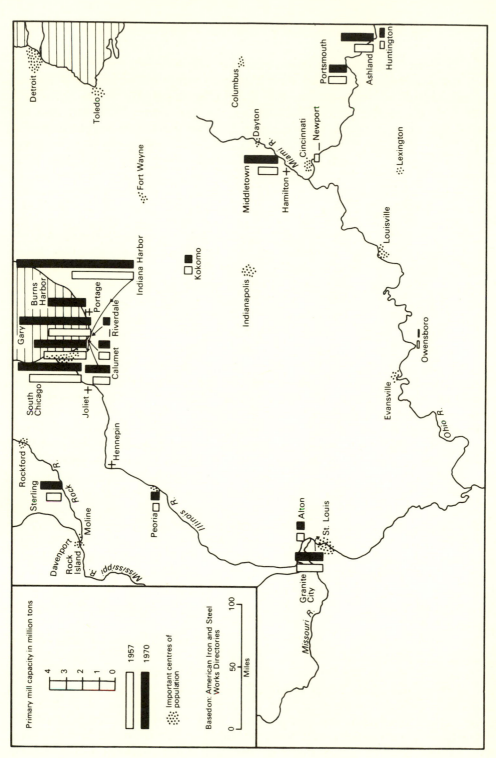

Primary mill capacity in million tons

4
3
2
1
0

□ 1957
■ 1970

.·.·. Important centres of population

Based on: American Iron and Steel Works Directories

0   50   100
Miles

38. The steel industry of the Chicago, St. Louis and Cincinnati districts 1957–1970

separate service facilities are taken into account. Yet coal can be assembled from the Allegheny valley and West Virginian mines by water and Wheeling has substantial advantages over the Valleys in shipments to the west and south-west down the Ohio. However, by the mid-sixties Wheeling was far behind the times in operating efficiency. In 1966 there were suggestions of a merger with Crucible Steel, whose Midland slabbing mill could have supplied some of Wheeling's needs. However, the link which was made was with Pittsburgh Steel. Pittsburgh's Monongahela valley works are further away, but are also accessible by water, and had the necessary surplus steel capacity to match Wheeling's excess in finishing capability.[8] Lower down the Ohio Newport was acquired in 1955 by Acme Steel of Chicago, in part to supply it with semi-finished steel, though it continued to supply outsiders with strip mill products as well. In 1950 Detroit Steel bought Portsmouth from Wheeling and by the early sixties had made it into an up-to-date integrated operation with a hot strip mill as its central unit. Modernization costs were reckoned only about one-third the investment needed for a new works.

Armco, with plants at Middletown, Ashland, and Butler is the largest company in the Ohio valley below Pittsburgh. In 1930, less than a decade after Ashland was acquired, and only 3 years after Butler's purchase, Middletown was the smallest plant and Butler the largest. By 1960 Middletown capacity had increased fivefold, Ashland had doubled in size and ranked second, and Butler was rather smaller than in 1930. This reflects their relative competitive situation.

Middletown is located away from the Ohio, and must therefore rail in its coal supplies, but it more than compensates for this by being in the midst of the big sheet markets of the industrialized Miami Valley. In 1954 metal fabrication took 620,000 tons of steel in the Hamilton—Middletown area, 390,000 tons in Cincinnati and 370,000 tons in Dayton, a total almost 70 per cent of that of the Pittsburgh area, but with, comparatively, very small steel capacity. By 1964 south-western Ohio was reckoned to consume 1·2 to 1·3 million tons of steel a year and by 1966 in the Hamilton—Middletown Standard Metropolitan Statistical Area (S.M.S.A.) primary metals employed 29·5 per cent of the manufacturing work force but 20·4 per cent were in fabricated metals, a highly favourable situation as compared with Steubenville—Weirton's 73·3 and 3·2 per cent or Johnstown's 50·4 and 1·5 per cent.[9]

By contrast Ashland is a high-cost producer from the point of view of raw material assembly, marketing, and in the early postwar years in processing costs as well. But in the ten years to 1963 $150 million was spent there, 80 per cent of the plant being rebuilt. Lack of scrap has been remedied by conversion to higher molten charges, cheaper now that pellets are used, and a large oxygen steel plant was completed in 1963.[10] Although Ashland's 1963 2·3 million tons capacity had been obtained at one-third, or perhaps even less, of the cost of a wholly new works, the development policy which

Armco has followed there is questionable. Middletown and Butler already give it a key position on the Ohio and it might instead have spent money on a new, better located works.

Butler has now been extended. It was provided with a new strip mill in the fifties and in the late sixties was converted wholly to electrical steel furnaces. It remains a cold metal works. This has been an important factor in its progressive concentration on quality rather than bulk steels.[11]

Having two plants far from lower lake ports, only one on navigable water and one a cold metal operation, Armco overall has high costs as compared with some of its major rivals. In 1959 it was estimated that its ingot costs were probably $10 a ton more than Bethlehem's.[12] Since then Armco has diversified its product line so that by 1967 it had reduced its dependence on the automotive industry to 15·4 per cent of total shipments. In 1971 the company's shareholders were told that the plant at Jervis Bay, New South Wales, in which Armco is a partner, will make steel at one-third of the cost at the Ohio works.[13]

Evidence on the value of an Ohio valley location is conflicting. In the postwar years of insistent demand and high-cost plant, expansion of existing plant was very attractive. New techniques such as oxygen steel-making and new blast furnace operations using pellets have improved operating costs, while further work on the Ohio navigation has opened a better route to south-western markets. Even in the sixties the rate of expansion of plant in the Ohio valley was impressive. Yet to the west and south-west new strip mill, wire, and tube capacity has been built. Ohio Valley plants will remain viable, but perhaps less so than ten or fifteen years ago.

LANDLOCKED CENTRES OF PRODUCTION BETWEEN THE OHIO AND THE GREAT LAKES

Steel-making and still more rolling mill plants are widely spread between the Lake shores and the Ohio river. Survival and growth in this belt is explained by the historical factor, the extension of demand, backward integration, and the advantages which accrue to existing locations. It shares much of the east-west nodality of the Lake shore, and is excellently placed to supply Lake Erie markets. In its smaller communities companies may sometimes obtain concessions on rating, water supply, or effluent disposal. Competition for labour may well be less keen, and scrap prices sometimes lower than in the major primary metal centres. Many steel firms have important local markets as consumers have built branch factories or complete plants for new processes away from the bigger industrial cities. On the other hand, as assembly costs for iron ore and coal are high there are usually no blast furnaces, and as iron costs and the operating costs of cold metal practice are increased, there has been an incentive to concentrate on higher-value lines.

The Mansfield steelworks and strip mill developed from an ordinary hand mill sheet plant. Increasingly it has had to specialize on high quality strip to meet the challenge of the bigger strip mills at integrated works. On a much larger scale, and over a longer period, ordinary sheet and steel bar firms in Canton and Massillon grew and merged into the so-called Central Alloy District of the Republic Steel Corporation. These works are exceptional, being fully integrated, and only about fifty-five miles from the ore docks at Cleveland. (Fig. 39) They are also closely linked with the Republic works at Warren at the head of the Mahoning Valley and in Cleveland, supplying slabs to strip mills there, and receiving hot rolled coils back for cold finishing

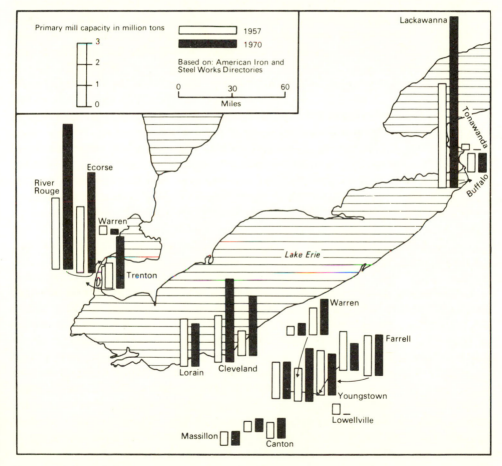

39. The steel industry on the Lake Erie shore, and in the Valleys and Canton-Massillon districts 1957–1970

stainless steel sheet. Elsewhere in this belt plants are operated by large steel consumers for whom slightly higher operating costs may be less serious than loss of control over supplies. Important examples are Babcock and Wilcox at Beaver Falls and Alliance, Timken at Canton, and Borg Warner at New Castle, Indiana, and at Franklin in north-western Pennsylvania.

However, in the whole landlocked belt west of Johnstown there are carbon steel works of 1 million tons capacity only in two areas, the Miami valley where Middletown is the only unit and in the Mahoning and Shenango valleys where in 1960, the last year when steel capacity figures were published, nine works had a capacity of 12·4 million tons, or only 500,000 tons less than in the whole nation west of the Mississippi. By 1970 primary mill capacity in the Valleys was a little under half that of the Chicago district. Forty years ago *Fortune* neatly summed up the dilemma of Valley producers in respect of iron manufacture: 'The Lake Company is the best situated. The River Company is second best. The Land Company is a very poor third.'[14] In spite of their very large capacity the Valleys are as disadvantaged now as then, and the effects are becoming steadily more painfully obvious.

Between 1947 and 1968, in spite of severe oscillations, U.S. ingot steel production went up by an average annual rate of 1·8 per cent. Of the eleven districts into which the country is divided by the American Iron and Steel Institute, ten recorded annual positive increases, though in the case of Pittsburgh the increase was only 0·2 per cent. In the Valleys there was an average annual decrease of 0·9 per cent. Valleys share of national output was 13·2 per cent in 1947, 10·8 per cent in 1955, and only 8·2 per cent in 1968. The Valley market for steel is small, and although the motor industry has increased employment in the area in the sixties, by 1970 over half the employment in manufacturing industry was in primary metals.

The three chief Valley companies are Youngstown Sheet and Tube, Republic Steel, and U.S. Steel in that order; all three have plants outside the area, which over the long run they have expanded more than their Valley works. (Tables 122 & 123) In 1955, shortly after it had proposed a merger with Youngstown Sheet and Tube, Bethlehem Steel announced the development programme which it would follow if the merger went through. This provided an interesting outside assessement of the Valleys. Indiana Harbor ingot capacity was to be increased 2 million tons while in the Valleys extensions would be only 1 million tons. Later, planned growth in the Valleys was cut to 600,000 tons, and by 1958 Bethelhem was proposing to spend a total of $90 million at Youngstown Sheet and Tube's Valley plants and $268 million at Indiana Harbor.[15]

Fuel economy has helped the Valleys but not to the extent that it benefited the lakeside plants, and in terms of the handling of a higher-grade iron burden, Pittsburgh and Ashland, with longer hauls than the Valleys, have gained more. (Table 124) The Valleys have by far the largest concentration of iron-making capacity for which all the raw materials must be delivered by rail. To this situation there are two remedies, one of minor value but practicable, the other of great importance but probably too costly.

Table 122.   *Ingot capacity of Valley and Lake-shore plants*
of Valley firms 1945, 1953, 1960
(million tons)

	1945	1953	1960
Republic Steel:			
Warren, Youngstown	3·30	3·14	3·64
Buffalo, Cleveland, Chicago	3·72	4·68	6·15
Youngstown Sheet and Tube:			
Brier Hill, Campbell	2·55	2·84	3·33
Indiana Harbor	1·44	2·10	3·42
United States Steel:			
Youngstown	2·34	2·73	2·71
Lorain, Chicago, Gary	12·17	13·92	16·24

Based on *A.I.S.I.*

Table 123. *Primary mill capacity of Valley and Lake-shore plants*
of Valley firms 1957 and 1970
(million tons)

	1957	1970
Warren, Youngstown	3·12	2·87
Buffalo, Cleveland, Chicago	4·21	6·58
Youngstown Sheet and Tube:		
Brier Hill, Campbell	3·36	4·28
Indiana Harbor	3·20	3·80
United States Steel:		
Youngstown	2·14	1·95
Lorain, Chicago, Gary	9·18	11·76

Based on A.I.S.I. *Works Directories.*

Table 124.   *Estimated assembly costs on iron ore and*
coking coal, Cleveland, Youngstown and
Pittsburgh, 1939, 1962
(dollars per ton of pig iron)

	Cleveland 1939 1962	Youngstown 1939 1962	Pittsburgh 1939 1962
Iron Ore	3·51 2·66	5·45 4·12	6·15 4·65
Coking Coal	3·13 2·46 (1962 as)	2·02 1·59 (1962 as)	0·73 0·57 (1962 as)
	(% of)	(% of)	(% of)
	(1939)	(1939)	(1939)
Total:	6·64 5·12 (77·1%)	7·47 5·71 (76·4%)	6·88 5·22 (75·7%)

Note: 1962 estimate is based on raw material economy corrections of the 1939 figure.
Based on: Haven 'The Manufacture of Pig Iron in America', *J.I.S.I.* 1940 1; *Bureau of Mines Yearbook,* 1962.

Shipping coke rather than coal reduces fuel haulage costs as the coke yield in ovens averages only about 69 to 70 per cent. Youngstown and Republic each have two coke plants in the Valleys, but Sharon's coke is made at Templeton sixty miles to the east in Armstrong County, and at Fairmont on the Monongahela in West Virginia, a small by-product coke plant. U.S. Steel supplies Youngstown with coke from Clairton plant. Works bringing coke from distant ovens have no coke oven gas (in the case of the Steel Corporation the gas can be used nearer to the ovens) but increased use of natural gas and fuel oil in open hearth furnaces or for power production has reduced the disadvantage which this might imply.

Another remedy for part of the trouble of the Valleys has been to barge coal down the Monongahela and Ohio and then transfer it to rail cars for the short haul to the furnaces. In the thirties Conway was an important transhipment point, and it was then estimated that 15 cents a ton or 10·5 per cent of the all-rail freight from Connellsville could be saved in this way.[16] Even so all-rail coal transport still predominates. There are pipe loading facilities at Conway, and at Colona, a little lower down, there are two river/rail transfer docks handling pipe and other bulk commodities. In 1958 it was estimated that only about 780,000 tons of coal for all purposes and destinations were transferred at Colona and about 200,000 tons of iron and steel were shipped from there.[17]

Solution to the problems of this major interior smelting centre might be obtained if ship canal navigation was established between the lakes and the Ohio. Such a project has been discussed for at least sixty years. In 1907 its cost was reckoned at $53 million. After a survey in the early thirties the U.S. Divisional Engineer for the Upper Mississippi valley rejected the scheme mainly on the grounds of inadequate water supply. By 1935 the Army Corps of Engineers was working on another survey. The Lake Erie and Ohio River Canal Board meanwhile estimated the costs for a 102 mile, 12 foot minimum depth canal from Ashtabula to the mouth of the Beaver River at $141 million.[18] By 1948 the Army Corps was recommending an 18 foot canal along the route Ashtabula to Rochester on the Ohio. Already standard Great Lakes ore carriers drew 21 feet, but on an 18 foot draft canal the limit would be about 16 feet, requiring special vessels, use of smaller ones, or standard carriers only partly loaded—all giving increased delivery costs. By 1948 the necessary investment was put at $439 million of which $23 million would be met by local communities, and the rest by the Federal Government. Carrying charges were reckoned $21 million a year—roughly 25 cents on every ton of ore and coal travelling between the Ohio and Lake Erie. 1964 estimates of cost were a startling $1,000 million.

The Valleys have had other processing cost disadvantages. One was the lack of large supplies of water. In the Second World War this problem was solved by construction, at Federal expense, of the Milton reservoir in the upper Mahoning valley and then of the much larger Mosquito Creek reservoir tributary to the main stream. A large Shenango valley reservoir was also built, and in the early sixties the Federal Government established a $33 million reservoir near Sharpesville, lower down the valley. Parts of the valley district have very congested narrow valleys, choked with plant, and hemmed in by railroads and urban sprawl. However, this applies especially to the lower Mahoning, and at the northern end around Warren the major plants of both Republic Steel and of Copperweld Steel still have ample flat sites and possibilities for expansion. Lack of steel consumption in the area implies a shortage of scrap so that its price there has long been above the Lake-shore level, though usually about the same as in Pittsburgh. This shortage has emphasized high molten metal charges and therefore has highlighted the Valley's problems in mineral assembly.

Lack of district demand is the basic problem, and the old empahsis on tubular and thin flat-rolled products is less helpful now that strip mill capacity is more widely spread, and tube and pipe plants are being expanded in the south-west. There is a Valley saying that when Detroit is down the Valleys are flat. U.S. 1966 steel output was 17 million tons higher than that of the good year 1955 but Valley output was 1 million tons or almost 8 per cent less. Significantly, there has been a long delay in installing oxygen steel plant in the Valleys, Youngstown Sheet and Tube and the Steel Corporation having none there yet.

In 1970 there were rumours that Youngstown Sheet and Tube would build an integrated works between New Orleans and Baton Rouge, which would send steel on for finishing in the Valleys where Youngstown Sheet and Tube iron and steel-making would cease. So far this has proved mere speculation, but the 1971 rehabilitation announced for Campbell works proved disappointingly small, merely a $9–10 million project. In the winter of 1971/2 the Steel Corporation Youngstown plant was shut down for 7 months. However extensive repairs and improvements were made at this time. The long-term future of the valleys remains in doubt. By spring 1972 Youngstown Sheet and Tube's president noted that 15,000–20,000 steel jobs had been lost in the Mahoning Valley alone since early postwar years. He went on to suggest that without a canal to the Ohio Youngstown would eventually be reduced to a light industry community.[19]

NOTES

[1] 'Regional Trends in Steel Production', Federal Reserve Bank of Cleveland, *Economic Review,* Oct. 1969, pp.10, 12.

[2] *I.A.,* 9 May 1946, p.129; see also *I.A.,* 16 May 1946, p.92, and *I.A.,* 5 Feb. 1942, p.121, 10 Mar. 1949, p.153.

[3] *I.A.,* 8 Aug. 1946, p.1048.

[4] *B.F.S.P.,* Jan. 1957, p.52.

[5] *B.F.S.P.,* Aug. 1948, p.972, Nov. 1948, p.1331; *I.A.,* 30 Sept. 1948, p.100.

[6] A. Longini, *Region of Opportunity,* Pittsburgh and Lake Erie Railroad Company, 1961, Table XXIV–50.

[7] Federal Reserve Bank of Cleveland, *Annual Report,* 1971, pp.10, 13.

[8] Wheeling Steel, *Annual Reports; Steel,* 24 Jan. 1966, p.27; *Fortune,* July 1967.

[9] *I.A.,* 23 July 1964, p.162; Federal Reserve Bank of Cleveland *Economic Review,* Oct. 1968, p.13.

[10] *I.S.Eng.,* Dec. 1963, p.133.

[11] *I.S.Eng.,* Aug. 1961.

[12] *Fortune* Nov. 1959, p.132.

[13] *Industry Week,* 5 July 1971, p.52.

[14] *Fortune* June 1932, p.86.

[15] *I.A.,* 2 June 1955, p.51, 26 Sept. 1958, p.47.

[16] C.E. McLaughlin, 'Probable Effects of an Ohio River to Lake Erie Canal on the competitive position of Pittsburgh Industry', *Pittsburgh Business Review,* 5, No.12, Dec. 1935, p.2.

[17] A. Longini, op. cit., p.VI–50.

[18] *I.A.,* 5 Sept. 1907, 11 Aug. 1932; McLaughlin, op. cit., p.1.

[19] *A.M.M.,* 29 June 1972, p.4.

CHAPTER 19

The Postwar Steel Industry IV.
The Manufacturing Belt:
The Lake Shore and the Mid West

In 1968 the four steel districts along the shores of the Great Lakes—Buffalo, Cleveland, Detroit, and Chicago—produced 50·9 million tons of steel, a total exceeded by only two steel-making nations, the U.S.S.R. and Japan. Between 1947 and 1968 the share of these districts in U.S. production rose from 33·6 to 38·8 per cent, an increase of 22·3 million tons out of a national increase of 46·2 million tons.

The largest part of the capacity is still in the Chicago district, but over the postwar period growth on Lake Erie has been much more impressive. In 1947 Chicago plants produced 20·2 per cent of U.S. steel, in 1960 20·9 per cent, and in 1968 20·4 per cent. For the Lake Erie districts, respective proportions were 13·4, 17·4, and 18·4 per cent, Lake Erie indeed has made in some ways the most impressive increase of any steel-producing area; even in the west and south the increase from 1947 to 1968 was only from 9·8 to 12·9 per cent of the national total.

THE LAKE ERIE SHORE

In 1968 steelworks on Lake Erie produced 24·1 million net tons of steel, a total which ranks the area high in the world league of producing districts. In the same year Ruhr works produced 28·2 million net tons. But whereas Ruhr output is from a district whose extreme dimensions are the less than 40 miles from Duisburg to Dortmund and in which, in spite of polarization at the eastern and western ends, production is still fairly widely spread, Erie shore works are in three distinct centres along over 300 miles of lake shore. In the postwar period the ranking of these three centres has been completely changed. (Table 125)

Table 125.   *Lake Erie centres: Ingot steel output and percentage of U.S. total 1947, 1960, 1968*

	1947 million tons	%	1960 million tons	%	1968 million tons	%
Buffalo	4·2	5·5	5·2	5·2	7·2	5·5
Cleveland	4·0	4·7	5·6	5·6	7·7	5·9
Detroit	3·1	3·7	6·5	6·6	9·2	7·0

Based on *A.I.S.I.*

THE SEAWAY, LAKE ONTARIO, AND BUFFALO

The 1954 decision to build the St. Lawrence Seaway, along with the opening of Labrador and other foreign orefields, seemed likely not only to benefit Great Lakes steel-making generally, but particularly its eastern centres. There was already some movement of ore by canal past the International Rapids, but it was small and costly. In 1951 206,000 tons of ore came through to the Great Lakes, 159,000 tons to be used at Buffalo—only 2·1 per cent of the crude ore tonnage terminating there—,47,000 tons for Detroit. On the basis of expected freight economies there were suggestions of new works on Lake Ontario using both Labrador and Adirondack ore. By 1956 New York State was pressing claims for a new steelworks at Oswego or, further west, at Sodus. Three years later a study of prospects in the St. Lawrence valley itself was made for Mohawk Power Corporation. There were ample resources for electrical reduction, good road, rail and, now, water transport. Yet the St. Lawrence valley was a hopeless proposition from a marketing point of view. To a large degree the same condition condemned the Lake Ontario projects. The industrial cities of the Mohawk Belt constituted only a small outlet for steel, none of them ranking among the first forty-seven state economic areas for steel consumption at the time of the 1954 census of manufactures. No large company seems to have shown the slightest interest in the St. Lawrence valley or Lake Ontario.[1]

Before the Seaway was opened it cost 22 per cent less to deliver Labrador ore in Buffalo than in Pittsburgh, but with the Seaway the advantage by 1961 had widened to 37 per cent.[2] On these grounds it appeared that among established steel firms Buffalo would benefit most from the new waterway. The area had much else to commend it—major established capacity, a key rail and road transport position between eastern and Great Lakes markets and rapid growth in local consumption. In fact the area's ore imports lagged, and its steel plants did not show the dynamic steel expansion expected. (Table 126)

The integrated works at Tonawanda, acquired by Colorado Fuel and Iron in an attempt to break into the eastern markets, proved especially vulnerable to foreign competition, largely because of specialization on rod and wire. It was abandoned in 1963. In 1962 another Tonawanda works, Buffalo Steel's bar rerolling plant, was closed. On the other hand the Roblin Steel Corporation at Dunkirk and North Tonawanda has had one of the notable though still rather small-scale successes of the 1960s. Since 1929 Republic Steel has operated the Buffalo plant built by the New York State Steel Company in 1907, specializing on bar production. Overall recent mill expansion has been much slower than at Chicago or Cleveland, the former of which especially has received a large part of the company's recent bar expansion. Republic's Buffalo works was noticeably almost completely absent from its $400 million capital development programme in the mid-sixties.

Lackawanna continues to dominate the area, but by the early 1970s it was in a crisis situation. For sixty years it was Bethlehem's best located plant to serve markets in the western part of the Manufacturing Belt and especially

Table 126.   *Blooming and slabbing mill capacity Buffalo,
Detroit, Cleveland and Lorain 1957, 1964, 1970
(million net tons)*

	1957	*1964*	*1970*
Buffalo area:			
Lackawanna	5·1	8·3	8·3
Colorado Fuel and Iron, Tonawanda	0·2	(scrapped 1963)	
Republic, Buffalo	0·8	0·8	0·8
Detroit[1]	7·8	9·4	14·1
Cleveland and Lorain[1]	4·4	6·1	4·8

[1] Because 1964 figures are not available Jones & Laughlin capacity figures have been omitted.
Based on A.I.S.I. *Works Directories,* 1957, 1964, 1970.

along the lake shores. In the late fifties it shipped 1 million tons of products annually to Detroit markets.[3] However, outside the Lake navigation season Buffalo has a long rail or truck haul compared with Cleveland or even Pittsburgh. Moreover, as competition became keener, Lackawanna's bridge-like location between eastern and western Manufacturing Belt markets became disadvantageous. In the twenties the plant had been a major supplier of structural steel to Chicago, but growth of local capacity pushed it out, so that by the late fifties it was only a marginal supplier, and its charges to Chicago fabricators were considerably higher than those of Inland Steel. By the late fifties Bethlehem Steel controlled 15 per cent of the U.S. steel market but shipped only about 1 per cent of its steel into the Chicago area.[4] These considerations led to the decision to build the new works at Burns Harbor, Indiana, and subsequently to problems in Lackawanna operations.

In the early sixties Lackawanna was provided with a new slabbing mill of over 3 million tons capacity, designed largely to supply the Burns Harbor plate and strip mill. In terms of its own finishing capacity Lackawanna made little headway, and indeed generally there was a slight writing down of its capacity. (Table 127) When complete integration was achieved at Burns Harbor the need for Buffalo semi-finished steel ceased, and Lackawanna was transformed from Bethlehem's western bridgehead into an uncomfortable middle status between Burns Harbor and Sparrows Point and Bethlehem, not so well placed for the biggest markets as them and more confined to the increasingly competitive Lake Erie markets east of Detroit. In late 1970 Bethlehem reduced Lackawanna steel capacity from 6·7 to 4·8 million tons and further scaling down was expected.

Table 127.   *Lackawanna steelworks: main finishing capacities*
1957, 1970
(thousand net tons)

	1957	*1970*
Billet and sheet bar	1,250	1,180
Rail and billet	750	715
Structural	1,150	1,485
Bar and structural	288	260
Bar	800	775
Joist	24	–
Hot strip	2,700	2,520
Cold strip	1,920	1,680
Cold finished sheet	2,193	2,100

Based on A.I.S.I. *Works Directories*, 1957, 1970.

CLEVELAND

Fuel economy has benefited Cleveland more than Buffalo, for on this score it was previously a more costly location. Cleveland's long-established advantage of major markets in a wide diversity of steel-using industries continues a major asset. At the end of the forties the area was reckoned to consume 29 per cent of Ohio's hot rolled sheet and strip, 44·4 per cent of its bars and 68·7 per cent of its wire and wire rods. By 1954 consumption of mill shapes and forms in the Cleveland economic area by all users except the primary metal firms themselves was about 69 per cent more than in Buffalo even though ingot capactiy was then 11 per cent less. By 1970 it was estimated that half the bar market of the United States was located within 300 miles of Cleveland.[5] Cleveland mills have a shorter haul to the Detroit market than the most easterly plants at the southern end of Lake Michigan.

Yet some Cleveland plants have suffered severely from foreign or domestic competition. U.S. Steel had 159 wire-drawing blocks and 10 continuous wire-drawing machines at its Cuyahoga works in 1957; ten years later there were 25 continuous machines but only 41 wire-drawing blocks. Lorain steel capactiy and primary mill capacity has been reduced. Tubular product capacity there has scarcely changed over the last fifteen years, but the plant has been made into a major Steel Corporation bar producer.

The two biggest Cleveland companies are Jones and Laughlin and Republic Steel. For them company structure is a factor which may have boosted Cleveland capacity more than a straight evaluation of costs might have suggested. Cleveland has been Jones and Laughlin's only foothold on the Lake shores for integrated operations, and even for Republic Cleveland is the only lake plant for flat-rolled products. Expansion of both works is still plagued by site difficulties.

Piecemeal extension of the Jones and Laughlin plant in a heavily developed area has resulted in the finishing mills being separated from the hot strip mill by the Baltimore and Ohio Railroad tracks. In 1962 it was decided to build a new hot strip mill, but, to make room, it was necessary to realign the Cuyahoga, straightening a 2,100 foot bend in the river, removing 600,000 cubic yards of earth—all for ten acres of land. Republic had similar troubles, but in 1964, on the eve of construction of a new cold reduction mill, acquired almost 200 acres of additional land in the valley. In the late sixties it built a major new strip mill there.[6]

DETROIT

In the long postwar boom years southern Michigan's growth as the nation's largest steel market continued. By the early fifties production was growing rapidly—1950 output was 48 per cent greater than in 1947 as compared with less than 15 per cent nationally—but was still less than half of consumption. (Table 128) F.O.B. pricing provided an additional irritant when customers had to make good the shortfall in local production by paying full freight on Ohio, Pennsylvania, or Chicago steel. As in the late twenties there were suggestions of a new plant on Saginaw Bay, but new

Table 128.   *Steel production and consumption in Pennsylvania and Michigan 1951 (million tons)*

	Pennsylvania	Michigan
Steel production	23;4	3·85
Steel consumption	6·8	9·19

Source: I.A. 21 Jan. 1954, p.41.

plant projects or major expansions were all much nearer the Detroit heart of the market area.

The major motor firms continued to follow completely different policies to ensure their supplies. Ford bought half its steel in the early postwar years, but strove to maintain its 'tapered integration' which included modernization of the Rouge operations at a cost of $54 million between

1947 and 1953. It still supplied only half its needs when it embarked on another $35 million steel investment programme in 1959. Replacement of the 1935 strip mill was long delayed. It was modernized in 1961 but not until ten years later were a new Rouge 68 inch strip mill and coke ovens announced. Presumably in the meantime Ford has borne the disadvantage of rather higher finished steel costs than those from the newer generation of strip mills.[7]

General Motors decided not to build its own steel capacity, but, whereas in the early thirties it considered acquiring Corrigan-McKinney of Cleveland or perhaps the local unintegrated operations of Newton Steel, in the post war period it made loans to steel firms. In 1950—2 General Motors lent Republic Steel $40 million for extensions, $15 million of Algoma Steel Corporation debentures were acquired and $28 million was loaned to Jones and Laughlin to provide the whole of the finance for melting shop and strip mill extensions, all of whose output it was to take.[8] In Detroit General Motors' finance was an important factor in an outstanding postwar success, the growth to major proportions of McLouth Steel. McLouth had operated hot and cold sheet and strip mills since 1936 using hot rolled coil from local firms. By 1949 its carbon and stainless steel capacity was just under 200,000 tons. It then built a 400,000 ton electrical steel plant, and a new hot strip mill at Trenton, with direct access to the Detroit river. Operating on a 100 per cent scrap burden proved costly, and in other respects too this development had the appearance of a cheese-paring development.[9] The Korean War provided opportunity for further expansion, and under the Defence Production Act of 1950 rapid amortization of part of the existing expansion programme was obtained. At this time General Motors found McLouth an attractive source of steel, near to many of its motor plants, and with only a small commitment in existing plant to stand in the way of new low-cost techniques. $25 million of McLouth preferred stock was bought and a contract to supply 5 per cent of General Motors' consumption of steel given. By 1963 half McLouth sales were to General Motors.[10] A blast furnace was built at Trenton in 1953, and by 1954 the reversing hot strip mill, built only six years before, had been sold to Spain, and replaced by a new hot mill with over five times its capacity. By building the first basic oxygen furnaces in the United States McLouth obtained steel extensions at much lower capital cost than with the open hearth extensions still being made by rivals. At the end of 1955 its 1·4 million tons steel capacity had cost only $100 per ton in a completely new plant on a greenfield site, about one-third of the generally accepted figure for such a development.

Other Detroit firms expanded, though rather less spectacularly than McLouth. Detroit Steel acquired the Portsmouth, Ohio, integrated works and finishing mills in Connecticut and Texas. A 1950 plan for an 800,000 tons integrated works at Gibralter on the Detroit River, a project with which Cyrus Eaton was associated, came to nothing. The postwar experiences of

National Steel Corporation at the Ecorse plant of its Great Lakes steel division show how other factors complicate the evaluation of the suitability of a production location. After the Second World War National's spectacular expansion continued, and the Korean War brought another spurt. The effects of the 1954 business recession on National Steel were serious, its operating rate falling from 93·5 per cent in 1953 to only 75·2 per cent in 1954, and net income was almost $19 million or 38·2 per cent less. To some extent this was, as before, the result of too close dependence on the motor industry. McLouth, still building at Trenton, suffered even more. Another factor was large freight absorption by rival strip mills firms, which chiselled away the differential which Detroit had been able to exact in times of shortage. Yet neither of these conditions explained the large size of National's declining activity as compared with that in the motor industry, or with the buoyancy of some rivals, such as Armco. A factor of key importance appears to have been the complacency which is always a danger in a superb location. This was linked to the decline in the health and judgement of George Fink, to whose vision and initiative National's earlier success in Detroit was largely due. In 1953 demand was over-estimated and inventories began to build up while at the same time quality and service fell away. In spring 1954 Fink was retired from management, though a year later he was still a director.[11] National Steel's Annual Report, issued in March 1955, spoke with remarkable frankness about the situation at Great Lakes Steel:

> ... we discovered that a number of practices had developed ... which were entirely out of harmony with the principles on which National Steel Corporation conducts its business. There was an excessive accumulation of inventory, insufficient maintenance, an undue amount of labor disturbance, and other deficiencies in operation. This had not come to notice prior to that time because operating conditions resulting from the lengthy boom in steel demand not only led to the gradual development of these faulty practices but also obscured them.

The remedy was also concisely summarized. 'The situation necessitated a thorough change in management personnel, policies and practices and a complete re-organization was put into effect.'[12] Great Lakes did badly in 1958, but this was probably due more directly to the recession in the motor industry. In the early sixties major extensions were made, partly to supply hot-rolled coil for the new Mid West Steel plant on Lake Michigan. By 1970 Ecorse was a 6·2 million ingot ton plant, some 80 per cent of its ingots going as finished products to the motor industry, and it was that industry's largest supplier of sheet and strip. The Great Lakes Steel president neatly summed up his firm as 'automotivated'.[13]

Even though its Weirton plant is a major and successful operation with an especially prominent position in 'packaging' steels—black plate, tinplate, and tin-free steel for cans—National Steel has expanded both rolling mills and steel capactiy there much less than at the Great Lakes division in Detroit. In 1930 the ingot capacity of the new Ecorse operation was 0·60 million tons, of Weirton 1·25 million. By 1970 their respective capacities were 6·2 million and 3·5 million tons. (Table 129)

Table 129.   *National steel corporation hot strip mill*
Capacity 1954, 1957, 1964, 1970
(million net tons)

	1954	*1957*	*1964*	*1970*
Great Lakes Steel, Ecorse	3·4	3·2	6·0	6·0
Weirton Steel, Weirton	1·8	2·5	3·0	3·1

Based on A.I.S.I. *Works Directories.*

In 1947 Detroit steel production was 38 per cent that of the combined output of the two other Lake Erie shore centres; by 1968 61 per cent. By 1970 its primary mill capacity was almost 80 per cent of the combined total for Buffalo and Cleveland.

THE MIDDLE WEST, ST. LOUIS, AND CHICAGO DISTRICTS

In 1947 the four industrialized mid-western states of Illinois, Indiana, Wisconsin, and Missouri consumed some 27·5 per cent of the steel taken by U.S. metal fabricating establishments; by 1963 27 per cent, a stability in their share in contrast to declining proportions of national consumption in all other Manufacturing Belt states except Ohio. Favourable regional demand conditions, a good location for access to expanding western and south-western markets, and developments in ore and coal procurement explain why the Mid-West has remained an area of impressive growth in steel since the Second World War. In 1947 output from the Chicago and St. Louis steel districts was just over 55 per cent that of Pittsburgh and Youngstown, by 1968 over 82 per cent as large. It is true that the district's lead over the Erie shore has narrowed greatly over this period—from output 63 per cent greater in 1947 to only 24 per cent more in 1968—but in the sixties hot strip and particularly cold reduced strip extensions in the Chicago area have been on an immense scale. (Table 130)

Away from Lake Michigan there are six considerable mid-western works extending from Kokomo, Indiana, west to Kansas City and south-west to St. Louis. Together by 1970 their primary mill capacity was 6 million tons or about 24 per cent that of Chicago mills. Some of these works benefit like those of the Lake Erie–Ohio River intermediate belt from operating in fairly small industrial agglomerations. They are also well placed to serve the important mid-western agricultural steel markets. Continental Steel at Kokomo produces concrete bars, wire, wire products, and galvanized sheet. Armco's Kansas City works with finishing capacity of over 800,000 tons operates in the wire, nail, and bar trade. At Peória another, long-established works integrated backwards from wire products to wire and later to rods and bars. Peoria itself is a large customer, and marketing south and south-

Table 130. *Primary, wide hot strip, and cold reduced strip mill capacity of Lake Erie and Chicago district integrated works 1957, 1964, 1970 (million tons)*

	1957			1964			1970		
	Primary	Hot strip	Cold reduced strip	Primary	Hot strip	Cold reduced strip	Primary	Hot strip	Cold reduced strip
Detroit	7·77	6·41	4·03	9·42	9·30	4·89	14·15	10·38	5·44
Cleveland	5·65	3·10	1·83	8·33	4·06	2·78	8·60	8·29	3·61
Buffalo	6·13	2·70	1·92	9·25	2·52	1·63	9·14	2·52	1·68
Total Lake Erie	19·55	12·21	7·78	27·00	15·88	9·30	31·89	21·19	10·73
Chicago	15·73	7·86	4·43	26·31	9·18	8·26	25·74	15·58	13·85

Based on A.I.S.I. *Works Directories.*

westwards is eased in this instance by the mill's location on the improved Illinois waterway. An even bigger operation is at Sterling on the Rock River. It too supplied barbed wire, fencing nails, and netting to western farms and ranches but later went into the bar, structural, and plate trades. The Moline—Rock Island—Davenport industrial area fifty miles away is a major market and source of scrap. In competing with Chicago plants, Sterling has a 100 mile advantage in reaching western customers.

In the past St. Louis was inferior to Chicago as a steel location. It has a longer rail haul on East Kentucky or West Virginia coal—though is more accessible by water—and Missouri ore production was of small importance. Its major raw material attraction was a large scrap supply, partly locally derived, partly from the wide area for which it is the chief rail junction. Scrap and good market access were the factors that gave rise to the un-integrated Laclede Steel Company at Alton. As an integrated plant, Granite City grew more rapidly, helped by its concentration on strip mill products, but also by improvement in its assembly conditions. At the end of the fifties over one-third of its deliveries were within the St. Louis metropolitan area, and 55 per cent to Missouri and Illinois.[14] For wider shipment to south and north St. Louis's rail and road junction position and barge shipment via the Mississippi and its connections are great assets. In 1971 Granite City was acquired by National Steel with the intention that its excess steel and hot strip capacity should supply the recently extended finishing capacity of Mid West Steel at Portage, Indiana.

By the fifties Granite City had largely achieved the substitution of Illinois coal for that from further east. By 1955 ovens there used 194,000 tons of Pocahontas and 474,000 tons of southern Illinois coal, the latter washed at the mines and mixed with the superior eastern coal before coking.[15] Local ore supplies are of growing importance. By the end of the fifties about one-third of the ore came from Iron Mountain, Missouri, the rest from Minnesota and Michigan, but since then Granite City has shared in ore developments on Pilot Knob 85 miles from the furnaces. Shipments began in 1968 and will eventually reach 1 million tons of 64 per cent Fe pellets annually.

CHICAGO

Between 1957 and 1970 the capacity of blooming and slabbing mills within 75 miles of Pittsburgh—an area which includes Johnstown, the Valleys (but not Canton—Massillon) and the Ohio to south of the Wheeling steel complex—went up from 38·5 to 40·3 million tons. Over the same period primary mill capacity within 75 miles of Chicago rose from 16·7 million to 28·5 million tons. Between these two dates the number of works operating primary mills in the Pittsburgh area fell from 28 to 25 and in the Chicago area increased from six to eight. Chicago works are not only far fewer but also much more concentrated, the most distant being well under 35 miles in a direct line from the centre of the city. (Installation of continuous casting units means that primary mill figures do not now fully represent the change in effective steel capacity but they provide the best guide to capacity available.)

This impressive difference in growth patterns—a 4·6 per cent growth over thirteen years for Pittsburgh and 70·9 per cent for Chicago—is by no means fully matched by changes in production, but even in this case the differences are striking enough, though unfortunately the statistical areas involved are rather different. In 1955 the Pittsburgh and Youngstown districts as recognized by the American Iron and Steel Institute produced 33·3 per cent of U.S. crude steel, by 1968 27·5 per cent; the share of the Chicago district increased from 20·2 to 20·4 per cent. Between 1965 and 1968 four major new hot strip mills were built in the Chicago area. A conservative estimate put their capacity at 12 million tons and potentially at 16 million.[16] In the Pittsburgh area—of 75 miles radius—two new strip mills were built in the second half of the sixties, at Sharon and Steubenville. By 1970 there were twelve producers of wide strip there are opposed to only four in the 75 miles surrounding Chicago, but whereas the Chicago area wide hot strip mill capacity rose from 9·18 to 15·58 million tons from 1964 to 1970 in the case of Pittsburgh the increase was only from 18·35 to 22·42 million tons.

As the major steel producer west of the Manufacturing Belt, and with the longest hauls on Appalachian coking coal, Chicago may be assumed to have gained most from postwar fuel economy. Assuming that the decline in coking coal consumption was at the same rate in all steel districts, then the charge for transporting coal per gross ton of pig iron would have fallen $0·157 in Pittsburgh between 1939 and 1962, $0·667 in Cleveland, and $0·743 in Chicago.[17] Keen competition between rail-including new unit trains—,rail—barge and rail—lake carriage of coking coal has cut delivery costs even more. Still more directly beneficial to Chicago has been the break-through in use of Illinois coal to make metallurgical coke. Granite City Steel has long used Illinois coal, but its employment in Chicago was made practicable by a combination of preparation and carbonization progress and a change in transport conditions.

In 1958 the introduction of multiple in place of the old single car rates cut freight charges on coal from Illinois and West Kentucky mines to Chicago by about 50 per cent. By 1965 Interlake Steel with long-term new contracts for this coal proved that coke costs could be cut, and Inland Steel was making major shipments from its expanding mining operations at Sesser in the Illinois coalfield 275 miles from Chicago three years later. By 1969 46 per cent of the 3·7 million tons of coal carbonized at Illinois ovens was from Illinois mines, 53 per cent from Kentucky and West Virginia. U.S. Steel and the new Bethlehem Steel plant do not yet carbonize Illinois coking coal. Iron ore and pellets come directly to lake-shore furnace stockyards mainly by water, but some rail transport is employed for direct delivery from Upper Lakes and Wisconsin mines, and there have been recent suggestions that growing demand may be met by transferring pellets from Mesabi by pipeline.[18]

The extensive and diverse metal fabricating trades of the Chicago metropolitan area and region have continued to grow rapidly. In 1954, when Chicago steel production still lagged behind that of Pittsburgh, metal fabricators in the Chicago economic area consumed 3·8 million tons of steel or almost twice as much as those in Pittsburgh. Milwaukee and the Gary-Hammond—East Chicago economic areas together took another 1·72 million tons.[19] In the wider Chicago region consumption is large, and until the mid 1960s was unsatisfied by local production. Some at least of the Chicago works have sold almost all their production in this area. From 1959 to 1964 about 65 per cent of Inland Steel's output was sold within a radius of 100 miles of the plant, 26 per cent within the next 300 miles, and only 8 to 9 per cent beyond. This compares with the situation in the Pittsburgh and upper Ohio districts where only 24 per cent of primary metal deliveries in 1963 were made within a distance of 100 miles and 15·2 per cent were made over distances greater than 400 miles. The proportion of Inland deliveries to distances under 100 miles is greater even than that for the Detroit—Toledo area.[20]

Since the late fifties all analysts have been agreed that consumption in the Chicago region would rise more rapidly than in the nation. In 1958 the Bethlehem Steel Commercial Research Department projected a 48 million ton increase in national capacity to 1970. Of this 24 million tons would be needed in the north-central states, 13 million tons of it in the Chicago district. Inland Steel, a little later, estimated that Chicago demand would increase 25 per cent over the next 10 years and for flat rolled steel the increase would be bigger still. A 1964—5 Illinois Institute of Technology study suggested a Chicago area steel consumption increase to 1980 of as much as 150 per cent requiring local production of about 50 million tons, or twice the high level of 1964.[21]   By the mid-sixties the mid-west area consumed 35 to 40 per cent of all steel used in the U.S.A. or 35 to 36 million tons a year; about one-fifth of this came from outside the area. This shortfall, along with the expected major growth in demand, has encouraged both expansion by existing firms and the arrival of new ones.

Three Chicago rerollers have integrated backwards to steel-making and in one instance even to iron manufacture. Borg-Warner in West Pullman and Ceco with a new plant at Lemont on the Illinois Waterway are small-scale instances. Acme Steel is larger and has achieved full integration. Beginning as a fabricator in the 1880s, it started to reroll bought steel in 1917 and in 1956 both acquired Newport steelworks on the Ohio to supply some of its semi-finished steel and began to make some of its own steel with small oxygen converters. In 1964 Acme merged with Interlake Iron and the following year molten iron movement from furnaces on the Calumet River 15 miles to the steelworks at Riverdale was begun.[22] Much bigger, long-established and fully integrated, the International Harvester steel plant has internal company outlets for about half its production. The other existing integrated concerns and major outsiders have responded more decisively to improving Chicago area prospects.

Republic Steel has more than doubled its Calumet billet and bar capacity in Chicago since 1957—during which time capacity for the same products at its Youngstown works has been almost halved. Alone among major companies Republic has no flat rolled capacity in Chicago. This is the product line in which capacity has expanded most. U.S. Steel has made much bigger extensions in the Chicago area than in Pittsburgh where its existing capacity was considerably larger. Homestead works on the Monongahela had a 1957 capacity for sheared and universal plate of 2·16 million tons compared with 1·30 million at the Steel Corporation's Chicago works (0·87 million at South works and 0·43 million at Gary). By 1970 the respective plate mill capacities were 1·93 and 2·38 (1·17 at South works and 1·20 at Gary). Even in structurals, though by no means so impressive, Chicago works have made more rapid progress. Clairton and Homestead structural mill capacity rose 1957—70 from 1·58 to 2·15 million tons; at South works capacity went up from 0·91 to 1·50 million tons. At its four

Manufacturing Belt wide strip mills—Irvin, Youngstown, Trenton and Gary—U.S. Steel increased nominal capacity only 340,000 tons 1957—70, but Gary's share of this was 280,000 tons.

Growth of Inland Steel at Indiana Harbor has not been merely an automatic response to market growth, but a process greatly helped by the traditional excellence of its management and careful cultivation of sales outlets. Especially in periods of bad trade these have given it operating rates higher than those of its rivals, even those of the Chicago area. In 1945 the Chicago Metropolitan area produced 19·8 per cent of U.S. crude steel; in 1964 18·8 per cent. Inland Steel over the same period increased its share of national output from 4·4 to 5 per cent.[23] Its ingot capacity at Indiana Harbor was 3·4 million tons in 1945, 6·5 million in 1960 and by 1969 7·7 million tons. By 1973 it will be 9·2 million tons, and at the end of the seventies Inland anticipates that it will be shipping one-third more steel than in 1971.[24] Youngstown Sheet and Tube, Inland's Indiana Harbor neighbour, has made considerable though smaller extensions. Both plants have built extensively on land reclaimed from the lake, which since the mid-sixties alone has comfortably accommodated a new 3 million ton oxygen steel plant and hot strip mill at the Youngstown Sheet and Tube works.

Three major outsiders have been attracted into Chicago since the late fifties. National Steel bought a 750—800 acre duneland site on the lake shore in Porter County, Indiana, in 1930. For almost thirty years the site was undeveloped. 1958 was a bad year for the automobile industry, and the Great Lakes Steel operating rate at Ecorse was only 42 per cent in the first nine months of that year. (Inland, less than half as dependent on the automobile market, managed an 81·3 per cent operating rate for the whole year.) By February 1959 National Steel had decided to build a plant at Portage to finish up to 840,000 tons a year of Ecorse hot rolled coil as cold reduced sheet, galvanized sheet, and tinplate. This would permit greater flexibility of Detroit area primary operations. By 1964, when Portage cold reduction capacity reached 960,000 tons, it was reckoned that half its shipments represented business which, without it, National Steel could not have obtained.[25] By 1970 finished product capacity there was 1·4 million tons. Eventually the works may be fully integrated.

In the mid-fifties Jones and Laughlin was associated with schemes for a Houston integrated works. Financing proved difficult, improvement of the Ohio made possible increasing pipe shipments from Aliquippa, and as the Chicago market expanded rapidly Jones and Laughlin began to think of development there. In 1963 it announced that, to meet competition from the expansion of existing works and new ones in the Chicago area, it too would build there. The location chosen was radically new. As the Jones and Laughlin Chairman remarked, 'we concluded that to locate a new Jones and Laughlin plant next door to our Chicago area competition would lack in imagination and economic sense. Accordingly, we conceived and developed a quite different approach to this competition problem—an offensive rather than defensive posture.' [26] The location was Hennepin on the Illinois River 100 miles west of Chicago where, in the 'long range', a fully integrated plant working on northern taconite pellets, or possibly using Southern Illinois coal in direct reduction processes, was planned. In addition to technology, marketing trends seemed to be moving in favour of this project.

The Illinois waterway has a ruling depth of nine feet, so that the plant has equally good navigation conditions and a shorter journey to southern markets than Pittsburgh works shipping via the Ohio. By Interstate route 80 Hennepin has speedy truck access to Chicago as compared with slow congested routes through the industrial and urban sprawl of Calumet. It is better poised for road and rail deliveries to western customers. Jones and Laughlin also acted imaginatively to improve their immediate outlets. The Hennepin site in the bend of the river was flat and covered almost 6,000 acres, over 10 times greater than the area of any existing Jones and Laughlin plant and more than that of any plant in the country until the building of U.S. Steel's Texas works. It was planned that much of the Hennepin site would be taken up by steel fabricators who would thereby become on-the-spot customers, benefitting along with the steel-maker from a key location in respect of mid-western markets.

Hennepin has worked in the main on hot rolled coil railed from Cleveland works at a freight charge which by 1968 was $4·70 a ton. It was equipped with cold rolled sheet mills and plant for galvanized sheet; by the late sixties capacity was 1 million tons. However by 1969/70 the works was operating at only 75 per cent of capacity. No major steel customers had occupied the large acreage of land assigned for a captive market, and it was realized that Jones and Laughlin had been too ambitious in moving so far out of the Chicago industrial area. It became known that company studies made before the location was chosen had in fact disclosed that most steel users who were considering relocation were planning to move less far towards the western extremity of the manufacturing belt, mostly preferring Kentucky or Indiana.[27] By 1972, in a strange reversal of old patterns of steel movement, but one which boosts its activity, Hennepin was contracting to cold reduced hot rolled coil derived from Kaiser at Fontana and designed for General Motors' Michigan and Ohio body shops.

The third new arrival, Bethlehem Steel, acted on a much bigger scale, but more conventionally. Bethlehem's proposed merger with Youngstown Sheet and Tube in 1930 and 1931 had been designed both to widen its market area and diversify its predominantly heavy emphasis. Private court action frustrated it. In 1954 the idea was revived. Bethlehem made it clear from the start that most of the subsequent expansion would be on Lake Michigan. It claimed it could make the planned extensions to the Youngstown Sheet and Tube Chicago and Youngstown plants for $135 a ton, but that costs on a greenfield site would be $300 a ton. It was inferred that Bethlehem could not afford this. Even so in 1956 it acquired a large tract of lake shore land in Porter County.[28]

The progress of the proposed merger before the Southern District Court of New York in 1958 threw much light on Chicago competitive conditions.[29] Discussion of the structural steel business, in which Bethlehem proposed Indiana Harbor extensions of 720,000 tons—half its total proposed mill extensions there and also just under half the Chicago area's current capacity—was decisive. The two existing producers made extensions which seemed to suggest Bethlehem's programme was unnecessary. Just before the court hearings began Inland Steel gave up rail manufacture and doubled its wide flanged beam capacity.[30] In September 1958, when the hearings were completed, but the judgement not yet decided, U.S. Steel announced it was installing new structural mills at South works. These had already been under construction for two years, but the public announcement at this time must, in effect if not in intention, have strongly suggested that no other major new structurals project was needed; and in November, rejecting the merger, the judge observed that plate and structural extensions by Inland and U.S. Steel would satisfy expanding Chicago area demand in these lines.[31] Lacking both financial resources and experience in heavy steel trades, Youngstown Sheet and Tube thereafter continued to expand in its well-established lines. Bethlehem Steel was left determined to gain access to the Chicago area.

In Porter County it still proved possible to obtain big unimpeded sites similar to those at Indiana Harbor and Gary almost sixty years earlier. 'The 3,300 acres site on the southern shore of Lake Michigan was devoid of industry: the steel facility was to be constructed from the ground up. Thus none of the usual problems existed; there were no power lines, buildings, roads, railroad tracks etc. in the way—it was virgin territory,' was how one commentator summed up the happy situation facing the plant engineers when Bethlehem announced, in December 1962, that it would build a plant at Burns Harbor, just east of National Steel's Portage works.[32] At this time it was reckoned that carrying charges on a completely new integrated works would be $25 or so a ton—$15 to almost $20 per ton more than at existing Chicago works.[33] Willingness to bear such heavy overheads emphasized the overwhelming attractiveness of a Chicago area development; ability to bear them was only possible in the case of a giant concern.

By mid-1965 Burns Harbor was rolling plate, cold reduced sheet and tinplate, and early in 1966 its 80 inch hot strip mill was operating. It worked on Lackawanna slabs shipped in large, New York Central unit trains at a freight cost of $5·25 a ton. Full integration was planned from the start, but by phasing out the project Bethlehem not only spaced out its capital investment programme but gained also on the technical and raw material account. Large capacity oxygen converters had proved their value for very big runs of common grade steel by the early sixties, so that, when it decided to build steel capacity, Bethlehem opted for 300 ton vessels. It steadily marshalled its mineral supplies. Appalachian plateau coking coal reserves were extended, and after 1964 a new eastern Kentucky mine was opened. In the late fifties Bethlehem had secured a half share in big ore developments on Pea Ridge in Washington County, Missouri, only 100 miles by rail from St. Louis. Shipments of 70 per cent Fe pellets began 1964, at which time Bethlehem's Johnstown works was building 800 100 ton rail hopper cars for the operations. Burns Harbor is 450 miles from the mine.[34] Full integration was announced in 1966 and by 1970 the works was complete with a finished product capactiy over 2 million tons. Overall costs had been $1,000 million, or a remarkable $50+ a ton annual standing charge. However, further expansion will be obtained at much lower cost for some units of plant have capacity well in excess of their present operating level. Eventual capacity of about 10 million tons is anticipated.

In spite of optimistic projections of Chicago area demand, the pace of recent expansion and especially the emphasis on hot strip and cold reduction mills may mean that for a time the market is over-supplied. In that case Chicago mills, as low-cost producers, may be expected to push on a bigger scale into the markets of southern Michigan and Ohio. Such a development would be at the expense of Valley or Pittsburgh mills.

NOTES

[1] *I.A.*, 23 Feb. 1956, p.56; *Newsweek* 15 Aug. 1955; *B.F.S.P.*, Dec. 1959, p.1320.

[2] Federal Reserve Bank of New York, *Monthly Review*, Aug. 1961, p.143.

[3] *New York Times*, 30 Apr. 1958, p.53.

[4] Ibid., 22 Apr. 1958, p.47, 3 May 1958, p.23.

[5] *I.A.*, 26 May 1949, p.72; *Census of Manufactures*, Bureau of the Census, 1954; *I.A.*, 27 Sept. 1956, pp.163–5; *A.M.M.*, 26 June 1970.

[6] *B.F.S.P.*, July 1962, p.666; Republic Steel, *Annual Reports*.

[7] *I.A.*, 9 Nov. 1947, p.125; *I.S. Eng.*, May 1953, p.114.

[8] *Administered Prices. Hearings Before the Sub-Committee on Anti-Trust and Monopoly of the Committee on the Judiciary*, U.S. Senate, 1958, part 6, *Automobiles*, evidence of H.H. Curtice, President of General Motors, p.2598; D.E. Lilienthal, *Big Business: A New Era*, 1952, p.85; *Fortune*, Jan. 1965, p.235.

[9] McLouth Steel Corporation, *Annual Report*, 1949; *B.F.S.P.*, Feb. 1949, p.239.

[10] *Fortune*, Jan. 1965, p.235.

[11] *Fortune*, Feb. 1964: 'National: a pace-setter in steel', p.208.

[12] National Steel, *Annual Report*, 1954, pp.8–9.

[13] *A.M.M.*, 23 Jan. 1970.

[14] Granite City Steel, *Post War Growth 1947–1959*, p.5.

[15] *B.F.S.P.*, Aug. 1957, p.860, May 1963, p.359.

[16] *A.M.M.*, 4 Mar. 1968, p.7.

[17] W.A. Haven, 'The Manufacture of Pig Iron in America', *J.I.S.I.*, 1940. 1, p.426, and U.S. Bureau of Mines, *Minerals Yearbook*, 1962.

[18] U.S. Bureau of Mines, *Circular 8512*, 1971.

[19] *1954 Census of Manufactures*.

[20] Inland Steel, *Transcript of meeting with New York Society of Security Analysts*, 1959 and, *Transcript of meeting with Los Angeles Society of Security Analysts*, 1961; 'Freight Transportation and Industrial Activity in the Fourth District', Federal Reserve Bank of Cleveland, *Economic Review*, Nov. 1968, pp.19, 21.

[21] *New York Times,* 9 May 1958, p.133; Inland Steel, *Transcript of meeting with New York Society of Security Analysts,* 24 May 1963; *Youngstown Bulletin,* Feb. 1965, p.18.

[22] *B.F.S.P.,* Sept. 1956, Oct. 1956, p.1207, Sept. 1957, p.1045, Dec. 1963; *I.A.,* 29 Oct. 1964, p.29.

[23] Chicago Association of Commerce, and Inland Steel company figures.

[24] Inland Steel, *Transcript of meeting with New York Society of Security Analysts,* 26 May 1972; *Fortune,* July 1958, 'Inland Steel does it again'.

[25] *Fortune,* Feb. 1964.

[26] Chairman's remarks Jones and Laughlin annual meeting 29 Apr. 1963, *Annual Report,* 1963.

[27] *Business Week,* 6 June 1970, pp.110–12.

[28] *New York Times,* 18 Apr. 1958, p.32, 25 Apr. 1958, p.46, 29 Apr. 1958, p.41; *I.A.,* 11 July 1957, p.66; Bethlehem Steel, *Annual Report,* 1956, p.6.

[29] The hearings were extensively reported in the *New York Times.*

[30] Inland Steel, *Annual Report,* 1957, p.16; *I.A.,* 12 Sept. 1957, p.94.

[31] *New York Times,* 12 Sept. 1958, p.35; *U.S. Steel Quarterly,* Feb. 1959, p.4; *New York Times,* 21 Nov. 1958, p.16.

[32] *I.S. Eng.,* Jan. 1972, p.48.

[33] *Fortune,* Apr. 1962, p.246.

[34] *B.F.S.P.,* June 1964, p.540, and Bethlehem Steel.

CHAPTER 20

The Postwar Steel Industry
The Manufacturing Belt since 1945:
The East

The steel centres east of the Alleghenies have made considerable headway since the Second World War. In 1947 their share of national output was 12·1 per cent but by 1968 13·8 per cent. However, improvement of the district's standing has been smaller than was widely anticipated by mining and plant engineers, government spokesmen, and acknowledged experts from the world of economic consulting and academic study in the early postwar years. Moreover its share of national production reached a peak between 1957 and 1960; since then there has been a slight fall. By 1970 fifteen eastern works operated a total mill capacity of 23·5 million tons, equal to 58 per cent the capacity at twenty-five works within 75 miles of Pittsburgh. Eastern steel-making has extreme dimensions of 450 miles, from Baltimore to Providence, Rhode Island and extending inland to Harrisburg and Williamsport, Pennsylvania, but less than 2 per cent of it is more than 95 miles from Philadelphia. The old-established eastern Pennsylvanian iron- and steelworks form its core. (Fig. 40)

In the late forties the wide expectation of a major eastwards shift of steel-making had every appearance of reasonableness. Both raw material assembly and marketing factors seemed to favour it. The life expectancy of Lake Superior direct shipping ores—which even in 1946/7 supplied 70 per cent of the ore for the Philadelphia area—was known to be short. As early as 1942 the War Production Board was informed that Mesabi might be worked out in eight to twelve years. Five years later open pit reserves were given a life of ten years. Mesabi ores by this time averaged about 48 per cent Fe while the ores being developed in Liberia were of 68 per cent iron. Large scale overseas mining and bulk ore carriers promised sharp cost reductions at tidewater, reductions which, because of the high grade of the ore, also involved fuel economy. One widely quoted commentator in 1947 summed up the problem facing the centres dependent on Great Lakes mineral flows: 'Costly new processes are going to have to be added to steel-making, and much of the industry is going to have to move to other parts of the country.'[1]

Eastern steel consumption grew rapidly in the war and seemed likely to expand still more, and again it appeared probable that the United States would become a much more important steel exporter. At the same time access to these markets from outside the region was becoming more difficult.

Blanket percentage rail freight rate increases above all penalized the long-distance shipper—in part because he was more dependent on rail transport than the producer with a short haul to market. The Pittsburgh to New York freight on finished steel rose from $7·20 in 1938 to $8·60 by early 1947 but from Sparrows Point only from $4·80 to $5·80. The differential widened still more after that.[2] Moreover, since 1938 Bethlehem and Sparrows Point had been the ruling basing points for a wide range of products delivered to eastern markets, so that as freight charges rose freight absorption by outsiders also had to increase. Following abandonment of basing point pricing in 1948 it was clear that, except in boom times when supplies were tight, major eastern consumers would choose to buy from eastern mills. Incentive to expand in the east or, alternatively, to build a new works there, was great.

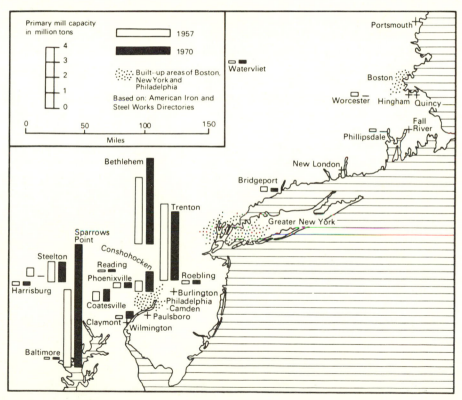

40. The steel industry in the east, 1957—1970

For purposes of analysis eastern works may be considered in three groups. Two, of not widely different size, are the dozen landbound works from Troy to Baltimore, along the Lehigh and Schuylkill, and at Harrisburg on the Susquehanna on the one hand, and the two tidewater works of Sparrows

Point and Trenton. Very much smaller, but deserving a disproportionate attention in this period is the New England steel industry (Table 131).

Table 131.   *Percentage of eastern capacity of various locations*
1945, 1957, 1970
(percentage)[1]

	1945 (ingots)	1957 (primary mill)	1970 (primary mill)
New England	4·9	3·4	1·2
Tidewater integrated	38·4	53·1	54·6
Other Eastern	56·7	43·5	44·2

[1] All figures refer only to works operating primary mills in 1957 and 1970. Based on A.I.S.I. *Works Directories.*

EASTERN LANDLOCKED PRODUCERS

The east retains a distinctive range of specializations. In 1957 it had 14·1 per cent of U.S. hot rolled product capacity (essentially in the area defined above except that the small works at Syracuse and Cortland are included). It had at this time only 10·6 per cent of U.S. hot rolled sheet capacity. In Buttweld tube eastern works had 21·3 per cent of U.S. capacity but only 7·8 per cent of Electric Resistance Weld (E.R.W.) tubing used especially for oil and gas operations. Plate has long been a regional specialism, and in 1957 capacity was 22·6 per cent of the national total for sheared plate and 26·3 per cent for universal plates. The largest independent eastern firms were in this trade. Lukens at Coatesville has suffered the competition of bigger firms and more recently of new Gulf coast plate capacity but has developed production of speciality steel plate with outlets in shipbuilding and particularly naval work. Lukens is unintegrated, but Alan Wood at Conshohocken operates a larger, integrated works whose record ingot output to date has been 970,000 tons. It too is still in the plate trade but since 1949 has widened its product range with a new 30 inch strip mill and later a cold reduction mill. Bethlehem Steel at Steelton has the only surviving rail mill in the region, but, although primary capacity there is over 1·3 million tons, the disadvantages of a landlocked operation have caused the plant to be converted to wholly cold metal operations. The Bethlehem works too is landbound but still a major unit—by 1970 its primary mill capacity was exceeded only by that at Sparrows Point, Lackawanna, and Inland Steel. Bethlehem remains the nation's largest producer of structural steel. Over a long period its pre-eminence in this line increased—from 17·7 per cent of the U.S. total in 1922 (structurals) to 26·9 per cent by 1957 (heavy structural shapes). At the latter date Bethlehem's capacity was more than twice that of the next works, U.S. Steel's South Chicago works, and almost three times that of Homestead. Since then Bethlehem's total structural capacity has not increased, but there has been large growth at both of these plants and also at Inland Steel. Perhaps even more indicative of the westward movement of structural demand, Bethlehem made considerable extensions to the Lackawanna structural mills. Bethlehem works uses not only foreign and lake ore but also material from modernized operations in Lebanon and Berks counties, whose ore was already mined when the plant was established in the 1850s. Phoenix Steel has structural capacity at Phoenixville, another of the very old eastern Pennsylvania locations, and plate mills at Claymont on the lower Delaware. Capacity at the former has been slightly written down over the last fifteen years, but considerable capital outlays have been made at Claymont.

TIDEWATER STEEL-MAKING

For over half a century Sparrows Point was unchallenged as the only major eastern tidewater plant. Within Bethlehem's domination of the regional markets it gradually increased its pre-eminence. In 1920 Sparrows Point capacity was only 58·8 per cent that of Bethlehem and Steelton; by 1930 73·9 per cent. Its 1945 capacity was 25·7 per cent greater than theirs and by 1960 51·8 per cent more. U.S. Steel was long credited with ambitions or even plans to break into the eastern markets; in the early postwar years it at last achieved this.

There were rumours of Steel Corporation interest in the East before the First World War. At the end of the twenties, when the existence of high-grade iron ore in eastern Venezuela had been proved, there were other schemes. The removal of many basing point differentials and establishment of new eastern basing points in 1938 was followed by more definite moves. By spring 1941 U.S. Steel seems to have made studies of the relative advantages of Wilmington, Richmond, and Norfolk as possible plant locations (indeed even in the early fifties, when it was building further north, U.S. Steel still seems to have kept a lingering interest in development possibilities in the Hampton Roads area). Immediately after the Second World War U.S. Steel sent out parties to search for and evaluate ore bodies in Africa, Canada, Mexico, Cuba, Nicaragua, and elsewhere. At the end of 1946, when the search was not going well, attention centred on Venezuela. At about the same time as the basing point system was abandoned, U.S. Steel proved the major orefields of Cerro Bolivar in the Llanos of Venezuela, a country eager to develop alternative exports to oil. New, liberal shipping policies introduced by the Truman administration in return for 1948 campaign support from the shipping interests improved eastern tidewater prospects still more. Plans for a U.S. Steel east coast mill were said to have been almost complete by late 1946, but it was not until the end of 1949 that the company bought 3,800 acres of land by the Delaware at Morrisville, near Trenton. At this stage it was intended that only finishing mills should be built, to be followed later by one blast furnace and five open hearth furnaces. The Korean War turned a long-term project into an urgent one. In October 1950 accelerated depreciation allowances were approved for defence projects and five months later ground was broken for a works, fully integrated from the start with two blast furnaces and nine open hearth furnaces.[3]

When this new Fairless works came into production in 1952 it was Sparrows Point's first large, tidewater competitor. For a time it was believed to have great advantages over its older rival. Yet it is difficult to avoid the conclusion that for many years Fairless was less successful than expected. Certainly modernization and, more significantly, expansion of U.S. Steel capacity went on more rapidly elsewhere, and the plant grew slowly even compared with Sparrows Point. A number of factors contributed to its relatively unspectacular growth.

In terms of marketing Fairless is certainly more centrally located in the east coast megalopolis than Sparrows Point. The latter, it is true, had its own shipyards and structural fabricating plants and long-established contacts with consumers, but some of them were delighted to find other sources of steel to break the near Bethlehem monopoly in certain lines.[4] New York and Philadelphia were each much nearer to Fairless than to Sparrows Point and ideally placed for truck delivery. By rail it was estimated in 1953 that New York customers could save $2·99 per ton on freight by buying from Fairless rather than Sparrows Point and Philadelphia buyers $2·42. However, this gain was reduced by the mill price differential which U.S. Steel introduced, nominally to cover part of the higher construction costs of the new works—$1 per net ton on cold reduced sheet by July 1957. In the keen competition of the early sixties this differential had to be removed.[5]

Fairless equipment was new, but this burdened it with much higher capital charges than Sparrows Point. Capacity cost an estimated $256 per ingot ton, whereas, for extensions which increased Sparrows Point capacity by almost as much, the investment was no more than $112 a ton. Allowing 12 per cent for interest and depreciation, Sparrows Point extensions gave it an operating cost advantage of some $18 a ton, an amount more than sufficient to cancel out any freight advantage of Fairless.[6] Between 1954 and 1960 Fairless was extended by a nominal 490,000 ingot tons, Sparrows Point by 2·45 million tons, by which time it was the nation's largest plant. Fairless had a much narrower product range, and though in some lines it was equipped with bigger units of plant notably in the case of its hot strip mill—much of the potential scale economy of this equipment was lost as a result of under-utilization. It was laid out on the grand scale on its almost 4,000 acre site, but such a layout, chosen to facilitate rational expansion, burdened it with high intraworks transfer costs.[7] A final weakness concerned water access. When production began navigation of the Delaware to the work's wharf was limited to vessels of less than 25 foot draught. In 1955 Congress was asked to authorize $91 million to dredge the river to 40 foot draft, but this motion was rejected.

By 1969 Fairless works was capable of 4 million tons of steel, still under half the capacity of Sparrows Point. The approaches through Chesapeake Bay to Sparrows Point have been deepened to allow ore cargo size to go up from 52,000 to 160,000 tons and there are now plans for even bigger terminal operations in the lower Delaware. U.S. Steel is credited with plans to expand Fairless to perhaps 10 million tons by 1980, at which time the full potential of its location will begin to be realized.

Other postwar plans to penetrate the east with integrated operations were abortive. National Steel bought a site at Paulsboro down river from Camden in 1951, but the imminent construction which was expected did not materialize and eight years later it decided to build on its Lake Michigan mill site rather than on its Delaware one. Colorado Fuel and Iron made a spirited bid to play an important role in the east in the late forties and fifties by acquiring steelworks and mills in Buffalo, at Claymont on the lower Delaware, and at Trenton, but by 1960, when it sold the Claymont works to Phoenix Steel, it was retracting. Finally Barium Steel of Canton, Ohio, acquired steelworks and plate mills at Harrisburg immediately after the war and then the Phoenixville structural mills. In 1956 it proposed an 800,000 ton integrated works on the east bank of the Delaware between Trenton and Burlington. Extensions would later bring this up to 2 million tons. Ingots were to be supplied to the mills at Harrisburg and Phoenixville. The certificate of necessity for a fast write-off which Barium requested was refused and the project foundered.[8]

NEW ENGLAND

Over a long period, as the metal fabricating trades of southern New England have expanded so the relative importance of New England metallurgical industries has declined. By the early twentieth century there were no integrated works, though as late as 1904 the section had three active blast furnaces and twenty-seven steelworks and rolling mills. Open hearth steel production rose from 57,000 tons in 1899 to 257,000 tons ten years later. By the mid-twenties there had been scarcely any increase on this figure though scrap shipments out of New England were already reaching high levels. For a time tariff duties had frustrated proposed coastal works using Nova Scotian coke—which was inferior—and foreign iron ore. When the tariff obstacle was removed the division of the substantial New England demand between a host of lines of rolled steel provided a more intractable problem. After the Second World War it seemed that New England ambitions for integrated steel operations would at last be fulfilled, but in the end they were dashed. The sharp growth of metal-working there in the war, freight rate increases, abolition of basing point pricing, and the general expectation of a major eastward shift of steel-making sent New England ambitions spiralling.

Between 1939 and 1947 the number of metal-working plants in New England rose from 2,970 to 5,236 and the number of workers in the industry increased from 331,000 to 579,000. Under 30 per cent of the section's employment was in metal-working in 1939, almost 40 per cent by 1950. New England was the area with the largest demand having no integrated works of its own. Total steel consumption by metal-working plants there by 1946/7 was variously estimated at between 1·6 and 1·9 million tons. In 1946 the New England Council began to urge the region's suitability for steel-making. The Econometric Institute Inc. was hired to make a study. Its report scotched the proposal, but the continuing rise in freight rates and the abolition of basing point pricing revived interest, so that the New England Council asked the Federal Reserve Bank of Boston to take up the study. The Federal Reserve Bank confirmed the high cost of outside supplies but pointed out that New England rolling mills already supplied between one-quarter and one-third of annual consumption. The remaining 1·2 million tons was divided between so many finished products that mill runs on any one would be small and costly. A smaller unit perhaps producing bars could be viable. Some of the new techniques already on the horizon, though by no means yet practicable processes, such as continuous casting, seemed to promise a reduction in the minimum profitable plant size.[9]

From the end of the war to about 1950 some three–quarters of the growth of New England employment was in metal-working, and this and news of the ore discoveries in Labrador, Venezuela, and elsewhere overseas provided new encouragement, so that the New England Council remained undeterred by these reports. It now contacted the big steel companies to try to induce them to give full consideration to a New England plant and meanwhile it went on with its own studies. In 1948 the Council set up a steel committee which in 1950 was given corporate status as the New England Steel Development Corporation, all of whose stock was held by members of the New England Council. The Steel Development Corporation commissioned feasability studies for a 1·25 million ingot ton $250 million integrated works—to produce 1 million tons of hot and cold rolled sheet, light plate, and hot rolled bars—first from Coverdale and Colpitts and then from H.A. Brassert. Both reported unfavourably. By this time also, after their individual studies, Jones and Laughlin, Republic Steel, and Youngstown Sheet and Tube had declined to take up an interest in the planned works. Bethlehem made its own analysis of the region at the end of the 1940s, and appears to have co-operated in the Coverdale and Colpitts study, but now decided not to take up an option to build. As its chairman remarked, U.S. Steel also gave 'thoughtful attention' to the possiblity of building its proposed east coast mill in New England. It was already one of the major factors in that region's steel business with a 250,000 ton melting shop at Worcester which supplied rod and wire mills there, a wire and rope plant at New Haven, and warehouses in Boston. However, it now proposed to build at Trenton and to supply some of New England's needs from there.[10] There was to be a long period of shrinking and occasionally reviving expectations but essentially this was the end of the New England integrated steel project.

The three sites which were under especial study were at Portsmouth, New Hampshire—where, after others turned the idea down, state officials still carried on negotiations with Allegheny Ludlum Steel—,at Hingham, Massachusetts, on the south eastern edge of the Boston conurbation, and at New London, Connecticut, at the eastern end of Long Island Sound. The latter became the favoured location. All of these locations were as well placed as other east coast ports to handle imported ore and had a shorter haul than any from Labrador. Conversely they would be disadvantaged by higher freight on coking coal even though it was anticipated that this would be shipped from Hampton Roads. The consultants of the time could not foresee the large fuel economy of the next twenty years which would substantially improve the fuel assembly situation of a New England Mill. On the scrap account too there were great advantages. In 1949 New England provided an estimated 4 per cent of all purchased scrap originating in the United States or about 830,000 tons; it consumed only 1·7 per cent of the total, about 383,000 tons. As a result scrap prices were low. Freight rates per ton of scrap to Pittsburgh were $11·08 at one point in the early fifties, and to Harrisburg $8·58, even from western Connecticut. A new plant of 1 million tons capacity could be expected to have scrap arisings from its own mill operations of about 250,000 tons, so that, even using a 60 per cent scrap charge, it would absorb only about 350,000 tons of the scrap which to this time had been annually shipped out of the region. It could therefore have retained the advantage of cheap scrap.[11]

On raw material grounds outside assessments of New England mill prospects may be seen with the benefit of hindsight, to have been too conservative. The trend of technology and raw material supply has improved the situation. But it is in relation to marketing that the true measure of the problem of successful major New England mill operations becomes so daunting. (Table 132) Essentially New England demand is still too small to give viable operations, and other parts of the eastern industrial belt are so much within the natural market area of other mills that a New England plant could not hope to control any large share of their business. Such mills have greater scale economies than would be available to New England mills. More efficient steel warehousing or steel finishing operations by companies whose major operations are outside the region has proved a more satisfactory way of meeting the region's demand than construction of major capacity there. In short, as compared with California and the West, once faced with some of these difficulties, New England remains at a disadvantage. Its market is smaller, though much more concentrated, demand has grown more slowly in the postwar years and New England is too near the great metal-making

Table 132.   *Estimated east coast raw material assembly and marketing costs per ton of steel 1952—1953 (dollars)*

| Location | Assembly Costs | | Freight to | |
	Ore	Coal	New York	Boston
New London	3·68	5·42	8·8	3·5
Sparrows Point	3·68	4·26	8·4	10·0
Bethlehem	5·56	5·06	5·8	8·2
Trenton	3·68	4·65	4·8	8·6

Based on W. Isard and R. Kuenne 1953, *Review of Economics and Statistics*, Nov. 1953, and separate estimates for delivery to Boston.

districts, so that a new plant there could not expect the price differential which for over fifteen years was a considerable boon to Fontana and the west coast cold metal plants. (Table 133) From 1950 onwards the New England mill project began to wither.

Table 133.   *New England and western steel production and consumption 1947, 1954, 1958, 1963 (percentage of U.S. total)*

| | New England | | West* | |
	Production	Consumption	Production	Consumption
1947	0·6	3·5	5·1	5·6
1954	0·4	3·0	6·1	7·3
1958	?	3·0	6·8	8·5
1963	?	2·7	6·2	7·8

*Pacific and Mountain States
Source Federal Reserve Bank of Cleveland *Economic Review*, Oct. 1969, p.10.

By the end of 1950 plans for Fairless Works had been announced but the New England Steel Development Corporation obtained a government certificate of necessity and assurance of a loan for the New London project, construction of which was expected to begin by the following midsummer. In March 1951 ground was broken for Fairless and enthusiasm for New London waned. The Federal Reserve Bank now commissioned a report from yet another firm of consultants, Arthur D. Little Inc., and this suggested a small speciality steel plant operating electrical furnaces and costing $14 million as opposed to the $250 million, 1 million ton integrated works. The certificates of necessity were allowed to lapse.[12] By 1953 the Steel Development Corporation had received authorization for rapid amortization of investment on a compromise project, a 300,000 ton electrical steel plant and mill, wholly scrap based.[13] After this New England efforts were to some extent divided. In 1954 the Northeastern Steel Corporation was formed to incorporate the Steel Development Corporation and with plans for the expansion of the long-existing open hearth steel plant at Bridgeport, Connecticut which not only had established market outlets but was less than half as far from New York as New London. From 1954 to 1956 the ingot capacity of the Bridgeport plant was increased from 190,000 to 300,000 tons, and there was a clear expectation that the expanded plant would become the nucleus of a major New England mill. Meanwhile in 1955 a Massachusetts legislative commission report recommended a 500,000 ton scrap-using continuous casting plant for Fall River. This was not built, and in 1957 Northeastern Steel became bankrupt.[14]

Defeated at last in its protracted bid for a large integrated works or indeed any large-scale operations, New England has become still more exclusively concerned with finishing steel made elsewhere. Bridgeport itself, as Northeastern Steel's successor, Carpenter Steel, put it, has become basically as extension of its mill at Reading in the Schuylkill Valley. In the fifties Jones and Laughlin and Detroit Steel built rerolling mills at Willimantic and Hamden, Connecticut, respectively, to be supplied from their bigger operations to the west in the case of Jones and Laughlin and with contracts for hot rolled coil from another eastern works in Detroit Steel's case. In 1958 U.S. Steel decided to dismantle the melting shop, blooming, and billet mills at Worcester, except for the briefly inflated Bridgeport works long the largest plant in New England. It had found that Fairless could deliver billets to the Worcester road and wire mill more cheaply. At the end of the sixties it decided to expand wire rod capacity and then wire capacity at Trenton and close the Worcester works. Only the small works at Bridgeport and Phillipsdale, Rhode Island survive in steel-making or primary mill operations, the remnants of an illustrious metal-making past, the relics of thwarted ambitions.

In summary, the experience of the east since the Second World War has confounded all predictions. New England has become more tributary than ever before to integrated mills east and west of the Appalachians. In spite of fuel economy, foreign ore, increasing freight rates and growth of the eastern megalopolis, the expected major eastward shift of steel-making to the mid-Atlantic region has failed to materialize.

NOTES

[1] H.E. Johnson, *World Iron Ore*, 1964; W.O. Hotchkiss, 'Iron Ore supply for the Future', *Economic Geology*, May 1947, pp.205—10; *Study in Monopoly Power*, Hearings before a Committee of the House of Representatives, 1950, part 4A, steel p.71; M. Barloon in *Harpers Monthly Magazine*, 1947, pp.154—5; G.S. Armstrong and Co., *An Engineering Interpretation of the Economic and Financial Aspects of American Industry*, X, 1952, pp.67—8.

[2] *I.A.*, 7 July 1938, p.846, 9 Jan. 1947, 7 Jan. 1954, p.248.

[3] *I.S. Eng.*, Feb. 1950, Dec. 1954, p.115; S.N. Whitney, *Antitrust Policies*, 1958, pp.277—8; *B.F.S.P.*, Feb. 1950, p.257 and other sources.

[4] House of Representatives, *Study in Monopoly Power*, 1950, 14, part 2A, p.847.

[5] *Hearings before the Senate Sub-Committee on Antitrust and Monopoly. Administered Prices*, part 2, Steel, 1957, p.615; W. Isard and R. Kuenne, 'The Impact of steel upon the Greater New York—Philadelphia Industrial Region: A Study in Agglomeration Projection', *Review of Economics and Statistics*, 35 4, 1953, pp.289, 301.

[6] *Fortune*, Mar. 1953, p.103.

[7] Artist's Impression of Works Layout in D.A. Fisher, *Steel Serves the Nation 1901—1951*, United States Steel Corporation, 1951, p.102.

[8] *New York Times*, 18 Dec. 1956, p.45.

[9] *I.A.*, 27 Jan. 1949, pp.109—11.

[10] *I.A.*, 1 Oct. 1953, p.38; *B.F.S.P.*, Mar. 1950, p.321 *Fortune*, Mar. 1953, p.104; *I.S. Eng.*, Jan. 1950, p.97.

[11] Department of the Interior, Materials Survey, *Iron and Steel Scrap*, Feb. 1953.

[12] *B.F.S.P.*, Jan. 1951, p.56, Feb. 1951, p.193; *I.S. Eng.*, Jan. 1953, p.127.

[13] *I.A.*, 18 Jan. 1951, p.79, 1 Oct. 1953, p.38; *I.S. Eng.*, Jan. 1954, p.125.

[14] *B.F.S.P.*, Feb. 1955, p.235, Mar. 1955, p.343, July 1955, p.791, Dec. 1957, p.1438, Jan. 1958, p.91; *I.S. Eng.*, Jan. 1956, p.121.

[15] Jones and Laughlin, *Annual Report*, 1956, p.4; Detroit Steel, *Prospectus*, Apr. 1964, pp.8, 11; *U.S. Steel Quarterly*, May 1958, p.3.

Review and Epilogue

The massive expansion of American iron and steel manufacture since the mid-nineteenth century has been accompanied by sweeping geographical changes. The most decisive locational influence has come from radical alterations in the distribution of demand. Growth of markets along the 'western waters' or in the woodlands or prairies between them pulled the growth centre of iron manufacture over the Appalachians in the middle years of the nineteenth century. In the decades around 1900 consumption west of the Mississippi provided a great boost for Chicago steelworks, and since 1940 the Pacific coast and the Gulf south-west have crossed consumption thresholds enabling these sections to support integrated works for the first time. These major areal shifts in steel consumption have not been matched in other leading industrial regions of the world. It is true that in the U.S.S.R. development of new metal districts has been compressed into a period only one-third as long and occurred over an even bigger area. Even so it is doubtful if changes in the distribution of demand have been as pronounced as those which have affected the various American metal markets. In Japan the eastern zone, from Tokyo to the western part of the Inland Sea, has been the focus of national population and economy for centuries. Important new industrial districts emerged in western Europe after the 1850s, but they did not involve the opening of wholly new sections of national territory.

Within the framework of the wide locational changes required by evolving patterns of settlement and economic growth, alterations in raw material supply and in the technology of processing have broadly influenced but never closely determined which shall be growth districts and which established centres shall decline. Until the 1850s dependence on charcoal fuel implied small-scale and generally scattered production, though swarming of producers occurred near the main orefields, especially those in the hinterland of the chief iron markets. The introduction of anthracite fuel increased both the scale of individual units and the possibility of localization in the industry, and resulted in the creation of the first great iron-making centres. The surge in coke smelting after the Civil War created another and soon even bigger concentration of iron capacity on the bituminous coalfield of the northern part of the Appalachian plateau. Small ore workings now had to be replaced by large-scale operations. The consequent development of the immense and long-distance water and rail movement of minerals, in the area between the head of the Great Lakes and the Ohio River, led to the domination of the supply situation by Lake Superior ore and Appalachian coal for more than three-quarters of the history of American iron and steel-making since 1850.

Beyond the Great Lakes basin long-established eastern metal districts, far from western markets except those of the Pacific coast states, declined, in some cases to the point of extinction. Some plants adjusted their product range and began to serve what were still major regional outlets, and from the late 1880s began to organize foreign ore supplies. In the years around 1900 Birmingham, Alabama, emerged as the only major focus of iron and steel-making outside the Manufacturing Belt. By this time the balance of locational advantage within the area dominated by Lake ore and Appalachian coal was shifting. Fuel economy in iron smelting, the substitution of Bessemer steel for puddled iron and the more rapid growth of consumption further west lessened the attractiveness of the districts located in the routes through the coalfields of western Pennsylvania and eastern Ohio.

In the twentieth century production became more diversified and the growth of consumer durable trades, especially the automobile industry, and the steel needs of urban expansion led to a concentration of consumption in the Manufacturing Belt. This was one factor which reduced the pace of locational change, but there were others. Works size had increased so that, by the early twentieth century, there were already a number of plants of over 500,000 tons capacity and some of over 1 million tons. It was more difficult to justify the abandonment of works of this size than the smaller ones of a few decades earlier. Growth in demand still encouraged the construction of new plants, old ones dropped out of the lists, but the pace of change in the geography of the industry was slowing, a certain maturity of the locational pattern was emerging. The dominant finished product of the last half of the nineteenth century, rails, declined to secondary status and subsequently dwindled almost to insignificance, while new demand lines rose to pre-eminence, as with tubular products, structurals, and particularly light, flat-rolled steels. These growth sectors helped to save or to bolster old areas which had chosen to specialize in them, such as the eastern district or the Valleys and Wheeling areas. They also helped build new producing districts, notably Detroit.

In and after the Second World War growth in their level of consumption encouraged a new western and south-western steel industry. This change in patterns of demand and of production, fuel economy, and a new access to richer, foreign ores led to the relative degeneration in the competitive position of the Birmingham district. Particularly after the mid-fifties major fuel economies were made in iron smelting as a result of better ore preparation, the use of iron ore pellets, the injection of other hydrocarbons to reduce the work of metallurgical coke, bigger blast furnaces and generally more efficient production. Coupled with increasing imports of ore and new patterns of market growth, this encouraged an already important centrifugal tendency in the distribution of locational advantages in the Manufacturing Belt, with relative and sometimes absolute decline in the old coal-based areas of the upper Ohio basin, stronger growth east of the Alleghenies and even more impressive expansion along the shores of Lake Erie and Lake Michigan. If there had not been a rapid inflation of capital cost, which widened the differential between the cost of extending existing works as compared with that for building new ones, locational change would have been more pronounced. In spite of this capital cost inflation, large outlays for oxygen steel-making plant, and to a smaller extent for continuous casting units and plant for producing and for using pellets of very high iron content were helping to speed locational change by the late sixties. The impetus for this large-scale re-equipment programme was now provided in large part by foreign competition of a severity not experienced for 100 years. Having operated to a considerable extent insulated from the rest of the metallurgical world by virtue of their massive home markets and low costs, made possible by rich mineral endowment, a superlative bulk transport system and large scales of operation, American steel-makers now had to face a reconstructed, extended, and very much more competitive foreign industry. The leading producers of the rest of the capitalist world have opened even richer orefields than those of the U.S. Upper Lakes district, have developed a new excellence in mineral movement, this time on a world as opposed to a subcontinental scale, and operate units technically at least as good as American plants, and in scale approaching and sometimes exceeding them.

Foreign markets are still of importance to American steel firms but, since the end of the fifties, exports have been exceeded by imports, and the imbalance has increased rapidly through the sixties. In 1971 consumption of foreign steel was at a record level of 18·3 million tons, or 18 per cent of the U.S. market; in the west the proportion supplied by imports was 31 per cent of the total. Foreign material has a still larger share of the special steel market. There have been some startling successes for foreign producers. By 1971/2 at least 5 per cent of General Motors' annual steel purchases came from overseas, and a group of Japanese companies has secured the contract for the 800 mile Alaskan oil pipeline. Since the mid-fifties American steel capacity and output has grown much less rapidly than that of some of its leading rivals. Temporarily at least it has lost its world leadership. In 1890 the United States passed Great Britain as the world's leading producer; in 1971 she in turn was first passed by the Soviet Union. In the pioneer days of the American trade heroic attempts were made to build up an industry while erecting barriers to hold out European and above all British iron; now a major established industry, with a host of plants, is striving to the extent of a $2,000 million annual outlay to reconstruct to meet a new flood of foreign material. It still seems that, in spite of these massive investments, this traffic can scarcely be checked let alone reversed. There are various accusations of past complacency and calls for present action. Yet, whatever its exact status or ranking, it is clear that American steel will remain a world leader. Its present state is the outcome of 150 years of unrivalled growth and success; its future propsects will be largely conditioned by that heritage, in organization, in business psychology, and in location.

INDEX